Recent Advances in Nonlinear Partial Differential Equations and Applications

Photograph courtesy of Dr. Ramon Escobedo

Peter D. Lax

Louis Nirenberg

Group photo at the Lorenzana Palace Courtyard

Proceedings of Symposia in
APPLIED MATHEMATICS

Volume 65

Recent Advances in Nonlinear Partial Differential Equations and Applications

Conference in Honor of Peter D. Lax and
Louis Nirenberg on Their 80th Birthdays
June 7–10, 2006
Universidad de Castilla-La Mancha
at Palacio Lorenzana, Toledo, Spain

L. L. Bonilla
A. Carpio
J. M. Vega
S. Venakides
Editors

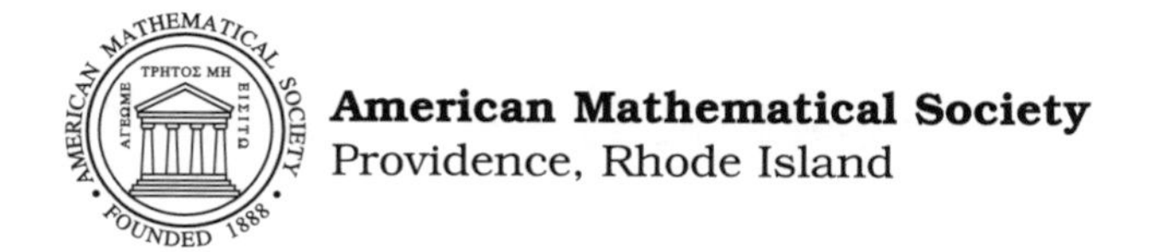

American Mathematical Society
Providence, Rhode Island

2000 *Mathematics Subject Classification.* Primary 35Lxx, 35Qxx, 37–XX, 44–XX, 70–XX, 74–XX, 76–XX, 82–XX, 83–XX.

Library of Congress Cataloging-in-Publication Data

Recent advances in nonlinear partial differential equations and applications : a conference in honor of Peter D. Lax and Louis Nirenberg on their 80th birthdays, June 7–10, 2006, Toledo, Spain / L. L. Bonilla . . . [et al.], editors.

p. cm. — (Proceedings of symposia in applied mathematics ; v. 65)

Includes bibliographical references.

ISBN 978-0-8218-4211-9 (alk. paper)

1. Differential equations, Nonlinear—Congresses. 2. Differential equations, Partial—Congresses. I. Lax, Peter D. II. Nirenberg, L. III. Bonilla, L. L. (Luis López), 1956– IV. Series.

QA377.R4237 2007
515′.353—dc22 2007060794

∞ The paper used in this book is acid-free and falls within the guidelines
established to ensure permanence and durability.
Visit the AMS home page at `http://www.ams.org/`

10 9 8 7 6 5 4 3 2 1 12 11 10 09 08 07

Contents

Preface

Peter D. Lax and Louis Nirenberg are two of the most distinguished mathematicians of our times. Their work has advanced the state of the art of partial differential equations throughout the last half-century dramatically and has profoundly influenced the course of mathematics. Their innovative approach to the subject has created new and active fields of mathematical research. The scientists they have trained are still carrying their ideas further. It is fair to say that the thinking of any researcher on any aspect of partial differential equations in the world today has been touched by their ideas.

Peter and Louis have been colleagues at the Courant Institute of Mathematical Sciences of New York University over their entire careers. They are best friends and they are of the same age. Their 80th birthdays offered an occasion for many mathematicians from all over the world who love and honor them to come to a conference on a subject that Peter and Louis have inspired so deeply. The conference took place in the scenic town of Toledo, in Spain, in the historic Lorenzana palace (which nowadays belongs to the Castilla–La Mancha University, one of the sponsors of the conference) on June 7-10, 2006, with the participation of a diverse group of over 180 scientists.

The scientific objective of the conference was to present the state of the art in some areas of the modern theory of nonlinear partial differential equations and their applications that have experienced great activity and the development of new tools in the last 5-10 years. The functional analytic approach to nonlinear PDE as well as topics such as conservation laws, transonic flows, hydrodynamics and vortical structures, turbulence, dispersive waves, combustion, materials science, brain spatiotemporal dynamics and others were presented by leaders in these areas in hour-long invited lectures. The present Proceedings volume records many of these lectures (see the following pages for a full list of invited speakers and titles). Abstracts of all invited talks and poster presentations are posted in the conference webpage http://www.mat.ucm.es/~ln06/, (alternatively, google search LN2006).

We wish to thank all those who have contributed to the success of the Conference. Among them the participants, the speakers, and, in particular, the members of the Local Organizing Committee chaired by Julio Muñoz (Universidad de Castilla La Mancha), who worked hard to run a smooth and successful conference. We owe particular thanks to the Comunidad de Castilla - La Mancha whose First Vice President, Fernando La Mata, co-presided the opening ceremony of the conference with the Vice President of the Universidad Castilla - La Mancha, Evangelina Aranda. Help from Pedro-Pablo Novillo, General Director of Coordination and Educational Policy (JCCM) and the Consejería de Educación y Ciencia is gratefully acknowledged.

It was possible for this conference to take place due to the financial support and the sponsorship by many Institutions both in Spain and in the United States. We acknowledge this support below by listing all of these Institutions; we extend to all of them our deepest thanks and the assurance that their support served a useful purpose. The conference was a forum for the dissemination of new scientific ideas and brought leading experts together with a large number of scientists, many of them at the early stages of their careers. At the same time it allowed the international mathematics community to honor two of its outstanding members. Unfortunately, Louis Nirenberg could not attend due to illness; his presence in spirit was very much felt in the lecture room. It was a pleasure for all participants, in particular the younger ones, to have a chance to converse with Peter Lax who has always been welcoming of such interactions.

On behalf of the American Mathematical Society, the publisher of this volume, and on behalf of all of us, we wish Peter and Louis many many more years of happiness and activity.

The editors

Sponsors

- United States:
  - National Science Foundation, Division of Mathematical Sciences, Grant DMS-0621292.
  - European Office of Aerospace Research and Development, Air Force Office of Scientific Research, US Air Force Research Laboratory.
  - Duke University, Office of the Vice Provost for International Affairs.
  - New York University, Office of the Provost.
  - American Mathematical Society (AMS).
  - Society for Industrial and Applied Mathematics (SIAM).
- Spain:
  - Ministerio de Educación y Ciencia, Grant MTM2005-25583-E.
  - Junta de Comunidades de Castilla-La Mancha. Consejería de Educación y Ciencia.
  - Comunidad Autónoma de Madrid, Grant S-0505/ENE/0229.
  - Universidad de Castilla-La Mancha.
  - Universidad Carlos III de Madrid.
  - Universidad Complutense de Madrid.
  - Universidad Politécnica de Madrid.
  - Universidad de Burgos.
  - Ayuntamiento de Toledo.
- European Union:
  - European Consortium for Mathematics in Industry (ECMI).

List of Speakers

- Luis L. Bonilla, Universidad Carlos III de Madrid, Spain, *Dynamics of defects in crystal lattices.*
- Haim Brézis, Université de Paris VI, France, *The degree of maps from S^1 into itself. An intriguing story which is not yet finished.*
- Alexandre Chorin, UC Berkeley, USA, *Averaging and memory.*
- Demetrios Christodoulou, ETH Zurich, Switzerland, *The formation of shocks in three-dimensional fluids.*
- Constantine M. Dafermos, Brown University, USA, *Entropy in hyperbolic conservation laws.*
- Athanassios S. Fokas, University of Cambridge, UK, *Lax pairs: Synthesis rather than separation of variables.*
- François Golse, Université Paris VII, France, *The incompressible Navier-Stokes limit of the Boltzmann equation.*
- F. Alberto Grünbaum, UC Berkeley, USA, *How often should Louis and Peter be simultaneously happy? Occupancy times for Brownian motion in a two dimensional wedge.*
- Javier Jiménez, Universidad Politécnica de Madrid, Spain, *Recent results from the direct simulation of turbulent flows.*
- Barbara L. Keyfitz, University of Houston, USA, and The Fields Institute, Canada, *Some interesting questions in multidimensional conservation laws.*
- Sergiu Klainerman, Princeton University, USA, *A break-down criterion for the Einstein equations in a vacuum.*
- C. David Levermore, University of Maryland, USA, *Fluid Dynamics from Boltzmann Equations.*
- Yanyan Li, Rutgers University, USA, *A geometric problem and the Hopf Lemma.*
- Amable Liñán, Universidad Politécnica de Madrid, Spain, *The role of the multiple scale character of the combustion phenomena in their mathematical analysis.*
- Andrew M. Majda, Courant Institute, New York University, USA, *New waves, PDE's, and coarse-grained stochastic lattice models for the tropics.*
- David W. McLaughlin, Courant Institute, New York University, USA, *Modeling the spatiotemporal cortical activity associated with the line-motion illusion in primary visual cortex.*
- Stephanos Venakides, Duke University, USA, *The focusing nonlinear Schrödinger equation: Rigorous semiclassical and long time asymptotics.*

List of Posters

(1) Optimal lower bounds on stress concentrations in random media, B. Alali, Louisiana State University, USA.
(2) An adaptative procedure for the spherical shallow water equations, B. Alonso, Universidad Politécnica de Madrid, Spain.
(3) High order TVD schemes using extrapolation, S. Amat, Universidad Politécnica de Cartagena, Spain.
(4) Propagation of fluctuations in biochemical systems, D. Anderson, Duke University, USA.
(5) Finite-difference adaptative mesh refinement for hyperbolic systems of conservation laws, A. Baeza, Université de Provence, France.
(6) A new reconstruction operator in central schemes for hyperbolic conservation laws, A. Balaguer, Universidad Politécnica de Valencia, Spain.
(7) Nonlinear Schrödinger equation with nonhomogeneous nonlinearities. Existence and orbital stability, J. Belmonte, Universidad de Castilla-La Mancha, Spain.
(8) A Meshless solution to the Hele-Shaw flow, F. Bernal, Universidad Carlos III, Spain.
(9) Reducibility and block diagonalizability problems for hydrodynamic type systems, O. Bogoyavlenskiij, Queen's University, Canada.
(10) Dispersive waves and turbulence interactions in the moderately small Rosby number limit, L. Bourouiba, McGill University, Canada.
(11) Long time asymptotics of the nonlinear Schrodinger equation shock problem, R. Buckingham, University of Michigan, USA.
(12) Navier's slip and NSE with temperature dependent viscosity, M. Bulicek, Charles University Prague, Czech Republic.
(13) Lyapunov inequalities for differential equations, A. Cañada, Universidad de Granada, Spain.
(14) An adaptative finite element method for the numerical simulation of lifted flames, J. Carpio, Universidad Politécnica de Madrid, Spain.
(15) The algebra of differential operators associated to a family of matrix values orthogonal polynomials, M. Castro, Universidad de Sevilla, Spain.
(16) Pattern formation in models of Faraday waves with non-smooth forcing, A. Catlla, Duke University, USA.
(17) Solutions to the bipolar Vlasov-Poisson-Boltzmann equations, E. Cebrián, Universidad de Burgos, Spain.
(18) Waves and shocks in integrodifferential conservation laws, A. Chesnokov, Laurentyev Institute of Hydrodynamics, Russia.

(19) Hyperbolic systems with memory, C. Christoforou, University of Northwestern, USA.
(20) Estimates of best constants for weighted Poincaré inequalities on convex domains, S.K. Chua, National University of Singapore, Singapore.
(21) Multidimensional detonation waves, N. Constanzino, Brown University, USA.
(22) Modulation theory for the Zakharov-Kuznetsov equation, G. Cruz, UNAM, México.
(23) Weak shocks for a one dimensional BGK kinetic model for conservation laws, C. Cuesta, University of Vienna, Austria.
(24) Homogeneization of a quasilinear parabolic equation with vanishing viscosity, A.L. Dalibard, Université Paris-Dauphine, France.
(25) Voltage switching and domain relocation in semiconductor superlattices, G. Dell' Acqua, Universidad Carlos III, Spain.
(26) Direct computation of critical flight conditions, I. de Mateo, Humboldt Universitaet zu Berlin, Germany.
(27) A new software to provide determining equations for Lie symmetries of PDE. Application to a 3+1 dimensional integrable system, J.M. Díaz, Universidad de Cádiz, Spain.
(28) Perturbing singular solutions of the Gelfand problem, L. Dupaigné, Université de Picardie, France.
(29) Algebraic approach to classification of Lax pairs, O. Efimovskaya, Lomonosov Moscow State University, Russia.
(30) Supersonic flow onto a solid wedge: exact solutions, V. Elling, Brown University, USA.
(31) Quantum drift-diffusion equations for semiconductor superlattices, R. Escobedo, Universidad Carlos III, Spain.
(32) Nonlinear stability of the Kuzmin disk, R. Firt, University of Bayreuth, Germany.
(33) High order finite volume schemes for solving hyperbolic systems with non conservative products. Applications to shallow water systems, J.M. Gallardo, Universidad de Málaga, Spain.
(34) A generalized finite difference method for nonlinear photonic soliton analysis, M.A. García March, Universidad Politécnica de Valencia, Spain.
(35) Searching for an existence result for a system involving a singular diffusion matrix, C. García Vázquez, Universidad de Cádiz, Spain.
(36) Numerical simulation of wave overtopping, M.T. González Cerverón, Universidad de Castellón, Spain.
(37) Some existence results to the termistor problem, M.T. González Montesinos, Universidad de Cádiz, Spain.
(38) Travelling fronts in porous media combustion, P. Gordon, New Jersey Institute of Technology, USA.
(39) Effect of tubular inhomogeneities on the tubuloglomerular feedback system, P.B. Grajdeanu, Ohio State University, USA.
(40) Symmetries in nonlinear PDE of slow dense granular flow, A. Grm, Fraunhofer Institut, Kaiserslauten, Germany.
(41) The binormal flow and cubic Schrodinger equations: Singularity formation phenomena, S. Gutiérrez, University of Nottingham, UK.

(42) Integrable nonlinear evolution equations in time dependent domains, I. Hitzazis, University of Patras, Greece.
(43) Transonic regular reflection for the isentropic gas dynamics equations, K. Jegdic, University of Houston, USA.
(44) Stability of vortices in two and three dimensional Bose-Einstein condensates, R. Kollar, University of Michigan, USA.
(45) Hamilton-Jacobi equation, transport system of equations and passing caustics for the aeroacustic hyperbolic system, O. Lafitte, Université Paris XIII, France.
(46) Parabolic behavior of a hyperbolic delay equation, T. Laurent, University of Duke.
(47) Probabilistic simulation of random fields for uncertain parameters used in the modelling of transport by groundwaters: application to the risk assessment analysis, S. Lazaar, Université A. Essaadi, Morocco.
(48) Blow-up and solitary waves for nonlinear equations describing Boson and Fermion stars, E. Lenzmann, ETH Zurich, Switzerland.
(49) Combinig finite volume methods and symplectic time numerical schemes for the shallow water equations with Coriolis terms, J.A. López, Universidad de Málaga, Spain.
(50) Singularities in a family of contour dynamics equations, A.M. Mancho, CSIC, Spain.
(51) Quasilinear problems depending on $|\nabla w|^2$ versus semilinear ones with blow-up at the boundary, P.J. Martínez Aparicio, Universidad de Granada, Spain.
(52) High order numerical simulation of shallow water equations, A. Martínez Gavara, Universidad de Valencia, Spain.
(53) Weak violation of the Lax's theorem on Rieman solutions, V. Matos, Universidade do Porto, Portugal.
(54) Unified description of electromagnetic nonlinear propagation in systems with translational invariance, L. Monreal, Universidad Politécnica de Valencia, Spain.
(55) New mathematical approaches for image reconstruction in medicine, M. Moscoso, Universidad Carlos III, Spain.
(56) A Lax-Wendroff type convergence result for nonconservative problems, ML Muñoz, Universidad de Málaga, Spain.
(57) A variational approach to propagation in infinite cylinders, C.B. Muratov, New Jersey Institute of Technology, USA.
(58) Numerical simulation of incompressible viscous flow with moving boundaries, A. Nanda Pati, University of Houston, USA.
(59) Very high order accurate well-balanced computation of river flows, S. Noelle, RWTH-Aachen, Germany.
(60) Existence of travelling solutions for Ginzburg-Landau type problems in infinite cylinders, M. Novaga, Universitá di Pisa, Italy.
(61) How we must change the hydrostatic approximation in an ocean with a general seabottom, F. Ortegón, Universidad de Cádiz, Spain.
(62) Metastability and dispersive shock waves in Fermi-Pasta-Ulam system, S. Paleari, Observatoire de la Côte d'Azur, France.

(63) Some results and open questions regarding higher-order inverse periodic spectral problems, V. Papanicolau, National Technical University of Athens, Greece.
(64) Complex analysis quaternions and boundary value problems, D.A. Pinotsis, University of Cambridge, UK.
(65) New error estimates for a viscosity-splitting scheme for the Navier-Stokes equations, M.V. Redondo, Universidad de Cádiz, Spain.
(66) Existence and uniqueness for Van Roosbroeck's system in Lebesgue spaces, J. Rehberg, Weierstrass Institute, Germany.
(67) Entire solutions for the Dirichlet problem and a characterization of ellipsoids, H. Render, Universidad de la Rioja, Spain.
(68) Approximating the 2D hydrostatic Navier-Stokes problem by rectangular fictitious domain, R. Rodríguez, Universidad de Cádiz, Spain.
(69) Collective coordinates for nonlinear Klein-Gordon equations, N. Rodríguez, Universidad de Sevilla, Spain.
(70) Blow up of smooth solutions to the compressible Navier-Stokes equations with the data highly decreasing at infinity, O. Rozanova, Moscow State University, Russia.
(71) Rearrangement inequalities and applications to isoperimetric problems for eigenvalues, E. Russ, Université Aix-Marseille, France.
(72) Existence and stability of static shells for the Vlasov-Poisson system, A. Schulze, University of Bayreuth, Germany.
(73) Delta and Delta shock type solutions of conservation laws: The Rankine Hugoniot conditions, geometrical and algebraical aspects, V. Shelkovitz, St Petersburg University, Russia.
(74) Elasticity and plasticity in plant growth, P. Shipman, Max Planck Institute, Germany.
(75) Open systems viewed through their conservative extensions, S. Shipman, Louisiana State University, USA.
(76) Local existence of dynamical and trapping horizons: stability of marginally trapped surfaces and existence of horizons, W. Simon, Universidad de Salamanca, Spain.
(77) Nonlinear waves in magma dynamics, G. Simpson, Columbia University, USA.
(78) Coupling two different nature methods for 3d fluid animation, N. Suárez, Universidad Politécnica de Cataluña, Spain.
(79) Universality in the large time behavior of the zero dispersion limits of the higher order KdV equation, Fei-Ran Tian, Ohio State University, USA.
(80) Solution of some boundary value problems for nonlinear Poisson equation with complex coefficients, I. Tralle, University of Rzeszow, Poland.
(81) Near equilibrium statistical solutions of NLS equations, B.E. Turkington, University of Massachusetts Amherst, USA.
(82) Asymptotic analysis and estimates of blow-up time for the radial symmetric semilinear heat equation in the one spectrum-case, D.E. Tzanetis, National Technical University of Athens, Greece.
(83) From Geroch reduction to quantized Einstein-Rosen waves, J.S. Villaseñor, CSIC, Spain.

(84) Phragmen-Lindelof principles for second order elliptic equations via weak Harnack inequality, A. Vitolo, University of Salerno, Italy (S1)
(85) On nonhomogeneous Neumann problem with weight and with critical nonlinearity in the boundary, H. Yazidi, Université Paris XII, France.
(86) Remarks on Gagliardo-Nirenberg type inequality with critical Sobolev space and BMO, H. Wadade, Tohoku University, Japan.

List of Participants

Alonso, B.
Universidad Politécnica de Madrid, Spain

Alvarez, J.D.
Universidad Carlos III de Madrid, Spain

Amat, S.
Universidad Politécnica de Cartagena, Spain

Anderson, D.
Duke University, USA

Angulo, P.
Universidad Complutense de Madrid, Spain

Arana, J.I.
Universidad Carlos III de Madrid, Spain

Arenas, A.
Universitat Rovira i Virgili, Spain

Azagra, D.
Universidad Complutense de Madrid, Spain

Bacim, A.
Louisiana State University, USA

Baeza, A.
Université de Provence, France

Balaguer-Beser, A.
Universidad Politécnica de Valencia, Spain

Belmonte, J.
Universidad de Castilla - La Mancha, Spain

Bermejo, R.
Universidad Politécnica de Madrid, Spain

Bernal, F.
Universidad Carlos III de Madrid, Spain

Bogoyavlenskij, O.
Queen's University, Canada

Bonfoh, AS.
ICTP Trieste, Italy

Bonilla, L.L.
Universidad Carlos III de Madrid, Spain

Boumediene, A.
Universidad Autónoma de Madrid, Spain

Bourouiba, L.
McGill University, Canada

Brauer, U.
Universidad Complutense de Madrid, Spain

Brézis, H.
Université Paris VI, France

Buckingham, R.
University of Michigan, USA

Bulicek, M.
Charles University Prague, Czech Republic

Cañada, A.
Universidad de Granada, Spain

Carpio, A.
Universidad Complutense de Madrid, Spain

Carpio, J.
Universidad Politécnica de Madrid, Spain

Carretero, M.
Universidad Carlos III de Madrid, Spain

Catlla, A.
Duke University, USA

Castello, X.
Universidad de Castellón, Spain

Castro, M.
Universidad de Sevilla, Spain

Cebrián, E.
Universidad de Burgos, Spain

Charro, F.
Universidad Autónoma de Madrid, Spain

Chesnokov, A.
Laurentiev Institute of hydrodynamics, Russia

Chorin, A.J.
UC Berkeley, USA

Christoforou, C.
Northwestern University, USA

Christodoulou, D.
ETH Zurich, Switzerland

Colorado, E.
Universidad de Granada, Spain

Constanzino, N.
Brown University, USA

Cruz, G.
UNAM, México

Cuesta, C.
University of Vienna, Austria

Dafermos, C.
Brown University, USA

Dalibard, A.L.
Université Paris IX, France

Davila, J.
Universidad de Chile, Chile

Dell'Acqua, G.
Universidad Carlos III de Madrid, Spain

Díaz, J.I.
Universidad Complutense de Madrid, Spain

Díaz, J.M.
Universidad de Cádiz, Spain

Donat, R.
Universidad de Valencia, Spain

Dupaigné, L.
Université Picardie, France

Efimovskaya, O.
Lomonosov University, Russia

Elling, V.
Brown University, USA

Escobedo, R.
Universidad Carlos III de Madrid, Spain

Firt, R.
University of Bayreuth, Germany

Fokas, A.S.
University of Cambridge, UK

Galán, P.
Universidad de Castilla - La Mancha, Spain

Gallardo, J.
Universidad de Málaga, Spain

Gancedo, F.
CSIC, Madrid, Spain

Garay, I.
CSIC, Madrid, Spain

García March, M.A.
Universidad Politécnica de Valencia, Spain

García Vázquez, C.
Universidad de Cádiz, Spain

Glavan, A.M.
Universidad Carlos III, Spain

Golse, F.
Université Paris VII, France

González Cerverón, M.T.
Universidad de Castellón, Spain

González Montesinos, M.T.
Universidad de Cádiz, Spain

Gordon, P.
New Jersey Institute of Technology, USA

Grajdeanu, P.
Ohio State University, USA

Grm, A.
Fraunhofer Institute, Kaiserslauten, Germany

Grunbaum, A.
UC Berkeley, USA

Gutiérrez, S.
University of Nottingham, UK

Hernando, P.J.
Universidad Carlos III de Madrid, Spain

Herrero, H.
Universidad de Castilla - La Mancha, Spain

Higuera, M.
Universidad Politécnica de Madrid, Spain

Hitzazis, I.
University of Patras, Greece

Jegdic, K.
University of Houston, USA

Jiménez, J.
Universidad Politécnica de Madrid, Spain

Kaplicky, P.
Charles University Prague, Czech Republic

Keyfitz, B.
Fields Institute (Canada) and Houston University (USA)

Kim, A.
UC La Merced, USA

Kindelan, M.
Universidad Carlos III de Madrid, Spain

Klainerman, S.
Princeton University, USA

Kollar, R.
University of Michigan, USA

Krupchyk, K.
University of Joensuu, Finland

Lafitte, O.
Laga, Université Paris XIII, France

Laurent, T.
Duke University, USA

Lazaar, S.
Université A. Essaadi, Morocco

Lax, P.
Courant Institute, NYU, USA

Lenzmann, E.
ETH Zurich, Switzerland

Levermore, C.D.
University of Maryland, USA

Li, Y.Y.
Rutgers University, USA

Liñán, A.
Universidad Politécnica de Madrid, Spain

López, J.A.
Universidad de Málaga, Spain

Lukaszewicz, G.
University of Warsaw, Poland

Maciá, F.
Universidad Complutense de Madrid, Spain

Majda, A.J.
Courant Institute, NYU, USA

Mancebo, F.J.
Universidad Politécnica de Madrid, Spain

Mancho, A.
CSIC, Madrid, Spain

Mande, J.V.
Universidad Politécnica de Cataluña, Spain

Marion, M
Ecole Centrale de Lyon, France

Martel, C.
Universidad Politécnica de Madrid, Spain

Martin, J.M.
CSIC, Madrid, Spain

Martínez Aparicio, P.J.
Universidad de Granada, Spain

Martínez Gavara, A.
Universitat de Valencia, Spain

Martínez, V.
Universidad de Castellón, Spain

Mateo, Ignacio de
Humboldt Universität zu Berlin, Germany

Matos, V.
Universidade do Porto, Portugal

McLaughlin, D.W.
Courant Institute, NYU, USA

Medina, P.
Universidad Carlos III de Madrid, Spain

Monreal, L.
Universidad Politécnica de Valencia, Spain

Morton, K.W.
University of Oxford, UK

Mosco, U.
Worcester Polytechnical Institute, USA

Moscoso, M.
Universidad Carlos III de Madrid, Spain

Mulet, P.
Universidad de Valencia, Spain

Muñoz, J.
Universidad de Castilla - La Mancha, Spain

Muñoz, M.L.
Universidad de Málaga, Spain

Muratov, C.B.
New Jersey Institute of Technology, USA

Nanda Pati, A.
University of Houston, USA

Noelle, S.
RWTH-Aachen, Germany

Novaga, M.
Universitá di Pisa, Italy

Ortegón, F.
Universidad de Cádiz, Spain

Paleari, S.
Observatoire CøCôtedd'Azur, Nice, France

Papanicolau, V.
Technical University of Athens, Greece

Parés, C.
Universidad de Málaga, Spain

Pelloni, B.
University of Reading, UK

Peral, I.
Universidad Complutense de Madrid, Spain

Pinotsis, D.A.
University of Cambridge, UK

Ponno, A.
Universitá di Milano, Italy

Porter, J.
Universidad Politécnica de Madrid, Spain

Prazak, D.
Charles University Prague, Czech Republic

Priego, M.
Universidad Complutense de Madrid, Spain

Primo, A.
Universidad Autónoma de Madrid, Spain

Redondo, V.
Universidad de Cádiz, Spain

Rehberg, J.
Weierstrass Institute, Germany

Render, H.
Universidad de la Rioja, Spain

Ricci, V.
Universitá di Palermo, Italy

Rodríguez, D.
Universidad de Granada, Spain

Rodríguez Galván, R.
Universidad de Cádiz, Spain

Rodríguez Quintero, N.
Universidad de Sevilla, Spain

Rozanova, O.
Moscow University, Russia

Ruiz Aguilar, D.
Universidad de Granada, Spain

Russ, E.
Université Paul Cezanne, France

Sadowski, W.
Warsaw University, Poland

Sanchez Villaseñor, E.J.
Universidad Carlos III de Madrid, Spain

Schlichtling, G.
Technische Universitat Muenchen, Germany

Schulze, A.
University of Bayreuth, Germany

Shelkovitz, V.
St Petersburg University, Russia

Shipman, S.
Louisiana State University, USA

Shipman, P.
Max Planck Institute, Germany

Simon, W.
Universidad de Salamanca, Spain

Simpson, G.
Columbia University, USA

Solomiak, M.
Weizmann Institute, Israel

Suarez, N.
Universidad Politécnica de Cataluña, Spain

Tapiador, B.
Universidad Complutense de Madrid, Spain

Tian, F.R.
Ohio State University, USA

Tralle, I.
University of Rzeszow, Poland

Turkington, B.E.
University of Massachusetts, USA

Tzanetis, D.
Technical University of Athens, Greece

Vega de Prada, J.M.
Universidad Politécnica de Madrid, Spain

Vela, M.
Universidad Complutense de Madrid, Spain

Venakides, S.
Duke University, USA

Villegas, R.M.
Universidad Carlos III, Spain

Villegas, S.
Universidad de Granada, Spain

Vitolo, A.
Universitá di Salerno, Italy

Wegner, D.
Humbolt Universitat, Germany

Wu, L.
INRIA, France

Yazidi, H.
Université Paris XII, France

Proceedings of Symposia in Applied Mathematics
Volume **65**, 2007

Null hypersurfaces with finite curvature flux and a breakdown criterion in General Relativity

Sergiu Klainerman

In honor of P.Lax and L. Nirenberg

1. Introduction

I report on recent work in collaboration with Igor Rodnianski concerning a *geometric* criterion for breakdown of solutions $(\mathbf{M}, \mathbf{g})$ of the vacuum Einstein equations,

$$\mathbf{R}_{\alpha\beta}(\mathbf{g}) = 0. \tag{1}$$

Here $\mathbf{R}_{\alpha\beta}$ denotes the Ricci curvature of the l 3+1 dimensional Lorentzian manifold $(\mathbf{M}, \mathbf{g})$. The main result discussed here is stated and proved in [**Kl-Ro6**]; the proof depends however on the results and methods of [**Kl-Ro1**], [**Kl-Ro2**], [**Kl-Ro3**] [**Kl-Ro4**] which establish a lower bound for the radius of injectivity of null hypersurfaces with finite curvature flux as well as [**Kl-Ro5**] in which we construct a Kirchoff-Sobolev type parametrix for solutions to covariant wave equations.

Assume that a part of space-time $\mathcal{M}_I \subset \mathbf{M}$ is foliated by the level hypersurfaces Σ_t of a time function t, monotonically increasing towards future in the interval $I \subset \mathbb{R}$, with lapse n and second fundamental form k defined by,

$$k(X,Y) = \mathbf{g}(\mathbf{D}_X \mathbf{T}, Y), \qquad n = \big(-\mathbf{g}(\mathbf{D}t, \mathbf{D}t)\big)^{-1/2} \tag{2}$$

where $\mathbf{T}$ is the future unit normal to Σ_t, $\mathbf{D}$ is the space-time covariant derivative associated with $\mathbf{g}$, and X, Y are tangent to Σ_t. Let Σ_0 be a fixed leaf of the t foliation, corresponding to $t = t_0 \in I$, which we consider the initial slice. We assume that the space-time region $\mathcal{M}_I$ is globally hyperbolic, i.e. every causal curve from a point $p \in \mathcal{M}_I$ intersects Σ_0 at precisely one point. Assume also that the initial slice verifies the assumption.

1991 *Mathematics Subject Classification.* 35J10.

The author is partially supported by NSF grant DMS-0070696.

A 1. There exists a finite covering of Σ_0 by a finite number of charts U such that for any fixed chart, the induced metric g verifies

$$\Delta_0^{-1}|\xi|^2 \leq g_{ij}(x)\xi_i\xi_j \leq \Delta_0|\xi|^2, \qquad \forall x \in U \tag{3}$$

with Δ_0 a fixed positive number.

Though our work in [**Kl-Ro6**] covers only the second of the following two situations below, it applies in principle to both.

(1) The surfaces Σ_t are asymptotically flat and maximal.

$$\mathrm{tr}k = 0.$$

(2) The surfaces Σ_t are compact, of Yamabe type -1, and of constant, negative mean curvature. They form what is called a (CMC) foliation .

$$trk = t, \qquad t < 0$$

We may assume in what follows that the region $\mathcal{M}_I$ corresponds to the time interval $I = [t_0, t_*)$ with $t_* < 0$. Without loss of generality we shall identify the entire space-time $\mathbf{M}$ with $\mathcal{M}_I$. We can also assume that the initial hypersurface Σ_0 corresponds to $t_0 = -1$.

Given a t-foliation of $\mathbf{M}$ and $p \in \mathbf{M}$ we can define a point-wise norm $|\Pi(p)|$ of any space-time tensor Π via the decomposition,

$$X = -X^0\mathbf{T} + \underline{X}, \quad X \in T\mathbf{M}, \quad \underline{X} \in T\Sigma_t.$$

We denote by $\|\Pi(t)\|_{L^p}$ the L^p norm of Π on Σ_t. More precisely,

$$\|\Pi(t)\|_{L^p} = \int_{\Sigma_t} |\Pi|^p dv_g$$

with dv_g the volume element of the metric g of Σ_t. The main result I want to report is the following,

THEOREM 1.1 (Klainerman-Rodnianski). *Let $(\mathbf{M}, \mathbf{g})$ be a globally hyperbolic development of Σ_0 foliated by the level hypersurfaces of a time function $t < 0$, verifying conditions (1) or (2) above, such that Σ_0 corresponds to the level surface $t = t_0$. Assume that Σ_0 verifies* **A1**. *Then the first time $T_* < 0$ of a breakdown is characterized by the condition*

$$\limsup_{t \to T_*^-} \left(\|k(t)\|_{L^\infty} + \|\nabla \log n(t)\|_{L^\infty} \right) = \infty. \tag{4}$$

More precisely the space-time together with the foliation Σ_t can be extended beyond any value $t_* < 0$ for which[1],

$$\sup_{t \in [t_0, t_*)} \|k(t)\|_{L^\infty} + \|\nabla \log n(t)\|_{L^\infty} = \Delta_0 < \infty. \tag{5}$$

[1]For simplicity we use below the same constant Δ_0 as in (3).

Condition (5) can be reformulated in terms of the deformation tensor of the future unit normal $\mathbf{T}$, $\pi = {}^{(\mathbf{T})}\pi = \mathcal{L}_{\mathbf{T}}\mathbf{g}$. By a simple calculation, expressed relative to an orthonormal frame $e_0 = \mathbf{T}, e_1, e_2, e_3$, we find,

$$\pi_{00} = 0, \quad \pi_{0i} = n^{-1}\nabla_i n, \quad \pi_{ij} = -2k_{ij}. \tag{6}$$

Thus condition (5) can be interpreted as the requirement that $\mathbf{T}$ is an approximate Killing vectorfield in the following sense,

A2. There exists a constant Δ_0 such that,

$$\sup_{t\in[t_0,t_*)} \|\pi(t)\|_{L^\infty} \le \Delta_0. \tag{7}$$

In addition to the constant Δ_0 in **A1**, **A2** the constant $\mathcal{R}_0$, which bounds the L^2 norm of the spacetime curvature tensor $\mathbf{R}$ on Σ_0, plays an essential role,

$$\|\mathbf{R}(t_0)\|_{L^2(\Sigma_0)} \le \mathcal{R}_0. \tag{8}$$

To prove the theorem we have to show that if assumptions **A1** and **A2** are satisfied then the space-time $\mathcal{M}_I$, $I = [t_0, t_*)$, $t_* < 0$ can be extended beyond t_*. We want to emphasize that theorem 1.1 is a large data result; indeed one need not make any smallness assumptions on the constants Δ_0 and $\mathcal{R}_0$.

Our theorem is connected and partially motivated by the following three earlier breakdown criteria results:

1. The first is a result of M. Andersson, [**And**], who showed that a breakdown can be tied to the condition that

$$\lim \sup_{t\to t_*^-} \|\mathbf{R}(t)\|_{L^\infty} = \infty. \tag{9}$$

It is clear that condition (4) is formally weaker than (9) as it requires one degree less of differentiability. Moreover a condition on the boundedness of the L^∞ norm of $\mathbf{R}$ covers all the dynamical degree of freedom of the equations. Indeed, once we know that $\|\mathbf{R}(t)\|_{L^\infty}$ is finite, one can find bounds for n, ∇n and k purely by elliptic estimates.

2. Our result can be also compared to the well known Beale-Kato-Majda criterion for breakdown of solutions of the incompressible Euler equation

$$\partial_t v + (v\cdot\nabla)v = -\nabla p, \qquad \text{div } v = 0,$$

with smooth initial data at $t = t_0$. A routine application of the energy estimates shows that a solution v blows up if and only if

$$\int_{t_0}^{t_*} \|\nabla v(t)\|_{L^\infty} dt = \infty. \tag{10}$$

The Beale-Kato-Majda criterion improves the blow up criterion by replacing it with the following condition on the vorticity $\omega = \text{curl } v$:

$$\int_{t_0}^{t_*} \|\omega(t)\|_{L^\infty} dt = \infty. \tag{11}$$

Similarly, in the case of the Einstein equations energy estimates, expressed relative to a special system of coordinates (such as wave coordinates), show that breakdown does not occur unless

$$\int_{t_0}^{t_*} \|\partial \mathbf{g}(t)\|_{L^\infty} dt = \infty.$$

This condition however is not geometric as it depends on the choice of a full coordinate system. Observe that both the spatial derivatives of the lapse ∇n and the components of the second fundamental form, $k_{ij} = -\frac{1}{2} n^{-1} \partial_t g_{ij}$, can be interpreted as components of $\partial \mathbf{g}$. Note however that after prescribing k and ∇n we are still left with many more degrees of freedom in determining $\partial \mathbf{g}$.

3. Finally, the result whose proof is closest in spirit to ours and which has played the main motivating role in developing our approach, is the proof of global regularity of solutions of the Yang-Mills equations in $\mathbb{R}^{3+1}$ by Eardley and Moncrief, see [**EM1**]. To explain the connection of their result to ours we review below its main ideas.

Recall that the curvature tensor $\mathbf{F}_{\alpha\beta} = \partial_\alpha \mathbf{A}_\beta - \partial_\beta \mathbf{A}_\alpha + [A_\alpha, A_\beta]$ of a Yang Mills connection $\mathbf{A}_\alpha dx^\alpha$, with values in the Lie algebra $su(N)$ is a critical point of the Yang-Mills functional

$$YM[\mathbf{F}] = \int_{\mathbb{R}^{3+1}} \mathrm{Tr}\left({}^\star\mathbf{F} \wedge \mathbf{F} \right)$$

and verifies the wave equation,

$$\square_{(\mathbf{A})} \mathbf{F} = \mathbf{F} \star \mathbf{F}, \tag{12}$$

where

$$\square_{(\mathbf{A})} = \mathbf{m}^{\mu\nu} \mathbf{D}_\mu^{(\mathbf{A})} \mathbf{D}_\nu^{(\mathbf{A})}$$

denotes the covariant wave operator, $\mathbf{D}_\mu^{(\mathbf{A})} = \partial_\mu + [A_\mu, \cdot]$. Thus, with $\square$ the usual D'Alembertian in $\mathbb{R}^{3+1}$,

$$\begin{aligned} \square_{(\mathbf{A})} &= \partial^\alpha\left(\partial_\alpha \mathbf{F} + [\mathbf{A}_\alpha, \mathbf{F}]\right) + [\mathbf{A}^\alpha, \partial_\alpha \mathbf{F} + [\mathbf{A}_\alpha, \mathbf{F}]] \\ &= \square \mathbf{F} + [\mathbf{A}, \partial \mathbf{F}] + [\partial \mathbf{A}, \mathbf{F}] + [\mathbf{A}, [\mathbf{A}, \mathbf{F}]]. \end{aligned}$$

Since the Minkowski space-time metric $\mathbf{m} = -dt^2 + \delta_{ij} dx^i dx^j$ is static (in particular $n = 1$ and $k = 0$) the energy of $\mathbf{F}$ associated with the energy-momentum tensor $\mathbf{Q}[\mathbf{F}]_{\alpha\beta} = \mathbf{F}_\alpha{}^\lambda \mathbf{F}_{\beta\lambda} + {}^\star\mathbf{F}_\alpha{}^\lambda\, {}^\star\mathbf{F}_{\beta\lambda}$ and vectorfield $\mathbf{T} = \partial_t$ is conserved. In particular, the flux of energy $\mathcal{F}_p$ through the null boundary $\mathcal{N}^-(p)$ of the domain of dependence $\mathcal{J}^-(p)$ of an arbitrary point p can be bounded by the energy of the initial data which we denote by I_0. We assume that smooth data for F is prescribed at $t = 0$ and restrict $\mathcal{J}^-(p)$ and $\mathcal{N}^-(p)$ to $t \geq 0$. We recall that the flux has the form,

$$\mathcal{F}_p = \Big(\int_{\mathcal{N}^-(p)} \mathbf{Q}[\mathbf{F}](\mathbf{L}, \mathbf{T}) \Big)^{1/2}$$

with $\mathbf{L}$ the null geodesic generator of $\mathcal{N}^-(p)$ normalized by $< \mathbf{L}, \mathbf{T} >= 1$.

The proof of the global regularity of solutions of the Yang-Mills equations is based on the boundedness of the flux $\mathcal{F}_p \leq I_0 < \infty$. Here is a summary of the main steps.

(1) Rewrite (12) in the form $\Box \mathbf{F} = \mathbf{F} \star \mathbf{F} - (\Box_{(\mathbf{A})} - \Box)\mathbf{F}$. Using the explicit representation, in $\mathbb{R}^{3+1}$, of solutions to the inhomogeneous wave equation, we deduce, for all points p, with $t > 0$,

$$\begin{aligned} \mathbf{F}(p) &= \mathbf{F}^{(0)}(p;\delta_0) + (4\pi)^{-1} \int_{\mathcal{N}^-(p;\delta_0)} r^{-1}\mathbf{F}\star\mathbf{F} \\ &- (4\pi)^{-1} \int_{\mathcal{N}^-(p;\delta_0)} r^{-1}(\Box_{(\mathbf{A})} - \Box)\mathbf{F}. \end{aligned} \tag{13}$$

Here $\mathcal{N}^-(p,\delta_0)$ represents the portion of the null cone $\mathcal{N}^-(p)$ included in the time slab $[t(p)-\delta_0, t(p)]$, with $t(p)$ the value of the time parameter at p. Also r is the distance, in euclidean sense, to the vertex p and $\mathbf{F}^{(0)}(p;\delta_0)$ represents the homogeneous solution to the wave equation whose initial data at $t = t(p) - \delta_0$ coincide with those of $\mathbf{F}$.

(2) Consider the second term on the right hand side of (13). Using the explicit form of the nonlinear term $\mathbf{F}\star\mathbf{F}$ one notices that at least one factor of the product can be estimated by the flux $\mathcal{F}_p$ through the null hypersurface $\mathcal{N}^-(p)$. Denoting $|\mathbf{F}| = \sum_{\alpha\beta} |\mathbf{F}_{\alpha\beta}|$, we have by a simple estimate[2],

$$\begin{aligned} |(\mathbf{F}(p) - \mathbf{F}^{(0)}(p;\delta_0)| &\lesssim \mathcal{F}_p\Big(\int_{\mathcal{N}^-(p;\delta_0)} r^{-2}\Big)^{1/2} \|\mathbf{F}\|_{L^\infty(\mathcal{J}^-(p;\delta_0))} \\ &\lesssim \delta_0^{1/2}\mathcal{F}_p\|\mathbf{F}\|_{L^\infty(\mathcal{J}^-(p,\delta_0))}, \end{aligned}$$

where $\|\mathbf{F}\|_{L^\infty(\mathcal{J}^-(p,\delta_0))}$ denotes the sup- norm of $|\mathbf{F}|$ for all points in the domain of dependence $\mathcal{J}^-(p)$ of p in the slab $[t(p)-\delta_0, t(p)]$ and $t_0 = t(p) > 0$ denotes the value of t at p. Therefore, we deduce the following,

LEMMA 1.2. *If $\delta_0^{1/2} \cdot \mathcal{F}_p$ is sufficiently small, then for any $t \geq 0$*

$$\|\mathbf{F}(t)\|_{L^\infty} \lesssim \|\mathbf{F}(t-\delta_0)\|_{L^\infty} + \|\mathbf{D}\mathbf{F}(t-\delta_0)\|_{L^\infty}. \tag{14}$$

(3) Arguing recursively and using the standard local existence theorem for the Yang-Mills system one can find bounds for all components of the curvature tensor $\mathbf{F}(p)$ depending only on the fact that $\mathcal{F}_p$ is uniformly bounded and the initial data $\mathbf{F}(0)$ is smooth.

(4) One can show that (14) remains true even as we take into consideration the presence of the third term in (13). This is done by choosing a specified gauge condition for A called the Crönstrom gauge. That is, one assumes that the connection 1-form A satisfies $(x-y)^\alpha A_\alpha = 0$, where x^α are the space-time coordinates of p and y^α those of a point $q \in \mathcal{N}^-(p)$. One can use this condition to derive uniform estimates of $\partial\mathbf{A}$ in terms of $\mathbf{F}$.

(5) In [**Kl-Ma**] the global regularity result was reproved by strengthening the classical local existence result to $A \in H^1(\mathbb{R}^3)$ and $E \in L^2(\mathbb{R}^3)$, which is at the same regularity level as the energy norm. That required, instead of the pointwise estimates (14), a new generation of L^4 type estimates, called bilinear. The premise of the [**Kl-Ma**] approach was the fact that, once we have a local existence result which depends only on the energy norm of the initial data, global existence can be easily derived by a simple continuation argument.

---

[2] Here and in what follows we denote by $A \lesssim B$ any inequality of the form $A \leq cB$, where $c > 0$ is a universal constant .

In what follows we give a short summary of how the mains ideas in the proof of the Eardley-Moncrief result for Yang-Mills can be adapted to General Relativity.

(1) One can easily show that the curvature tensor $\mathbf{R}$ of a $3+1$ dimensional vacuum spacetime $(\mathbf{M}, \mathbf{g})$, see (1), verifies a wave equation of the form,

$$\square_{\mathbf{g}} \mathbf{R} = \mathbf{R} \star \mathbf{R} \tag{15}$$

where $\square_{\mathbf{g}}$ denotes the covariant wave operator $\square_{\mathbf{g}} = \mathbf{D}^\alpha \mathbf{D}_\alpha$.

(2) Recall that the Bel-Robinson energy-momentum tensor has the form

$$\mathbf{Q}[\mathbf{R}]_{\alpha\beta\gamma\delta} = \mathbf{R}_{\alpha\lambda\gamma\mu} \mathbf{R}_{\beta\ \delta}^{\ \lambda\ \mu} + {}^\star\mathbf{R}_{\alpha\lambda\gamma\mu}\, {}^\star\mathbf{R}_{\beta\ \delta}^{\ \lambda\ \mu}.$$

and verifies, $\mathbf{D}^\delta \mathbf{Q}_{\alpha\beta\gamma\delta} = 0$. It can thus be used to derive energy and flux estimates for the curvature tensor $\mathbf{R}$. As opposed to the case of the Yang-Mills theory, however, in General Relativity the background metric is dynamic and thus does not admit, in general, Killing fields (and in particular a time-like Killing field). This means that we can not associate conserved quantities to a divergence free Bel-Robinson tensor. It is at this point where we need crucially our bounded deformation tensor condition **A2**. Indeed that condition suffices to derive bounds for both energy and flux associated to the curvature tensor $\mathbf{R}$. Using the Bel-Robinson energy momentum tensor $\mathbf{Q}$ the energy associated to a slice Σ_t is defined by the integral

$$\mathcal{E}(t) = \int_{\Sigma_t} \mathbf{Q}[\mathbf{R}](\mathbf{T}, \mathbf{T}, \mathbf{T}, \mathbf{T}), \tag{16}$$

while the flux, through the null boundary $\mathcal{N}^-(p)$ of the domain of dependence (or causal past) $\mathcal{J}^-(p)$ of a point p, is given by the integral

$$\mathcal{F}^-(p) = \Big(\int_{\mathcal{N}^-(p)} \mathbf{Q}[\mathbf{R}](\mathbf{L}, \mathbf{T}, \mathbf{T}, \mathbf{T}), \Big)^{\frac{1}{2}} \tag{17}$$

where $\mathbf{L}$ is the null geodesic generator of $\mathcal{N}^-(p)$ normalized at the vertex p by $< \mathbf{L}, \mathbf{T} >= 1$.

As in the case of the Yang-Mills equations, it is precisely the boundedness of the flux of curvature that plays a crucial role in our analysis. In General Relativity the flux takes on even more fundamental role as it is also needed to control the geometry of the very object it is defined on, i.e. the boundary of the causal past of p. This boundary, unlike in the case of Minkowski space, are not determined a-priori but depend in fact on the space-time we are trying to control.

(3) In the construction of a parametrix for (15) we cannot, in any meaningful way, approximate $\square_{\mathbf{g}}$ by the flat D'Alembertian $\square$. To deduce a formula analogous to (14) one might try to proceed by the geometric optics construction of parametrices for $\square_{\mathbf{g}}$, as developed in [**Fried**]. Such an approach would require additional bounds on the background geometry, determined by the metric $\mathbf{g}$, incompatible with the limited assumption **A2** and the implied finiteness of the curvature flux. We rely instead on a geometric version, which we develop in [**Kl-Ro5**], of the Kirchoff-Sobolev formula, similar to that used by Sobolev in [**Sob**] and Y. C. Bruhat in [**Br**], see also [**M**]. Roughly, this can be obtained by applying to (15)

the measure $\mathbf{A}\delta(u)$, where u is an optical function[3] whose level set $u = 0$ coincides with $\mathcal{N}^-(p)$ and $\mathbf{A}$ is a 4-covariant 4-contravariant tensor defined as a solution of a transport equation along $\mathcal{N}^-(p)$ with appropriate (blowing-up) initial data at the vertex p. After a careful integration by parts we arrive at the following analogue of the formula (14):

$$\mathbf{R}(p) = \mathbf{R}^0(p;\delta_0) - \int_{\mathcal{N}^-(p;\delta_0)} \mathbf{A}\cdot(\mathbf{R}\star\mathbf{R}) + \int_{\mathcal{N}^-(p;\delta_0)} \mathrm{Err}\cdot\mathbf{R}, \tag{18}$$

where $\mathcal{N}^-(p;\delta_0)$ denotes the portion of the null boundary $\mathcal{N}^-(p)$ in the time interval $[t(p)-\delta_0, t(p)]$ and the error term Err depends only on the extrinsic geometry of $\mathcal{N}^-(p;\delta_0)$. The term $\mathbf{R}^0(p;\delta_0)$ is completely determined by the initial data on the hypersurface $\Sigma_{t(p)-\delta_0}$. As in the flat case [4], one can prove bounds for the sup-norm of $\mathbf{R}^0(p;\delta_0)$ which depend only on uniform bounds for $\mathbf{R}$ and its first covariant derivatives at values of $t' \le t(p) \le t-\delta$.

(4) As in the case of Yang-Mills, the structure of the term $\mathbf{R}\star\mathbf{R}$ allows us to estimate one of the curvature terms by the flux of curvature:

$$\begin{aligned} \Big|\int_{\mathcal{N}^-(p;\delta_0)} \mathbf{A}\cdot(\mathbf{R}\star\mathbf{R})\Big| &\lesssim \mathcal{F}^-(p)\cdot\|\mathbf{R}\|_{L^\infty(\mathcal{N}^-(p;\delta_0))}\cdot\|\mathbf{A}\|_{L^2(\mathcal{N}^-(p;\delta_0))} \\ &\lesssim \delta_0^{1/2}\cdot\mathcal{F}^-(p)\cdot\|\mathbf{R}\|_{L^\infty(\mathcal{N}^-(p;\delta_0))}, \end{aligned} \tag{19}$$

provided that,

$$\|\mathbf{A}\|_{L^2(\mathcal{N}^-(p;\delta_0))} \lesssim \delta_0^{1/2}. \tag{20}$$

Neglecting, for a moment, the third integral in (18) one may thus expect to prove a result analogous to that of Lemma 1.2.

THEOREM 1.3. *There exists a sufficiently small $\delta_0 > 0$ and a large constant C, depending only on Δ_0 in assumptions* **A1** *and* **A2** *as well as $\mathcal{R}_0$ in (8) such that for all $t_0 \le t < t_*$,*

$$\|\mathbf{R}(t)\|_{L^\infty} \lesssim C \sup_{t\le t-\delta}\big(\|\mathbf{R}(t)\|_{L^\infty} + \|\mathbf{DR}(t)\|_{L^\infty}\big). \tag{21}$$

(5) The proof of (19) depends on verifying (20). In addition, to estimate the third term in (18), we need to provide estimates for tangential derivatives of $\mathbf{A}$ and other geometric quantities associated to the null hypersurfaces $\mathcal{N}^-(p)$. In particular, it requires showing that $\mathcal{N}^-(p)$ remains a *smooth* (not merely Lipschitz) hypersurface in the time slab $(t(p)-\delta_0, t(p)]$ for some with dependent only on the constants Δ_0 and $\mathcal{R}_0$. Thus to prove the desired theorem we have to show that all geometric quantities, arising in the parametrix construction, can be estimated only in terms of the flux of the curvature $\mathcal{F}_p^-$ along $\mathcal{N}^-(p)$ and our main assumption **A1**. Yet, to start with, it is not even clear that we can provide a lower bound for the radius of injectivity of $\mathcal{N}^-(p)$. In other words the congruence of null geodesics, initiating at p, may not be controllable[5] only in terms of the

[3] That is $\mathbf{g}^{\alpha\beta}\partial_a u\partial_\beta u = 0$.

[4] This is by no means obvious as we need to rely once more on the Kirchoff-Sobolev formula.

[5] Different null geodesics of the congruence may intersect, or the congruence itself may have conjugate points, arbitrarily close to p.

curvature flux. Typically, in fact, lower bounds for the radius of conjugacy of a null hypersurface in a Lorentzian manifold are only available in terms of the sup-norm of the curvature tensor $\mathbf{R}$ along the hypersurface, while the problem of short, intersecting, null geodesics appears not to be fully understood even in that context. The situation is similar to that in Riemannian geometry, exemplified by the Cheeger's theorem, where pointwise bounds on sectional curvature are sufficient to control the radius of conjugacy but to prevent the occurrence of short geodesic loops one needs to assume in addition an upper bound on the diameter and a lower bound on the volume of the manifold.

In a sequence of papers, [**Kl-Ro1**]–[**Kl-Ro3**], we have been able to prove a lower bound, depending essentially only on the curvature flux, for the radius of conjugacy of null hypersurfaces[6] in a Lorentzian spacetime which verifies the Einstein vacuum equations. The methods used in these articles can be adapted to provide all the desired estimates, except a lower bound on the "size" of intersecting null geodesics which needs a separate argument. The lower bound on the radius of injectivity of the null hypersurfaces $\mathcal{N}^-(p)$, needed here, is discussed in [**Kl-Ro4**]

(6) As in the case of Yang-Mills equations one can use the result of Lemma (1.3), together with the classical local existence result for the Einstein equations to show that solutions can be extended as long the bounds on ${}^{(\mathbf{T})}\pi$ hold true.

(7) Finally we observe that our construction is gauge invariant, as it does not require a gauge condition such as that of Eardley-Moncrief, but depends instead on an appropriate tensorial definition of the transport term $\mathbf{A}$ along $\mathcal{N}^-(p)$. This in fact suggest a possibility of reproving the Eardley-Moncrief result itself with the help of a carefully constructed *covariant* Hadamard parametrix for $\square_A$. This was in fact implemented in [**Kl-Ro5**].

The proof of Theorem 1.1 contains the following main steps.

1. Curvature L^2-bound . One shows, based on condition **A2**, that the flux of curvature $\mathcal{R}_0$, through the null boundary $\mathcal{N}^-(p)$ of the causal past $\mathcal{J}^-(p)$ of any point p to the future of Σ_0, must remain bounded by a universal constant. This follows quite easily from energy estimates obtained with the help of the Bel-Robinson tensor $\mathbf{Q}$ defined above.

2. Higher derivatives bounds. One shows how to derive L^2 bounds for higher covariant derivatives of $\mathbf{R}$ by using condition **A2** as well as an auxiliary condition on the L^∞ norm of the curvature tensor $\mathbf{R}$. The proof, though quite involved, is rather standard. It is based on higher energy estimates and L^2 elliptic theory along the leaves Σ_t of the time foliation.

3. Curvature L^∞-bound This is the main step in the proof. One shows how to transform the L^2 bound of the curvature flux established in Step 1 into a pointwise bound. This require a geometric, approximate, representation formula for solutions to the tensorial equation (15). The representation formula (18), with a precise

[6]together with many other estimates of various geometric quantities associated to $\mathcal{N}^-(p)$.

expression for the error term Err, was developed in [**Kl-Ro5**]. In applying formula (18) to prove Theorem 1.3 one needs to have good control on the regularity of the null hypersurfaces $\mathcal{N}^-(p)$. More precisely we need,

3a. Lower bound for the radius of injectivity of the null hypersurfaces $\mathcal{N}^-(p)$. Null hypersurfaces can develop singularities, called caustics, even in flat space. The presence of caustics is detected by an important invariant, called the null second fundamental form,

$$\chi(X,Y) = \mathbf{g}(\mathbf{D}_X\mathbf{L}, Y) \tag{22}$$

Here X, Y are arbitrary vectorfields tangent to the null hypersurface while $\mathbf{L}$ is the null geodesic generator of the hypersurface. According to the Jacobi identity we have,

$$\mathbf{D}_{\mathbf{L}}\chi + \chi\cdot\chi = \mathbf{R}(\cdot,\mathbf{L},\cdot,\mathbf{L}) \tag{23}$$

The trace $\mathrm{tr}\chi$ of χ verifies the well known Raychadhouri equations,

$$\mathbf{L}(\mathrm{tr}\chi) + \frac{1}{2}\mathrm{tr}\chi^2 + |\hat{\chi}|^2 = 0 \tag{24}$$

where $\hat{\chi}$ is the traceless part of χ, see [**Kl-Ro1**] for precise definitions. From (24) it is easy to see that if $\mathrm{tr}\chi$ is negative at some point then it must become infinite along the null geodesic passing through that point, independently on the size of the curvature tensor. This corresponds to a conjugate point. To avoid conjugate points along $\mathcal{N}^-(p)$ we need to show that $\mathrm{tr}\chi$ remains close to its flat value $\frac{2}{s}$, where s is the afine parameter such that $s(p) = 0$, for a uniform interval $s \in [0, \delta]$, with δ independent of p. To do this we need to be able to get a uniform bound for the integrals along past null geodesics from p, $\int_0^\delta |\hat{\chi}|^2$. Observe that equation (23) does not allow one to obtain such an estimate since it seems to require pointwise bounds for $\mathbf{R}$. In the next section we shall discuss our main ideas for avoiding this difficulty.

3b. Higher derivative estimates for $\mathcal{N}^-(p)$. It is not enough to control $\mathrm{tr}\chi$ in the uniform norm we also need to control up to 2 derivatives of χ tangential to $\mathcal{N}^-(p)$. This can be done by using an extension of the ideas discussed in the next section.

To finish this section we would like to point out possible refinements of our main Theorem 1.1. We expect that one should be able to replace the pointwise condition **A2** with the integral condition,

$$\int_{t_0}^{t_*} \|\pi(t)\|^2_{L^\infty} dt < \infty \tag{25}$$

Moreover it may be possible to improve the result even further by eliminating the term $\nabla \log n$ in (4) or (25) and requiring instead only a pointwise bound on n.

2. Lower bound for the radius of injectivity of the null hypersurfaces $\mathcal{N}^-(p)$

Null hypersurfaces in a Lorentzian space can be viewed as level hypersurfaces of an optical function u, i.e. a solution to the Eikonal equations

$$\mathbf{g}^{\alpha\beta}\partial_\alpha u \partial_\beta u = 0. \tag{26}$$

Let $L = -\mathbf{g}^{\alpha\beta}\partial_a u\, \partial_b$ be the corresponding null generator vectorfield and s its affine parameter, i.e. $L(s) = 1, \quad s|_{S_0} = 0$ where $S_0 \subset \Sigma$ is a given, initial, two dimensional compact surface which we assume to be diffeomorphic to the standard sphere. The level surfaces S_s of s generates the geodesic foliation on $\mathcal{H}$. We denote by ∇ the covariant differentiation on S_s and by ∇_L the projection to S_s of the covariant derivative with respect to L. At every point of $p \in \mathcal{H}$ we denote by $\underline{L}$ to be the unique null vector which is orthogonal to the S_s sphere passing through p and such that $\mathbf{g}(L, \underline{L}) = -2$. We introduce the total curvature flux[7] along $\mathcal{H}$ to be the integral,

$$\mathcal{R}_0 = \left(\|\alpha\|^2_{L^2(\mathcal{H})} + \|\beta\|^2_{L^2(\mathcal{H})} + \|\rho\|^2_{L^2(\mathcal{H})} + \dots \right)^{\frac{1}{2}} \tag{27}$$

with $\alpha, \beta, \rho, \dots$ null components of the curvature tensor $\mathbf{R}$. More precisely, for any vector-fields X, Y on $\mathcal{H}$ tangent to the 2-surfaces S_s

(28)

$$\alpha(X,Y) = \mathbf{R}(X, L, Y, L), \quad \beta(X) = \frac{1}{2}\mathbf{R}(X, L, \underline{L}, L), \quad \rho = \frac{1}{4}\mathbf{R}(L, \underline{L}, L, \underline{L}) \dots$$

The geometry of $\mathcal{H}$ depends in particular on the null second fundamental form

$$\chi(X,Y) = < D_X L \,, Y > \tag{29}$$

with X, Y arbitrary vector-fields tangent to the s-foliation. We denote by $\mathrm{tr}\chi$ the trace of χ, i.e. $\mathrm{tr}\chi = \delta^{ab}\chi_{ab}$ where χ_{ab} are the components of χ relative to an orthonormal frame $(e_a)_{a=1,2}$ on S_s. In view of the Einstein equations we have,

$$\frac{d}{ds}\mathrm{tr}\chi + \frac{1}{2}(\mathrm{tr}\chi)^2 = -|\hat{\chi}|^2 \tag{30}$$

with $\hat{\chi}_{ab} = \chi_{ab} - \frac{1}{2}\mathrm{tr}\chi\delta_{ab}$ the traceless part of χ.

2.1. Conjugate points.

Even in Minkowski space one cannot adequately control the geometry of null hypersurfaces without a uniform bound on, at least, $\mathrm{tr}\chi$. Indeed one can show by an explicit calculation that if the L^∞ norm of $\mathrm{tr}\chi$ is allowed to become arbitrarily large bad singularities, called caustics, occur. To avoid them we are led to ask whether we can bound the L^∞ norm of $\mathrm{tr}\chi$ simply in terms of an initial data norm[8] $\mathcal{I}_0$ and total curvature flux $\mathcal{R}_0$. At first glance this seems impossible. Indeed, by integrating (30), in order to control $\|\mathrm{tr}\chi\|_{L^\infty}$ we need

[7] The justification for this quantity can be found in [**C-K**] in connection to the Bell Robinson tensor and the Bianchi identities.

[8] Which contains informations about S_0.

to control uniformly the integrals $\int_\Gamma |\hat{\chi}|^2$, along all null geodesic generators Γ of $\mathcal{H}$. On the other hand $\hat{\chi}$ verifies a transport equation of the form,

$$\frac{d}{ds}\hat{\chi} + \frac{1}{2}\mathrm{tr}\chi \cdot \hat{\chi} = -\alpha. \tag{31}$$

where α, the null curvature component defined by (28), is only L^2 integrable on $\mathcal{H}$. Thus, unless there is a miraculous gain of a spatial derivative(along the surfaces of the s- foliation) of the integrals of α along the null geodesic generators Γ, we have no chance to control the L^∞ norm of $\mathrm{tr}\chi$. Fortunately this cancellation occurs and it is best seen by observing that $\hat{\chi}$ verifies a Codazzi type equation of the form,

$$\mathrm{div}\ \hat{\chi} = -\beta + \frac{1}{2}\nabla\mathrm{tr}\chi + \ldots \tag{32}$$

with β as defined above. One can show that div $\hat{\chi}$ defines an elliptic system on S_s and therefore we expect that $\hat{\chi}$ behaves like $\mathcal{D}^{-1}(-\beta + \frac{1}{2}\nabla\mathrm{tr}\chi)$ with $\mathcal{D}^{-1}$ a pseudo-differential operator of order -1. We are thus led to control, uniformly, the integrals

$$I_1 = \int_\Gamma |\mathcal{D}^{-1}\beta|^2, \quad I_2 = \int_\Gamma |\mathcal{D}^{-1}\cdot\nabla\mathrm{tr}\chi|^2. \tag{33}$$

Consider first I_1. Can we estimate it in terms of the curvature component $\|\beta\|_{L^2(\mathcal{H})}$?. The idea is to interpret this as as a restriction problem. This makes sense dimensionally and it corresponds to a trace type inequality of the type $\|U\|_{L^2(\Gamma)} \lesssim \|\nabla U\|_{L^2(\mathcal{H})}$, applied to the tensor $U = \mathcal{D}^{-1}\beta$. Unfortunately it is well known that this type of trace theorems, just as the sharp Sobolev embedding into L^∞, fail to be true. We might be able to overcome this difficulty if we could write $\beta = \nabla_L Q$ with Q a tensor verifying[9] $\|\bar{\nabla}^2 Q\|_{L^2(\mathcal{H})} \lesssim \mathcal{R}_0$. Indeed such a sharp trace theorem holds true in flat space[10]. The breakthrough, which allows us to make use of an appropriately adapted version of the sharp trace theorem mentioned above, is provided by the Bianchi identities which, expressed relative to our null pair L, $\underline{L}$, takes the form of a system of first order equations connecting the L derivatives of the null components $\alpha, \beta, \rho, \sigma \ldots$ to their spatial derivatives ∇. In particular we have, ignoring the quadratic and higher order terms[11] ,

$$\mathrm{div}\ \beta = L(\rho) + \ldots, \quad \mathrm{curl}\ \beta = -L(\sigma) + \ldots \tag{34}$$

Once more this is an elliptic Hodge system in β and we can write, formally

$$\mathcal{D}^{-1}\beta = \mathcal{D}^{-2}L(\rho, -\sigma) + \ldots = \nabla_L\big(\mathcal{D}^{-2}(\rho, -\sigma)\big) + [\mathcal{D}^{-2}, \nabla_L](\rho, -\sigma) + \ldots,$$

with $\mathcal{D}^{-2}$ a pseudo-differential operator of order -2. Ignoring for the moment the commutator $[\mathcal{D}^{-2}, \nabla_L](\rho, -\sigma)$ and the other error terms, we have indeed

$$\mathcal{D}^{-1}\beta = \nabla_L Q + \ldots, \qquad Q = \mathcal{D}^{-2}(\rho, -\sigma), \tag{35}$$

and one can check, with the help of L^2 elliptic estimates, that $\|\bar{\nabla}^2 Q\|_{L^2(\mathcal{H})}$ can indeed be bounded by the curvature flux $\mathcal{R}_0$.

This circle of ideas seem to take care of the integral I_1 in (33). Unfortunately we hit another serious obstacle with I_2. The problem is that the operator $\mathcal{D}^{-1}\cdot\nabla$

[9]We denote by $\bar{\nabla} = (\nabla, \nabla_L)$ all first derivatives on $\mathcal{H}$ and by $\bar{\nabla}^2 Q$ all second derivatives.

[10]That is the estimate $\big(\int_\Gamma |\nabla_L Q|^2\big)^{\frac{1}{2}} \lesssim \|\bar{\nabla}^2 Q\|_{L^2(\mathcal{H})} + \|Q\|_{L^2(\mathcal{H})}$ holds true.

[11]It turns out in fact that the ignored terms are not so easy to treat as they required various renormalizations and delicate estimates.

is a nonlocal operator of order zero and therefore does not map L^∞ into L^∞. To overcome this difficulty we were forced to try to prove a stronger estimate for $\mathrm{tr}\chi$. The idea is to try to prove the boundedness of $\mathrm{tr}\chi$ not only in L^∞ but rather in a Besov space of type $B^1_{2,1}(S_s)$ which both imbeds in $L^\infty(S_s)$ and is stable relative to operators of order zero. This simple, unavoidable fact, forces us to work with spaces defined by Littlewood-Paley projections which adds a lot of technical difficulties[12]. Moreover the standard LP- theory, based on Fourier transform, would require the use of local coordinates on $\mathcal{H}$. The most natural coordinates, those transported by the hamiltonian flow generated by L, are not sufficiently regular. Thus we are forced to rely on an invariant, geometric, version of the LP-theory which we develop, together with an appropriate paradifferential calculus, in [**Kl-Ro2**] by following a heat flow approach. An informal introduction to the sharp trace type theorems in Besov spaces needed in our work and the geometric LP theory on which they rely is given in [**Kl-Ro3**].

2.2. Cut Locus points. The circle of ideas discussed above will only give us a lower bound for the radius of conjugacy of the null hypersurfaces $\mathcal{N}^-(p)$. It is however possible that the radius of conjugacy of the null congruence is bounded from below and yet there are past null geodesics from a point p intersecting again at points arbitrarily close to p. Indeed, this can happen on a Lorentzian manifold with a thin neck, i.e. which looks locally like $\mathbf{M} = \mathbb{S}^1 \times \mathbb{R}^2 \times \mathbb{R}$. Clearly on such a flat Lorentzian manifold $\mathbf{M}$ there can be no conjugate points for the congruence of past or future null geodesics from a point and yet there are plenty of distinct null geodesics from a point p in $\mathbf{M}$ which intersect on a time scale proportional to the size of the neck. There can be thus no lower bounds on the null radius of injectivity expressed only in terms of bounds for the curvature tensor $\mathbf{R}$. This problem occurs, of course, also in Riemannian geometry where we can control the radius of conjugacy in terms of uniform bounds for the curvature tensor, yet, in order to control the radius of injectivity we need to make other geometric assumptions such as, in the case of compact Riemannian manifolds, lower bounds on volume and upper bounds for its diameter. It should thus come as no surprise that we also need, in addition to bounds for the curvature flux, other assumptions on the geometry of solutions to the Einstein equations in order to ensure control on the null cut-locus of points in $\mathbf{M}$ and obtain lower bounds for the null radius of injectivity. This is discussed in [**Kl-Ro4**]

3. Gauge Invariant Yang-Mills theory

The methods developed in connection with theorem 1.1 can be used to give a straightforward gauge invariant proof of the global existence result of Eardley-Moncrief mentioned above. Assume that $\mathbf{A}$ is a Yang-Mills connection in the 4-dimensional Minkowski space, i.e., it verifies the equations

$$\mathbf{D}^\beta_{(\mathbf{A})}\mathbf{F}_{\alpha\beta} = 0. \tag{36}$$

where $\mathbf{F}$ denotes the curvature of the connection.

[12] It requires, in particular, a sharp trace theorem in Besov spaces in a rough background.

As mentioned earlier the approach of Eardley-Moncrief was based on the fundamental solution for a scalar wave equation in Minkowski space (Kirchoff formula) and made use of Cronström gauges. In [**Kl-Ro5**] we present a *gauge independent* proof which we review below. The main new ingredient is the use of a gauge covariant first order Kirchoff-Sobolev parametrix. Recall that $\mathbf{F}$ verifies the gauge invariant wave equation,

$$\square^{(\mathbf{A})}\mathbf{F}_{\alpha\beta} = 2[\mathbf{F}^{\sigma}{}_{\alpha}\,,\,\mathbf{F}_{\sigma\beta}] \tag{37}$$

The proof is based on the following Kirchoff-Sobolev parametrix formula, see [**Kl-Ro5**],

$$\begin{aligned} 4\pi < \mathbf{F}(p), \mathbf{J} > &= -2\int_{\mathcal{N}^-(p;\delta_0)} < \mathbf{W}, [\mathbf{F},\mathbf{F}] > -\frac{1}{2}\int_{\mathcal{N}^-(p;\delta_0)} < [\mathbf{F}_{\underline{\mathbf{L}}\mathbf{L}}, \mathbf{W}], \mathbf{F} > \\ &+ \int_{\mathcal{N}^-(p;\delta_0)} < \Delta^{(\mathbf{A})}\mathbf{W}, \mathbf{F} > +\mathcal{I}_0 \end{aligned} \tag{38}$$

where

$$\mathcal{I}_0 = \int_{\mathcal{N}^-(p)\cap\Sigma_{t(p)-\delta_0}} (< \mathbf{W}, \mathbf{D}_{\mathbf{T}}\mathbf{F} > - < \mathbf{D}_{\mathbf{T}}\mathbf{A}, \mathbf{F} >) \tag{39}$$

depends only on initial conditions on $\Sigma_{t(p)-\delta_0}$. Here $\mathbf{J}$ is an arbitrary $\mathcal{G}$ valued anti-symmetric 2-tensor on $\mathbb{R}^{3+1}$, $\mathbf{W}$ is a $\mathcal{G}$ valued 2-form on $\mathbb{R}^{3+1}$ verifying the gauge invariant transport equation [13]

$$\mathbf{D}_{\mathbf{L}}\mathbf{W} + r^{-1}\mathbf{W} = 0, \qquad (r\mathbf{W})|_{r=0} = \mathbf{J}, \tag{40}$$

$<,>$ denotes the positive definite invariant scalar product on the Lie algebra and $\Delta^{(\mathbf{A})}$ denotes the tangential gauge invariant Laplacian. The first term in (38)

$$I = \int_{\mathcal{N}^-(p;\delta_0)} < \mathbf{W}, [\mathbf{F},\mathbf{F}] >,$$

can be estimated precisely as before in terms of the flux $\mathcal{F}_p$ and the sup norm of $\mathbf{F}$,

$$|I| \;\le\; \delta_0^{1/2}\mathcal{F}_p\|\mathbf{F}\|_{L^\infty(\mathcal{J}^-(p,\delta_0))}.$$

The second and third terms

$$II = \int_{\mathcal{N}^-(p;\delta_0)} < [\mathbf{F}_{\underline{\mathbf{L}}\mathbf{L}}, \mathbf{W}], \mathbf{F} >, \qquad III = \int_{\mathcal{N}^-(p;\delta_0)} < \Delta^{(\mathbf{A})}\mathbf{W}, \mathbf{F} >$$

can be estimated by,

$$\begin{aligned} |II| &\lesssim \mathcal{F}_p\|\mathbf{W}\|_{L^2(\mathcal{J}^-(p,\delta_0))}\|\mathbf{F}\|_{L^\infty(\mathcal{J}^-(p,\delta_0))}, \\ III &\lesssim \|\mathbf{F}\|_{L^\infty(\mathcal{J}^-(p,\delta_0)}\|\Delta^{(\mathbf{A})}\mathbf{W}\|_{L^1(\mathcal{J}^-(p,\delta_0)}. \end{aligned}$$

We estimate $\|\mathbf{W}\|_{L^2(\mathcal{J}^-(p,\delta_0))}$ and $\|\Delta^{(\mathbf{A})}\mathbf{W}\|_{L^1(\mathcal{J}^-(p,\delta_0)}$ using the transport equation (40). In fact we show the following[14],

$$\|\mathbf{W}\|_{L^2(\mathcal{J}^-(p,\delta_0))} \;\lesssim\; \delta_0^{1/2}|\mathbf{J}| \tag{41}$$

$$\|\mathbf{W}\|_{L^1(\mathcal{J}^-(p,\delta_0)} \;\lesssim\; \delta_0^{1/2}|\mathbf{J}|\mathcal{F}_p \tag{42}$$

---

[13]Here $\mathbf{L} = \partial_r - \partial_t$, $\mathbf{D}_{\mathbf{L}} = \partial_r - \partial_t + [\mathbf{A}_{\mathbf{L}}, \cdot]$ and $r = |y|$.

[14]In fact the estimate for $\Delta^{(\mathbf{A})}\mathbf{W}$ is a lot more complicated as it leads to a logarithmic divergence, but we should ignore this here.

The second estimate is particularly dangerous since it seems to depend on derivatives of $\mathbf{F}$. Indeed, introducing $\bar{\mathbf{W}} = r\mathbf{W}$, we rewrite (40) in the form,

$$\mathbf{D}_{\mathbf{L}}^{(\mathbf{A})}\bar{\mathbf{W}} = 0, \qquad \bar{\mathbf{W}}|_{r=0} = \mathbf{J}.$$

Commuting the transport equation $\mathbf{D}_{\mathbf{L}}^{(\mathbf{A})}\bar{\mathbf{W}} = 0$ with $r^2\Delta^{(\mathbf{A})}$ we obtain,

$$\mathbf{D}_{\mathbf{L}}^{(\mathbf{A})}(r^2\Delta^{(\mathbf{A})}\bar{\mathbf{W}}) = r^2\,[\mathbf{F}_{\mathbf{L}}{}^{a}, \nabla_a^{(\mathbf{A})}\bar{\mathbf{W}}] + r^2\,\nabla_a^{(\mathbf{A})}[\Lambda_{\mathbf{L}}{}^{a}, \bar{\mathbf{W}}].$$

We also have the equation

$$\mathbf{D}_{\mathbf{L}}^{(\mathbf{A})}(r\nabla_a^{(\mathbf{A})}\bar{\mathbf{W}}) = r\,[\mathbf{F}_{\mathbf{L}a}, \bar{\mathbf{W}}].$$

We combine these equations into the system:

$$\begin{aligned}
&\mathbf{D}_{\mathbf{L}}^{(\mathbf{A})}\bar{\mathbf{W}} = 0,\\
&\mathbf{D}_{\mathbf{L}}^{(\mathbf{A})}(r\nabla_a^{(\mathbf{A})}\bar{\mathbf{W}}) = r\,[\mathbf{F}_{\mathbf{L}a}, \bar{\mathbf{W}}],\\
&\mathbf{D}_{\mathbf{L}}^{(\mathbf{A})}(r^2\Delta^{(\mathbf{A})}\bar{\mathbf{W}}) = 2r^2\,[\mathbf{F}_{\mathbf{L}}{}^{a}, \nabla_a^{(\lambda)}\bar{\mathbf{W}}] + r^2\,[\nabla_a^{(\mathbf{A})}\mathbf{F}_{\mathbf{L}}{}^{a}, \bar{\mathbf{W}}].
\end{aligned} \tag{43}$$

Observe indeed that the last term on the right hand side of (43) contains covariant derivatives of $\mathbf{F}$. Thus estimate (42) seems at first glance impossible. To avoid this problem we need to take into account the fact that the integral along a null geodesic of $r^2\,[\nabla_a^{(\mathbf{A})}\mathbf{F}_{\mathbf{L}}{}^{a}, \bar{\mathbf{W}}]$ compensates for the loss of derivative. This can only be done by using once more the Bianchi identities.

References

[And] M. Andersson, *Regularity for Lorentz metrics under curvature bounds*,arXiv:gr-qc/020907 v1, Sept 20, 2002.

[Br] Y. Choquét-Bruhat, *Theoreme d'Existence pour certains systemes d'equations aux derivees partielles nonlineaires.*, Acta Math. **88** (1952), 141-225.

[C-K] D. Christodoulou, S. Klainerman, *The global nonlinear stability of the Minkowski space*, Princeton Math. Series **41**, 1993.

[EM1] D. Eardley, V. Moncrief, *The global existence of Yang-Mills-Higgs fields in 4-dimensional Minkowski space. I. Local existence and smoothness properties.* Comm. Math. Phys.**83** (1982), no. 2, 171–191.

[EM2] D. Eardley, V. Moncrief, *The global existence of Yang-Mills-Higgs fields in 4-dimensional Minkowski space. II. Completion of proof.* Comm. Math. Phys.**83** (1982), no. 2, 193–212.

[Fried] H.G. Friedlander *The Wave Equation on a Curved Space-time*, Cambridge University Press, 1976.

[HE] Hawking, S. W. & Ellis, G. F. R. *The Large Scale Structure of Space-time*, Cambridge: Cambridge University Press, 1973

[HKM] Hughes, T. J. R., T. Kato and J. E. Marsden *Well-posed quasi-linear second-order hyperbolic systems with applications to nonlinear elastodynamics and general relativity*, Arch. Rational Mech. Anal. 63, 1977, 273-394

[Kl] S. Klainerman. *PDE as a unified subject* Special Volume GAFA 2000, 279-315

[Kl-Ma] S. Klainerman and M. Machedon, *Finite Energy Solutions for the Yang-Mills Equations in $\mathbb{R}^{1+3}$*, Annals of Math., Vol. 142, (1995), 39-119.

[Kl-Ro1] S. Klainerman and I. Rodnianski, *Causal geometry of Einstein-Vacuum spacetimes with finite curvature flux* Inventiones Math. 2005, vol 159, No 3, pgs. 437-529.

[Kl-Ro2] S. Klainerman and I. Rodnianski, *A geometric approach to Littlewood-Paley theory*, to appear in GAFA(Geom. and Funct. Anal)

[Kl-Ro3] S. Klainerman and I. Rodnianski, *Sharp trace theorems for null hypersurfaces on Einstein metrics with finite curvature flux*, to appear in GAFA

[Kl-Ro4] S. Klainerman and I. Rodnianski, *Lower bounds for the radius of injectivity of null hypersurfaces*, preprint

[Kl-Ro5] S. Klainerman and I. Rodnianski, *A Kirchoff-Sobolev parametrix for the wave equations in a curved space-time.* preprint.

[Kl-Ro6] S. Klainerman and I. Rodnianski, *A breakdown criterion in General Relativity* preprint.

[M] V. Moncrief, Personal communication.

[Sob] S. Sobolev, *Methodes nouvelle a resoudre le probleme de Cauchy pour les equations lineaires hyperboliques normales*, Matematicheskii Sbornik, vol 1 (43) 1936, 31 -79.

DEPARTMENT OF MATHEMATICS, PRINCETON UNIVERSITY, PRINCETON NJ 08544

E-mail address: `seri@math.princeton.edu`

Proceedings of Symposia in Applied Mathematics
Volume **65**, 2007

The Formation of Shocks in 3-Dimensional Fluids

Demetrios Christodoulou

This lecture is dedicated to Peter Lax and Louis Nireberg, two mathematicians whom I love and admire greatly, on the occasion of their 80th birthdays.

I am grateful to Peter Lax, Cathleen Morawetz and Andrew Majda for introducing me to the wonderful field of fluid mechanics some fifteen years ago. I have concentrated my efforts to this field for the past seven years and the fruit of this work is the monograph [Ch1], the contents of which I would like to summarize in the present lecture. All the material which I shall report here is expounded in the monograph [Ch1].

The monograph considers the relativistic Euler equations in three space dimensions for a perfect fluid with an arbitrary equation of state. We consider regular initial data on a spacelike hyperplane Σ_0 in Minkowski spacetime which outside a sphere coincide with the data corresponding to a constant state. We consider the restriction of the initial data to the exterior of a concentric sphere in Σ_0 and we consider the maximal classical development of this data. Then, under a suitable restriction on the size of the departure of the initial data from those of the constant state, we prove certain theorems which give a complete description of the maximal classical development. In particular, the theorems give a detailed description of the geometry of the boundary of the domain of the maximal classical development and a detailed analysis of the behavior of the solution at this boundary. A complete picture of shock formation in three-dimensional fluids is thereby obtained. Also, sharp sufficient conditions on the initial data for the formation of a shock in the evolution are established and sharp lower and upper bounds for the temporal extent of the domain of the maximal solution are derived.

The reason why we consider only the maximal development of the restriction of the initial data to the exterior of a sphere is in order to avoid having to treat the long time evolution of the portion of the fluid which is initially contained in the interior of this sphere. For, we have no method at present to control the long time behavior of the pointwise magnitude of the vorticity of a fluid portion, the vorticity satisfying a transport equation along the fluid flow lines. Our approach to the general problem is the following. We show that given arbitary regular initial data which coincide with the data of a constant state outside a sphere, if the size of the initial departure from

---

1991 *Mathematics Subject Classification.* Primary 35L67, 76L05; Secondary 35L65, 35L70, 58J45, 76N15, 76Y05.

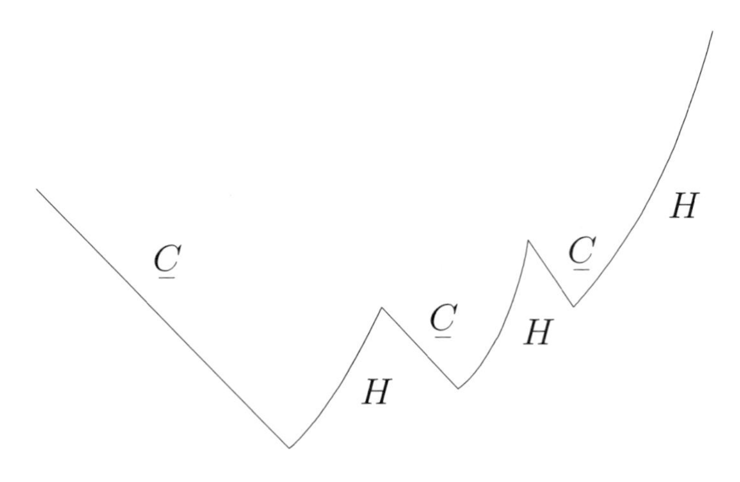

FIGURE 1

the constant state is suitably small, we can control the solution for a time interval of order $1/\eta_0$, where η_0 is the sound speed in the surrounding constant state. We then show that at the end of this interval a thick annular region has formed, bounded by concentric spheres, where the flow is irrotational and isentropic, the constant state holding outside the outer sphere. We then study the maximal classical development of the restriction of the data at this time to the exterior of the inner sphere. We should emphasize here that if we were to restrict ourselves from the begining to the irrotational isentropic case, we would have no problem extending the treatment to the interior region, thereby treating the maximal solution corresponding to the data on the complete initial hyperplane Σ_0. In fact, it is well known that sound waves decay in time faster in the interior region and our constructions can readily be extended to cover this region. It is only our present inability achieve long time control of the magnitude of the vorticity along the flow lines of the fluid, that prevents us from treating the interior region in the general case.

The boundary of the domain of the maximal solution consists of a regular part and a singular part. Each component of the regular part $\underline{C}$ is an incoming characteristic hypersurface with a singular past boundary. The singular part of the boundary of the domain of the maximal solution is the locus of points where the inverse density of the wave fronts vanishes. It is the union $\partial_- H \bigcup H$, where each component of $\partial_- H$ is a smooth embedded surface in Minkowski spacetime, the tangent plane to which at each point is contained in the exterior of the sound cone at that point. On the other hand each component of H is a smooth embedded hypersurface in Minkowski spacetime, the tangent hyperplane to which at each point is contained in the exterior of the sound cone at that point, with the exception of a single generator of the sound cone, which lies on the hyperplane itself. The past boundary of a component of H is the corresponding component of $\partial_- H$. The latter is at the same time the past boundary of a component of $\underline{C}$. (See Fig. 1).

The maximal classical solution is the physical solution of the problem up to $\underline{C} \bigcup \partial_- H$, but not up to H. The problem of the physical continuation of the solution is then set up as the shock development problem. This problem is associated to each component of $\partial_- H$ and its solution requires the construction of a hypersurface of

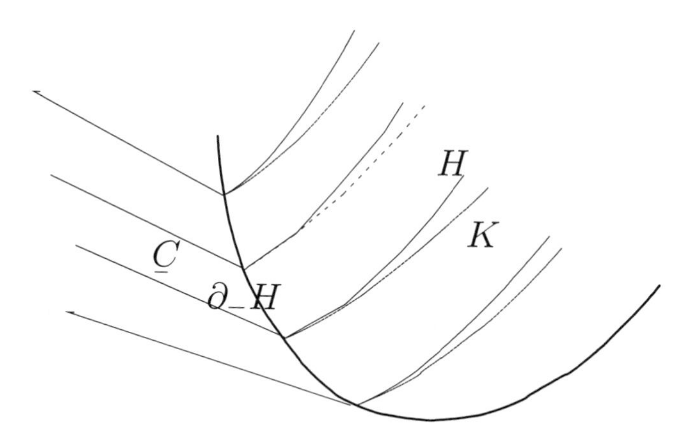

FIGURE 2

discontinuity K, lying in the past of the corresponding component of H, but having the same past boundary as the latter, namely the given component of $\partial_- H$. The maximal classical solution provides the right boundary conditions at $\underline{C} \bigcup \partial_- H$, as well as a barrier at H. (See Fig. 2).

The actual treatment of the shock development problem and the subsequent shock interactions shall be the subject of a subsequent monograph. The present monograph concludes with a derivation of a formula for the jump in vorticity across K, which shows that while the flow is irrotational ahead of the shock, it acquires vorticity immediately behind, which is tangential to the shock front and is associated to the gradient along the shock front of the entropy jump.

One reason for working with the relativistic equations is that there is a substantial gain in geometric insight because of the spacetime geometry viewpoint of special relativity. As an example we give the following equation:

$$i_u \omega = -\theta ds \tag{1}$$

Here ω is the vorticity 2-form:

$$\omega = d\beta \tag{2}$$

where β is the 1-form defined, relative to an arbitary system of coordinates, by:

$$\beta_\mu = -\sqrt{\sigma} u_\mu, \qquad u_\mu = g_{\mu\nu} u^\nu \tag{3}$$

$\sqrt{\sigma}$ being the relativistic enthalpy per particle, u^μ the fluid velocity and $g_{\mu\nu}$ the Minkowski metric. In 1, θ is the temperature and s the entropy per particle, while i_u denotes contraction on the left by the vectorfield u. Equation 1 is arguably the simplest explicit form of the energy-momentum equations. The 1-form β plays a fundamental role in the monograph. In the irrotational isentropic case it is given by $\beta = d\phi$, where ϕ is the wave function.

Moreover, no special care is needed to extract information on the non-relativistic limit. This is due to the fact that the non-relativistic limit is a regular limit, obtained by letting the speed of light in conventional units tend to infinity, while keeping the sound speed fixed. To allow the results in the non-relativistic limit to be extracted from our treatment in a straightforward manner, we have chosen to avoid summing quantities having different physical dimensions when such sums would make sense only when a unit of velocity has been chosen.

We have:

$$\sqrt{\sigma} = e + pv \tag{4}$$

where e is the relativistic energy per particle, p the pressure and v is the volume per particle. Let H be the function defined by:

$$1 - \eta^2 = \sigma H \tag{5}$$

where η is the sound speed. The derivative of H with respect to σ at constant s plays a central role in shock theory . This quantity is expressed by:

$$\left(\frac{dH}{d\sigma}\right)_s = -a\left\{\left(\frac{d^2v}{dp^2}\right)_s + \frac{3v}{\sqrt{\sigma}}\left(\frac{dv}{dp}\right)_s\right\} \tag{6}$$

where a is the positive function:

$$a = \frac{\eta^4}{2\sigma v^3}$$

The sign of $(dH/d\sigma)_s$ in the state ahead of a shock determines the sign of the jump in pressure in crossing the shock to the state behind. The jump in pressure is positive if this quantity is negative, the reverse otherwise. The value of $(dH/d\sigma)_s$ in the surrounding constant state is denoted by ℓ. This constant determines the character of the shocks for small initial departures from the constant state. In particular when $\ell = 0$, no shocks form and the domain of the maximal classical solution is complete. Consider the function $(dH/d\sigma)_s$ as a function of the thermodynamic variables p and s. Suppose that we have an equation of state such that at some value s_0 of s the function $(dH/d\sigma)_s$ vanishes everywhere along the adiabat $s = s_0$. We show that in this case the irrotational isentropic fluid equations corresponding to the value s_0 of the entropy are equivalent to the minimal surface equation, the wave function ϕ defining a minimal graph in a Minkowski spacetime of one more spatial dimension. Thus the minimal surface equation defines a dividing line between two different types of shock behavior.

The general concept of variation, or variation through solutions, is a basic concept on which our treatment not only of the irrotational isentropic case but also of the general equations of motion is based. This concept has been discussed in the general context of Euler-Lagrange equations, that is, systems of partial differential equations arising from an action principle, in a previous monograph [Ch2]. To a variation is associated a linearized Langangian, on the basis of which energy currents are constructed. It is through energy currents and their associated integral identities that the estimates essential to our approach are derived. Here the first order variations correspond to the one-parameter subgroups of the Poincaré group, the isometry group of Minkowski spacetime, extended by the one-parameter scaling or dilation group, which leave the surrounding constant state invariant. The higher order variations correspond to the one-parameter groups of diffeomorphisms generated by a set of vectorfields, the commutation fields. The construction of an energy current requires a multiplier vectorfield which at each point belongs to the closure of the positive component of the inner characteristic core in the tangent space at that point.

In the irrotational isentropic case the characteristic in the tangent space at a point consists only of the sound cone at that point and this requirement becomes the requirement that the multiplier vectorfield be non-spacelike and future directed with respect to the acoustical metric:

$$h_{\mu\nu} = g_{\mu\nu} + (1-\eta^2)u_\mu u_\nu, \qquad u_\mu = g_{\mu\nu}u^\nu, \tag{7}$$

This is a Lorentzian metric on spacetime the null cones of which are the sound cones. We use two multiplier vectorfields in our analysis of the isentropic irrotational problem. The first multiplier field is the vectorfield K_0:

$$K_0 = (\eta_0^{-1} + \alpha^{-1}\kappa)L + \underline{L}, \qquad \underline{L} = \alpha^{-1}\kappa L + 2T \tag{8}$$

Here, α is the inverse density of the hyperplanes Σ_t corresponding to the constant values of the time coordinate t and κ is the inverse spatial density of the wave fronts, both with respect to the acoustical metric. The vectorfield L is the tangent vectorfield of the bicharacteristic generators, parametrized by t, of a family of outgoing characteristic hypersurfaces C_u, the level sets of an acoustical function u. The wave fronts $S_{t,u}$ are the surfaces of intersection $C_u \bigcap \Sigma_t$. The vectorfield T defines a flow on each of the Σ_t, taking each wave front onto another wave front, the normal, relative to the induced acoustical metric $\overline{h}$, flow of the foliation of Σ_t by the surfaces $S_{t,u}$.

The second multiplier field is the vectorfield K_1 defined by:

$$K_1 = (\omega/\nu)L \tag{9}$$

Here ν is the mean curvature of the wave fronts $S_{t,u}$ relative to their characteristic normal L. However ν is defined not relative to the acoustical metric $h_{\mu\nu}$ but rather relative to a conformally related metric $\tilde{h}_{\mu\nu}$:

$$\tilde{h}_{\mu\nu} = \Omega h_{\mu\nu} \tag{10}$$

It turns out that there is a choice of conformal factor Ω such that in the isentropic irrotational case a first order variation $\dot{\phi}$ of the wave function ϕ satisfies the wave equation relative to the metric $\tilde{h}_{\mu\nu}$. This choice defines Ω and the definition makes Ω the ratio of a function of σ to the value of this function in the surrounding constant state, thus Ω is equal to unity in the constant state. It turns out moreover that Ω is bounded above and below by positive constants. The function ω appearing in 10 is required to satisfy certain conditions and it is shown that the function $\omega = 2\eta_0(1+t)$ does satisfy these requirements. The multiplier field K_1 corresponds to the generator of inverted time tranlsations, which are proper conformal tranformations of the Minkowski spacetime with its Minkowskian metric $g_{\mu\nu}$. The latter was first used by Morawetz [M] to study the decay of solutions of the initial-boundary value problem for the classical wave equation outside an obstacle. The vectorfield K_1 is an analogue of the multiplier field of Morawetz for the acoustical spacetime which is the same underlying manifold but equipped with the acoustical metric $h_{\mu\nu}$. To each variation ψ, of any order, there are energy currents associated to ψ and to K_0 and K_1 respectively. These currents define the energies $\mathcal{E}_0^u[\psi](t)$, $\mathcal{E}_1'^u[\psi](t)$, and fluxes $\mathcal{F}_0^t[\psi](u)$, $\mathcal{F}_1'^t[\psi](u)$. For given t and u the energies are integrals over the exterior of the surface $S_{t,u}$ in the hyperplane Σ_t, while the fluxes are integrals over the part of the outgoing characteristic hypersurface C_u between the hyperplanes Σ_0 and Σ_t. It is these energy and flux integrals, together with a spacetime integral $K[\psi](t,u)$ associated to K_1, to be discussed below, which are used to control the solution.

Evidently, the means by which the solution is controlled depend on the choice of the acoustical function u, the level sets of which are the outgoing characteristic hypersurfaces C_u. The function u is determined by its restriction to the initial hyperplane Σ_0. The divergence of the energy currents, which determines the growth of the energies and fluxes, itself depends on ${}^{(K_0)}\tilde{\pi}$, in the case of the energy current associated to K_0, and ${}^{(K_1)}\tilde{\pi}$, in the case of the energy current associated to K_1. Here for any vectorfield X in spacetime, we denote by ${}^{(X)}\tilde{\pi}$ the Lie derivative of the conformal acoustical metric $\tilde{h}$ with respect to X. We call ${}^{(X)}\tilde{\pi}$ the deformation tensor corresponding to X. In the case of higher order variations, the divergences of the energy currents depend also on the ${}^{(Y)}\tilde{\pi}$, for each of the commutation fields Y to be discussed below.

All these deformation tensors ultimately depend on the acoustical function u, or, what is the same, on the geometry of the foliation of spacetime by the outgoing characteristic hypersurfaces C_u, the level sets of u. The most important geometric property of this foliation from the point of view of the study of shock formation is the density of the packing of its leaves C_u. One measure of this density is the inverse spatial density of the wave fronts, that is, the inverse density of the foliation of each spatial hyperplane Σ_t by the surfaces $S_{t,u}$. This is the function κ which appears in 8 and is given in arbitrary coordinates on Σ_t by:

$$\kappa^{-2} = (\overline{h}^{-1})^{ij}\partial_i u \partial_j u \tag{11}$$

where $\overline{h}_{ij}$ is the induced acoustical metric on Σ_t. Another measure is the inverse temporal density of the wave fronts, the function μ given in arbitrary coordinates in spacetime by:

$$\frac{1}{\mu} = -(h^{-1})^{\mu\nu}\partial_\mu t \partial_\nu u \tag{12}$$

The two measures are related by:

$$\mu = \alpha\kappa \tag{13}$$

where α is the inverse density, with respect to the acoustical metric, of the foliation of spacetime by the hyperplanes Σ_t. The function α also appears in 8 and is given in arbitrary coordinates in spacetime by:

$$\alpha^{-2} = -(h^{-1})^{\mu\nu}\partial_\mu t \partial_\nu t \tag{14}$$

It is expressed directly in terms of the 1-form β. It turns out moreover, that it is bounded above and below by positive constants. Consequently μ and κ are equivalent measures of the density of the packing of the leaves of the foliation of spacetime by the C_u. Shock formation is characterized by the blow up of this density or equivalently by the vanishing of κ or μ.

The other entity, besides κ or μ which describes the geometry of the foliation by the C_u is the second fundamental form of the C_u. Since the C_u are null hypersurfaces with respect to the acoustical metric h their tangent hyperplane at a point is the set of all vectors at that point which are h-orthogonal to the generator L, and L itself belongs to the tangent hyperplane, being h-orthogonal to itself. Thus the second fundamental form χ of C_u is intrinsic to C_u and in terms of the metric $\not{h}$ induced by the acoustical metric on the $S_{t,u}$ sections of C_u, it is given by:

$$\not{\mathcal{L}}_L \not{h} = 2\chi \tag{15}$$

where $\not{\mathcal{L}}_X \vartheta$ for a covariant $S_{t,u}$ tensorfield ϑ denotes the restriction of $\mathcal{L}_X \vartheta$ to $TS_{t,u}$.

The acoustical structure equations are:

The propagation equation for χ along the generators of C_u.

The Codazzi equation which expresses $\text{d}\!\!/\text{iv}\,\chi$, the divergence of χ intrinsic to $S_{t,u}$, in terms of $\text{d}\!\!/\text{tr}\chi$, the differential on $S_{t,u}$ of $\text{tr}\chi$, and a component of the acoustical curvature and of k, the second fundamental form of the Σ_t relative to h.

The Gauss equation which expresses the Gauss curvature of $(S_{t,u}, h\!\!\!/)$ in terms of χ and a component of the acoustical curvature and of k.

An equation which expresses $\mathcal{L}\!\!\!/_T\chi$ in terms of the Hessian of the restriction of μ to $S_{t,u}$ and another component of the acoustical curvature and of k.

These acoustical structure equations seem at first sight to contain terms which blow up as κ or μ tend to zero. The analysis of the acoustical curvature then shows that the terms which blow up as κ or μ tend to zero cancel.

The most important acoustical structure equation from the point of view of the formation of shocks is the propagation equation for μ along the generators of C_u:

$$L\mu = m + \mu e \tag{16}$$

where the function m given by:

$$m = \frac{1}{2}(\beta_L)^2\left(\frac{dH}{d\sigma}\right)_s (T\sigma) \tag{17}$$

and the function e depends only on the derivatives of the β_α, the rectangular components of β, tangential to the C_u. It is the function m which determines shock formation, when being negative, causing μ to decrease to zero.

The path we have followed in attacking the problem of shock formation in 3-dimensional fluids illustrates the following approach in regard to quasilinear hyperbolic systems of partial differential equations. That the quantities which are used to control the solution must be defined using the causal, or characteristic, structure of spacetime determined by the solution itself, not an artificial background structure. The original system of equations must then be considered in conjunction with the system of equations which this structure obeys, and it is only through the study of the interaction of the two systems that results are obtained. The work with Klainerman [C-K] on the stability of the Minkowski space in the framework of general relativity was the first illustation of this approach. In the present case however, the structure, which is here the acoustical structure, degenerates as shocks begin to form, and the precise way in which this degeneracy occurs must be guessed beforehand and established in the course of the argument of the mathematical proof. The fact that the underlying structure degenerates implies that our estimates are no longer even locally equivalent to standard energy estimates, which would of necessity have to fail when shocks appear.

We first establish a theorem, the fundamental energy estimate, which applies to a solution of the homogeneous wave equation in the acoustical spacetime, in particular to any first order variation. The proof of this theorem relies on certain bootstrap assumptions on the acoustical entities. The most crucial of these assumptions concern the behavior of the function μ. These assumptions are established later on on the basis of the final set of bootstrap assumptions, which consists only of pointwise estimates for the variations up to certain order. To give an idea of the nature of these assumptions, one of the assumptions required to obtain the fundamental energy estimate up to time s is:

$$\mu^{-1}(T\mu)_+ \leq B_s(t) \quad : \text{for all } t \in [0,s] \tag{18}$$

where $B_s(t)$ is a function such that:

$$\int_0^s (1+t)^{-2}[1+\log(1+t)]^4 B_s(t)dt \leq C \tag{19}$$

with C a constant independent of s. Here T is the vectorfield defined above and we denote by f_+ and f_-, respectively the positive and negative parts of an arbitrary function f. This assumption is then established by a certain proposition with $B_s(t)$ the following function:

$$B_s(t) = C\sqrt{\delta_0}\frac{(1+\tau)}{\sqrt{\sigma-\tau}} + C\delta_0(1+\tau) \tag{20}$$

where $\tau = \log(1+t)$, $\sigma = \log(1+s)$, and δ_0 is a small positive constant appearing in the final set of bootstrap assumptions.

The spacetime integral $K[\psi](t,u)$ mentioned above, is essentially the integral of

$$-\frac{1}{2}(\omega/\nu)(L\mu)_-|\not{d}\psi|^2$$

in the spacetime exterior to C_u and bounded by Σ_0 and Σ_t. Another assumption states that there is a positive constant C independent of s such that in the region below Σ_s where $\mu < \eta_0/4$ we have:

$$L\mu \leq -C^{-1}(1+t)^{-1}[1+\log(1+t)]^{-1} \tag{21}$$

In view of this assumption, the integral $K[\psi](t,u)$ gives effective control of the derivatives of the variations tangential to the wave fronts in the region where shocks are to form. The same assumption, which is then established by a certain proposition, also plays an essential role in the study of the singular boundary.

The final stage of the proof of of the fundamental energy estimate is the analysis of system of integral inequalities in two variables t and u satisfied by the five quantities $\mathcal{E}_0^u[\psi](t)$, $\mathcal{E}_1^{\prime u}[\psi](t)$, $\mathcal{F}_0^t[\psi](u)$, $\mathcal{F}_1^{\prime t}[\psi](u)$, and $K[\psi](t,u)$.

After this, the commutation fields Y, which generate the higher order variations, are defined. They are five: the vectorfield T which is tranversal to the C_u, the field $Q = (1+t)L$ along the generators of the C_u and the three rotation fields $R_i : i = 1,2,3$ which are tangential to the $S_{t,u}$ sections. The latter are defined to be $\Pi \overset{\circ}{R}_i : i = 1,2,3$, where the $\overset{\circ}{R}_i$ $i = 1,2,3$ are the generators of spatial rotations associated to the background Minkowskian structure, while Π is the h-orthogonal projection to the $S_{t,u}$. Expressions for the deformation tensors ${}^{(T)}\tilde{\pi}$, ${}^{(Q)}\tilde{\pi}$, and ${}^{(R_i)}\tilde{\pi}$ $: i = 1,2,3$ are then derived, which show that these depend on the acoustical entities μ and χ. The last however depend in addition on the derivatives of the restrictions to the surfaces $S_{t,u}$ of the spatial rectangular coordinates x^i $: i = 1,2,3$, as well as on the derivatives of the x^i with respect to T and L, that is, on the rectangular components T^i and L^i of the vectorfields T and L.

The higher order variations satisfy inhomogeneous wave equations in the acoustical spacetime, the source functions depending on the deformation tensors of the commutation fields. These source functions give rise to error integrals, that is to spacetime integrals of contributions to the divergence of the energy currents.

The expressions for the source functions and the associated error integrals show that the error integrals corresponding to the energies of the $n+1$st order variations contain the nth order derivatives of the deformation tensors, which in turn contain the nth order derivatives of χ and $n+1$st order derivatives of μ. Thus to achieve closure, we must obtain estimates for the latter in terms of the energies of up to the $n+1$st order variations. Now, the propagation equations for χ and μ give appropriate expressions for $\not{\mathcal{L}}_L\chi$ and $L\mu$. However, if these propagation equations, which may be thought of as ordinary differential equations along the generators of the C_u, are integrated with respect to t to obtain the acoustical entities χ and μ themselves, and their spatial derivatives are then taken, a loss of one degree of differentiability would result and closure would fail. We overcome this difficulty in the case of χ by considering the propagation equation for $\mu\mathrm{tr}\chi$. We show that, by virtue of a wave equation for σ, which follows from the wave equations satisfied by the first variations corresponding to the spacetime translations, the principal part on the right hand side of this propagation equation can be put into the form $-L\check{f}$ of a derivative of a function $-\check{f}$ with respect to L. This function is then brought to the left hand side and we obtain a propagation equation for $\mu\mathrm{tr}\chi+\check{f}$. In this equation $\hat{\chi}$, the trace-free part of χ enters, but the propagation equation in question is considered in conjuction with the Codazzi equation, which constitutes an elliptic system on each $S_{t,u}$ for $\hat{\chi}$, given $\mathrm{tr}\chi$. We thus have an ordinary differential equation along the generators of C_u coupled to an elliptic system on the $S_{t,u}$ sections. More precisely, the propagation equation which is considered at the same level as the Codazzi equation is a propagation equation for the $S_{t,u}$ 1-form $\mu\not{d}\mathrm{tr}\chi+\not{d}\check{f}$, which is a consequence of the equation just discussed. To obtain estimates for the angular derivatives of χ of order l we similarly consider a propagation equation for the $S_{t,u}$ 1-form:

$$^{(i_1\ldots i_l)}x_l=\mu\not{d}(R_{i_l}\ldots R_{i_1}\mathrm{tr}\chi)+\not{d}(R_{i_l}\ldots R_{i_1}\check{f})$$

In the case of μ the aforementioned difficulty is overcome by considering the propagation equation for $\mu\not{\Delta}\mu$, where $\not{\Delta}\mu$ is the Laplacian of the restriction of μ to the $S_{t,u}$. We show that by virtue of a wave equation for $T\sigma$, which is a differential consequence of the wave equation for σ, the principal part on the right hand side of this propagation equation can again be put into the form $L\check{f}'$ of a derivative of a function $\check{f}'$ with respect to L. This function is then likewise brought to the left hand side and we obtain a propagation equation for $\mu\not{\Delta}\mu-\check{f}'$. In this equation $\hat{\not{D}}^2\mu$, the trace-free part of the Hessian of the restriction of μ to the $S_{t,u}$ enters, but the propagation equation in question is considered in conjuction with the elliptic equation on each $S_{t,u}$ for μ, which the specification of $\not{\Delta}\mu$ constitutes. Again we have an ordinary differential equation along the generators of C_u coupled to an elliptic equation on the $S_{t,u}$ sections. To obtain estimates of the spatial derivatives of μ of order $l+2$ of which m are derivatives with respect to T we similarly consider a propagation equation for the function:

$$^{(i_1\ldots i_{l-m})}x'_{m,l-m}=\mu R_{i_{l-m}}\ldots R_{i_1}(T)^m\not{\Delta}\mu-R_{i_{l-m}}\ldots R_{i_1}(T)^m\check{f}'$$

This allows us to obtain estimates for the top order spatial derivatives of μ of which at least two are angular derivatives. A remarkable fact is that the missing top order spatial derivatives do not enter the source functions, hence do not contribute to the error integrals. In fact it is shown that the only top order spatial derivatives of the acoustical entities entering the source functions are those in the 1-forms $^{(i_1\ldots i_l)}x_l$ and the functions $^{(i_1\ldots i_{l-m})}x'_{m,l-m}$.

The paradigm of an ordinary differential equation along the generators of a characteristic hypersurface coupled to an elliptic system on the sections of the hypersurface as the means to control the regularity of the entities describing the geometry of the characteristic hypersurface and the stacking of such hypersurfaces in a foliation, was first encountered in the work [C-K] on the stability of the Minkowski space. It is interesting to note that this paradigm does not appear in space dimension less than three. In the case of the work on the stability of the Minkowski space however, in contrast to the present case, the gain of regularity achieved in this treatment is not essential for obtaining closure, because there is room of one degree of differentiability. This is due to the fact that the Einstein equations arise from a Lagrangian which is quadratic in the derivatives of the unknown functions, in contrast to the equations of fluid mechanics, or more generally of continuum mechanics, which in the Lagrangian picture are equations for a mapping of spacetime into the material manifold, the Lagrangian not depending quadratically on the differential of this mapping (see [Ch2]). As a consequence, the metric determining the causal structure depends in continuum mechanics on the derivatives of the unknowns, rather than only on the unknowns themselves.

In the present work, the appearence of the factor of μ, which vanishes where shocks originate, in front of $\not{d}R_{i_l}...R_{i_1}\mathrm{tr}\chi$ and $R_{i_{l-m}}...R_{i_1}(T)^m\not{\Delta}\mu$ in the definitions of ${}^{(i_1...i_l)}x_l$ and ${}^{(i_1...i_{l-m})}x'_{m,l-m}$ above, makes the analysis far more delicate. This is compounded with the difficulty of the slower decay in time which the addition of the terms $-\not{d}R_{i_l}...R_{i_1}\check{f}$ and $R_{i_{l-m}}...R_{i_1}(T)^m\check{f}'$ forces. The analysis requires a precise description of the behavior of μ itself, given by certain propositions, and a separate treatment of the condensation regions, where shocks are to form, from the rarefaction regions, the terms refering not to the fluid density but rather to the density of the stacking of the wave fronts. To overcome the difficulties the following weight function is introduced:

$$\overline{\mu}_{m,u}(t) = \min\left\{\frac{\mu_{m,u}(t)}{\eta_0}, 1\right\}, \quad \mu_{m,u}(t) = \min_{\Sigma_t^u}\mu \tag{22}$$

where Σ_t^u is the exterior of $S_{t,u}$ in Σ_t, and the quantities $\mathcal{E}_0^u[\psi](t)$, $\mathcal{E}_1'^u[\psi](t)$, $\mathcal{F}_0^t[\psi](u)$, $\mathcal{F}_1'^t[\psi](u)$, and $K[\psi](t,u)$ corresponding to the highest order variations are weighted with a power, $2a$, of this weight function. The following lemma then plays a crucial role here as well as in the proof of the main theorem where everything comes together. Let:

$$M_u(t) = \max_{\Sigma_t^u}\left\{-\mu^{-1}(L\mu)_-\right\}, \quad I_{a,u} = \int_0^t \overline{\mu}_{m,u}^{-a}(t')M_u(t')dt' \tag{23}$$

Then under certain bootstrap assumptions in the past of Σ_s, for any constant $a \geq 2$, there is a positive constant C *independent of* s, u *and* a such that for all $t \in [0,s]$ we have:

$$I_{a,u}(t) \leq Ca^{-1}\overline{\mu}_{m,u}^{-a}(t) \tag{24}$$

Now, estimates for the derivatives of the spatial rectangular coordinates x^i with respect to the commutation fields must also be obtained, the derivatives of the x^i with respect to the vectorfields $\hat{T}$ and L being the spatial rectangular components $\hat{T}^i$ and L^i of these vectorfields. Here $\hat{T} = \kappa^{-1}T$, is the vectorfield of unit magnitute with respect to h corresponding to T. Thus, although the argument depends mainly on the causal structure of the acoustical spacetime, the underlying Minkowskian structure, to which the rectangular coordinates belong, has a role to play as well, and it is the estimates in question which analyze the mutual relationship of the two structures. The derivation of these estimates occupies a considerable part of the work. The required estimates for the deformation tensors of the commutation fields in terms of the acoustical entities are then obtained.

After this, the acoustical assumptions on which the previous results depend are established, using the method of continuity, on the basis of the final set of bootstrap assumptions, which consists only of pointwise estimates for the variations up to certain order. Then, the estimates for up to the next to the top order angular derivatives of χ and spatial derivatives of μ are derived. These, when substituted in the estimates established earlier give control of all quantities involved in terms of estimates for the variations. A fundamental role is played by the propositions which establish the coercivity hypotheses on which the previous results depend. These propositions roughly speaking show that for any covariant $S_{t,u}$ tensorfield ϑ, the sum $\sum_i |\not{\mathcal{L}}_{R_i}\vartheta|^2$ bounds pointwise $|\not{D}\vartheta|^2$, and that if X is any $S_{t,u}$-tangential vectorfield and ϑ any covariant $S_{t,u}$ tensorfield then we can bound pointwise $\not{\mathcal{L}}_X\vartheta$ in terms of the $\not{\mathcal{L}}_{R_i}\vartheta$ and the $\not{\mathcal{L}}_{R_i}X = [R_i, X]$.

We then analyze the structure of the terms containing the top order spatial derivatives of the acoustical entities, showing that these can be expressed in terms of the 1-forms ${}^{(i_1\dots i_l)}x_l$ and the functions ${}^{(i_1\dots i_{l-m})}x'_{m,l-m}$. These terms are shown to contribute *borderline error integrals*, the treatment of which is the main source of difficulties in the problem. These borderline integrals are all proportional to the constant ℓ mentioned above, hence are absent in the case $\ell = 0$. We should make clear here that the only variations which are considered up to this point are the variations arising from the first order variations corresponding to the group of spacetime translations. In particular the final bootstrap assumption involves only variations of this type, and each of the five quantities $\mathcal{E}^u_{0,[n]}(t)$, $\mathcal{F}^t_{0,[n]}(u)$, $\mathcal{E}'^u_{1,[n]}(t)$, $\mathcal{F}'^t_{1,[n]}(u)$, and $K_{[n]}(t,u)$, which together control the solution, is defined to be the sum of the corresponding quantity $\mathcal{E}^u_0[\psi](t)$, $\mathcal{F}^t_0[\psi](u)$, $\mathcal{E}'^u_1[\psi](t)$, $\mathcal{F}'^t_1[\psi](u)$, and $K[\psi](t,u)$, over all variations ψ of this type, up to order n. To estimate the borderline integrals however, we introduce an additional assumption which concerns the first order variations corresponding to the scaling or dilation group and to the rotation group, and the second order variations arising from these by applying the commutation field T. This assumption is later established through energy estimates of order 4 arising from these first order variations and derived on the basis of the final bootstrap assumption, just before the recovery of the final bootstrap assumption itself. It turns out that the borderline integrals all contain the factor $T\psi_\alpha$, where $\psi_\alpha : \alpha = 0,1,2,3$ are the first variations corresponding to spacetime translations and the additional assumption is used to obtain an estimate for $\sup_{\Sigma^u_t}\left(\mu^{-1}|T\psi_\alpha|\right)$ in terms of $\sup_{\Sigma^u_t}\left(\mu^{-1}|L\mu|\right)$, which involves on the right the factor $|\ell|^{-1}$. Upon substituting this estimate in the borderline integrals, the factors involving ℓ cancel, and the integrals are estimated using the inequality 24. The above is an outline of the main steps in the estimation of the borderline integrals associated to the vectorfield K_0. The estimation of the borderline integrals associated to the vectorfield K_1, is however still more delicate. In this case we first perform an integration by parts on the outgoing characteristic hypersurfaces C_u, obtaining hypersurface integrals over Σ^u_t and Σ^u_0 and another spacetime volume integral. In this integration by parts the terms, including those of lower order, must be carefully chosen to obtain appropriate estimates, because here the long time behavior, as well as the behavior as μ tends to zero, is critical. Another integration by parts, this time on the surfaces $S_{t,u}$, is then performed to reduce these integrals to a form which can be estimated. The estimates of the hypersurface integrals over Σ_t^u are the most delicate (the hypersurface integrals over Σ_0^u only involve the initial data) and require separate treatment of the condensation and rarefaction regions, in which the properties of the function μ, established by the previous propositions, all come into play.

In proceeding to derive the energy estimates of top order, $n = l + 2$, the power $2a$ of the weight $\overline{\mu}_{m,u}(t)$ is chosen suitably large to allow us to transfer the terms contributed by the borderline integrals to the left hand side of the inequalities resulting from the integral identities associated to the multiplier fields K_0 and K_1. The argument then proceeds along the lines of that of the fundamental energy estimate, but is more complex because here we are dealing with weighted quantities. Once the top order energy estimates are established, we revisit the lower order energy estimates using at each order the energy estimates of the next order in estimating the error integrals contributed by the highest spatial derivatives of the acoustical entities at that order. We then establish a descent scheme, which yields, after finitely many steps, estimates for the five quantities $\mathcal{E}^u_{0,[n]}(t)$, $\mathcal{F}^t_{0,[n]}(u)$, $\mathcal{E}'^u_{1,[n]}(t)$, $\mathcal{F}'^t_{1,[n]}(u)$, and $K_{[n]}(t,u)$, for $n = l + 1 - [a]$, where $[a]$ is the integral part of a, in which weights no longer appear.

It is these unweighted estimates which are used to close the bootstrap argument by recovering the final bootstrap assumption. This is accomplished by the method of continuity through the use of the isoperimetric inequality on the wave fronts $S_{t,u}$, and leads to the main theorem. This theorem shows that there is another differential structure, that defined by the acoustical coordinates t, u, ϑ, the $\vartheta = const.$ coordinate lines corresponding to the bicharacteristic generators of each C_u, such that relative to this structure the maximal classical solution extends smoothly to the boundary of its domain. This boundary contains however a singular part where the function μ vanishes, hence, in these coordinates, the acoustical metric h degenerates. With respect to the standard differential structure induced by the rectangular coordinates x^α in Minkowski spacetime, the solution is continuous but not differentiable on the singular part of the boundary, the derivative $\hat{T}^\mu \hat{T}^\nu \partial_\mu \beta_\nu$ blowing up as we approach the singular boundary. Thus, with respect to the standard differential structure, the acoustical metric h is everywhere in the closure of the domain of the maximal solution non-degenerate and continuous, but it is not differentiable on the singular part of the boundary of this domain, while with respect to the differential structure induced by the acoustical coordinates h is everywhere smooth, but it is degenerate on the singular part of the boundary.

After the proof of the main theorem, we establish a general theorem which gives sharp sufficient conditions on the initial data for the formation of a shock in the evolution. The proof of this theorem is through the propositions describing the properties of the function μ and is based on the study of the evolution with respect to t of the mean value on the sections $S_{t,u}$ of each outgoing characteristic hypersurface C_u of the quantity:

$$\underline{\tau}' = (1 - u + \eta_0 t)\underline{i} - v_0(p - p_0) \tag{25}$$

where v_0 and p_0 are respectively the volume per particle and pressure in the surrounding constant state. Here i and $\underline{i}$ are the functions:

$$i = L^\mu \xi_\mu, \quad \underline{i} = \underline{L}^\mu \xi_\mu \tag{26}$$

and ξ the 1-form:

$$\xi_\mu = \dot{\beta}_\mu + \theta \dot{s} u_\mu \tag{27}$$

corresponding to any first order variation $(\dot{p}, \dot{s}, \dot{u})$ of a general solution (p, s, u) of the equations of motion. We consider in particular the variation corresponding to time translations. The proof of the theorem uses the estimate provided by the spacetime integral $K(t, u)$ associated to this variation. Certain crucial integrations by parts on the $S_{t,u}$ sections as well as on C_u itself are performed, in which the structure of C_u as a characteristic hypersurface comes into play. The theorem also gives a sharp upper bound on the time interval required for the onset of shock formation.

The last part of the work is concerned with the structure of the boundary of the domain of the maximal classical solution and the behavior of the solution at this boundary. We first establish a proposition which describes the singular part of the boundary of the domain of the maximal classical solution from the point of view of the acoustical spacetime. This singular part has the intrinsic geometry of a regular null hypersurface in a regular spacetime and, like the latter, is ruled by invariant curves of vanishing arc length. On the other hand, the extrinsic geometry of the singular boundary is that of a spacelike hypersurface which becomes null at its past boundary. The main result of the last part of the work is the trichotomy theorem. This theorem shows that at each point q of the singular boundary, the past sound cone in the cotangent space at q degenerates into two hyperplanes intersecting in a 2-dimensional plane. We thus have a trichotomy of the bicharacteristics, or null geodesics of the acoustical metric, ending at q, into the set of outgoing null geodesics ending at q, which corresponds to one of the hyperplanes, the set of incoming null geodesics ending at q, which corresponds to the other hyperplane, and the set of the remaining null geodesics ending at q, which corresponds to the 2-dimensional plane. The intersection of the past characteristic cone of q with any Σ_t in the past of q similarly splits into three parts, the parts corresponding to the outgoing and to the incomings sets of null geodesics ending at q being embedded discs with a common boundary, an embedded circle, which corresponds to the set of the remaining null geodesics ending at q. *All outgoing null geodesics ending at q have the same tangent vector at q.* This vector is then an invariant characteristic vector associated to the singular point q. (See Fig. 3).

This striking result is in fact the reason why the considerable freedom in the choice of the acoustical function does not matter in the end. For, considering the transformation from one acoustical function to another, we show that the foliations corresponding to different families of outgoing characteristic hypersurfaces have equivalent geometric properties and degenerate in precisely the same way on the same singular boundary. Finally, the last proposition gives a detailed description of the boundary of the domain of the maximal classical solution from the point of view of Minkowski spacetime.

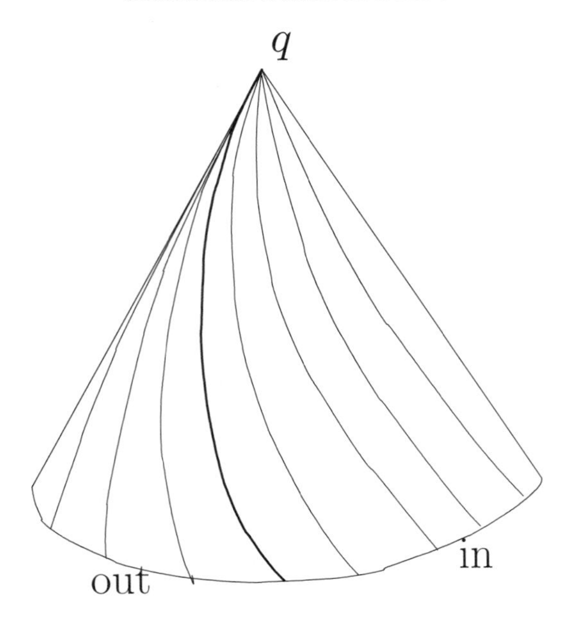

FIGURE 3

References

[Ch1] D. Christodoulou, *The Formation of Shocks in 3-Dimensional Fluids*, EMS Monographs in Mathematics, EMS Publishing House, ISBN 978-3-03719-031-9, 2007.

[Ch2] D. Christodoulou, *The Action Principle and Partial Differential Equations*, Ann. Math. Stud., vol. 146, Princeton University Press, 2000.

[C-K] D. Christodoulou and S. Klainerman, *The Global Nonlinear Stability of the Minkowski Space*, Princeton Mathematical Series, vol. 41, Princeton University Press, 1993.

[M] C. Morawetz, *The decay of solutions of the exterior initial-boundary value problem for the wave equation*, Comm. Pure & Appl. Math., **14** (1961), 561–568.

DEPARTMENT OF MATHEMATICS, ETH-ZENTRUM, 8092 ZÜRICH, SWITZERLAND
E-mail address: demetri@math.ethz.ch

Proceedings of Symposia in Applied Mathematics
Volume **65**, 2007

Occupation time for two dimensional Brownian motion in a wedge

F. Alberto Grünbaum and Caroline McGrouther

Abstract. We study the distribution of the random variable that measures the total time, within the interval $(0,1)$, that a Brownian particle spends in a wedge anchored at the origin. This is related to the quest for a two dimensional version of the famous arc-sine law of P. Levy. This should help answer the question: How often should Peter and Louis be simultaneously happy?

1. Introduction

This paper deals with a probabilistic question which could be solved by solving a boundary value problem for a partial differential equation. In this paper, we do not solve the problem analytically, but give a precise formulation of what needs to be done to get such a solution. We also give very accurate numerical simulations which could eventually be compared to an analytic solution.

This connection between Brownian motion and boundary value problems is certainly not new and everyone knows that the interplay between these two fields has been a very fruitful one.

The title of this paper could be : How often should Peter and Louis be simultaneously happy? Ths question refers to the following situation: Louis and Peter go to the casino and spend a long night repeatedly playing a coin game not against each other but rather they play independent games of *heads or tails* against the house. We assume that they each use a fair coin, and we also assume that a) they have enough capital to allow for heavy loses, b) the casino will not ask them to leave before the end of the night even if they keep winning. Several of these assumptions are, of course, unrealistic.

If, on a given night, you plot their net fortunes as functions of time you may get something like what is shown in Fig 1 below.

If we were dealing with the case of just one player, let us say Peter, and we wanted to know how often he should be happy during this long night at the casino, it turns out that the answer to the question is both known and interesting.

2000 *Mathematics Subject Classification.* 60J65, 35J05, 35J25.
Key words and phrases. Two dimensional Brownian motion, Occupation time.
The first author was supported in part by NSF Grant # DMS0603901.

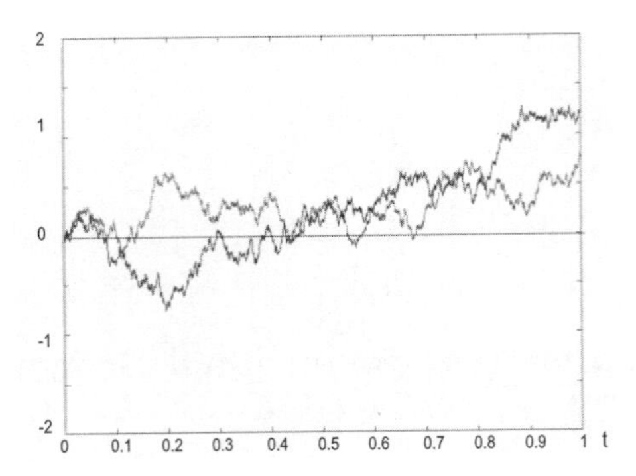

FIGURE 1. The case of two players

In this case Peter's fortune as a function of time may look like what is given in Fig 2 below.

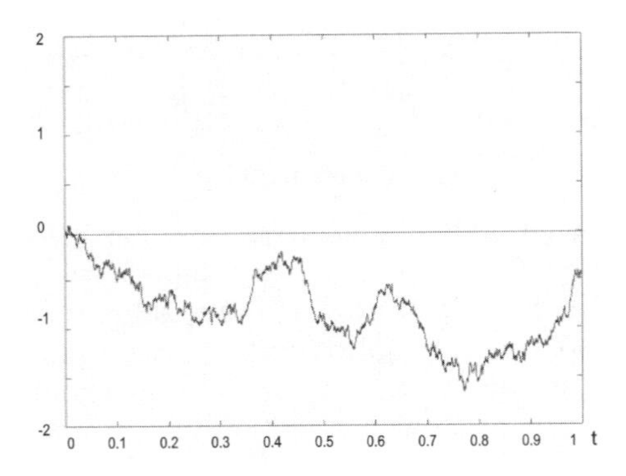

FIGURE 2. The case of one player

To deal with the case of an actual coin game is not so simple, and it is easier to deal with the limiting case when you replace the corresponding one dimensional random walk by one dimensional Brownian motion. This means that Peter (or Louis) plays for a finite length of time but he plays so fast and so many times that we can approximate random walk by Brownian motion. We normalize the length of the night to be 1. For a lovely treatment of the case of the coin, the reader should consult chapter 3 of [**F**].

If we denote by $\tau(w)$ the length of time, in the interval $(0, 1)$, that a Brownian particle labelled by w spends on the positive side of the real axis (i.e. the net fortune of our player is positive), then Paul Levy , [**L**] proved a long time ago that

$$
\begin{aligned}
Pr(\omega, \tau(\omega) \leq x) &= \frac{1}{\pi} \int_0^x \frac{dt}{\sqrt{t(1-t)}} \\
&= \frac{2}{\pi} \arcsin \sqrt{x}
\end{aligned}
$$

A plot of the cumulative distribution function of τ and its density is given below in Fig 3 and Fig 4 respectivly.

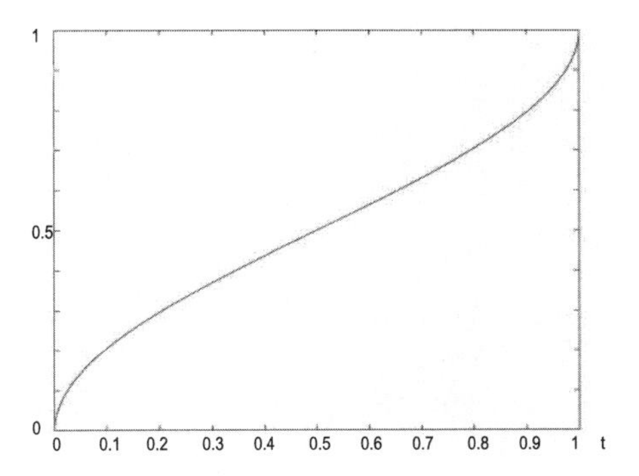

FIGURE 3. The cumulative distribution of τ

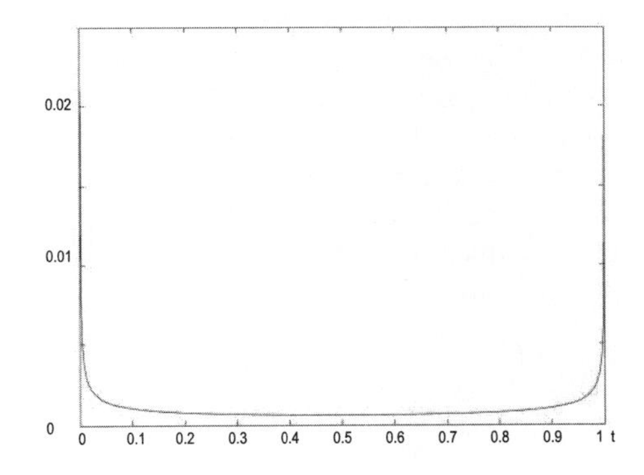

FIGURE 4. The density of the distribution of τ

This result shocks most people, and it should. It proves, for instance, that the probability of having a well balanced night is much smaller that that of having a wild night. A well balanced night is one when Peter is ahead more or less half of the time and a wild night is one when he is ahead (or behind) almost all the time.

Notice that we have not defined happiness very carefully, so here it is: Peter is happy when he is ahead of the casino, and when Louis and Peter are both playing they are simultaneously happy when they are both ahead in their games. Someone remarked to me that Louis, being such a nice guy, would be uncomfortable when he is winning and that he might prefer to have Peter winning while he is himself losing. It is one of these facts of life that the random variable that keeps track of the time when this is happening has the same distribution as the one we are considering here.

Being a bit more quantitative in the case of one player, we can see from the expression above that it is about *four times* more likely that Peter will be happy either less that 5 percent of the time or more than 95 percent of the time than he will be happy somewhere between 45 and 55 percent of the time.

Explicitly:

$$\begin{aligned} Pr(\omega, .45 \le \tau(\omega) \le .55) &\cong .06377 \\ Pr(\omega, \tau(\omega) \le .5 \text{ or } \tau(\omega) \ge .95) &\cong .2871 \\ \text{ratio} &\cong 4.5 \end{aligned}$$

In each case we have a 10 percent window, and extreme behaviour is about four times more likely than very balanced behaviour on the part of the coin. As the window becomes narrower the effect is even more pronounced.

Needless to say, all the above holds if Louis is the lone player.

Most people find this very surprising and a contradiction to common sense.

Returning to the case of two players:

We will be interested only in the total time , in the interval $(0, 1)$, when both Louis' and Peter's fortunes, which are two independent one dimensional Brownian motions, are positive. It is rather natural to describe the joint state of their net fortunes at time t, by a two dimensional Brownian motion.

We are clearly interested in the length of time, within the interval $(0, 1)$, that this Brownian particle spends on the positive quadrant. The case of one player being ahead while the other is behind amounts to looking at the second or fourth quadrant, and the rotational invariance of Brownian motion insures that the distribution of the occupation time for all these cases coincide.

The time evolution of Louis's and Peter's net fortunes can be, as mentioned above be seen by looking at a plot of two dimensional Brownian motion stopped at time 1, as given in Fig 5.

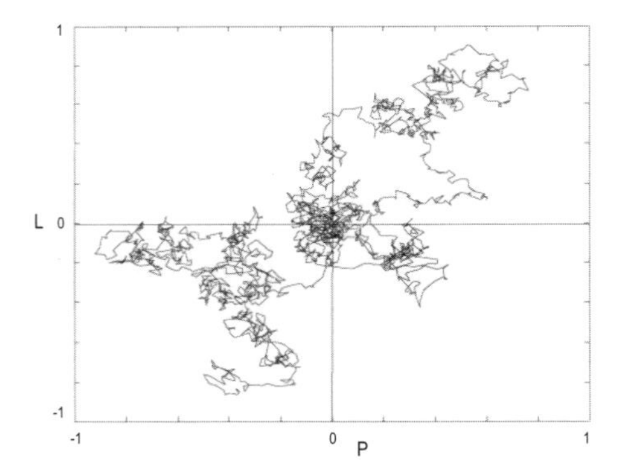

FIGURE 5. The joint net fortunes in the case of two players

We are ready to consider the quantity of interest: if we denote with $\mu(w)$ the length of time, in (0,1), that a two dimensional Brownian path w spends in the positive quadrant, what is

$$Pr(\omega, \mu(\omega) \le x)$$

A detailed, but neither analytical nor explicit, solution to the question at hand is contained in the following figure, Fig 6, that plots the (cumulative) distribution function of the random variable $\mu(w)$.

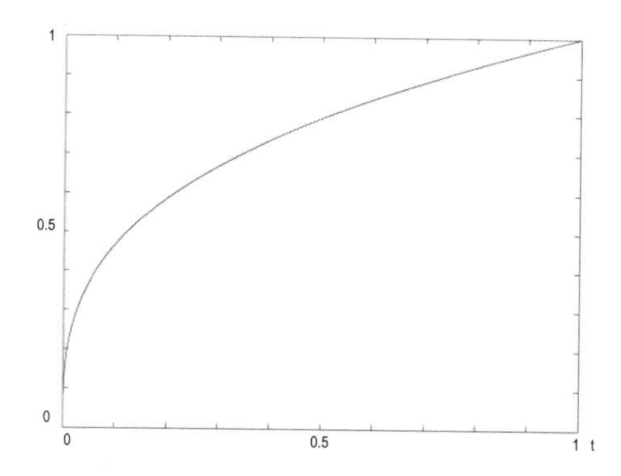

FIGURE 6. The cumulative distribution of μ

From the graph in Fig 6 one can get a graph of its density which is shown as Fig 7.

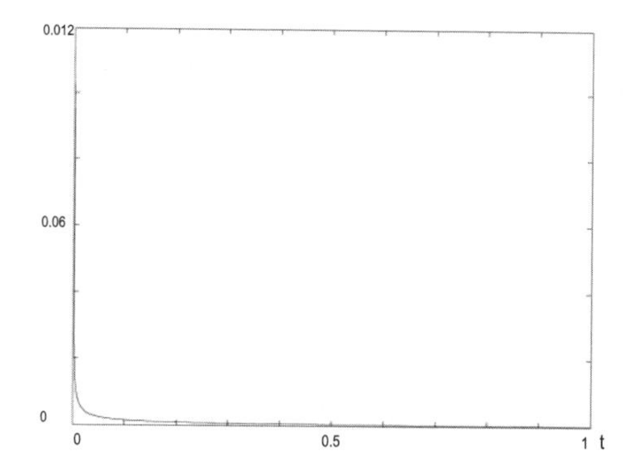

FIGURE 7. The density of the distribution of μ

This already shows that as long as we are interested in the time that Louis and Peter are simultaneously happy we can trust our common sense. By looking at the last two plots we see that now nothing dramatic happens. This is quite different from the case of one player, where common sense proves to be misleading.

If Louis and Peter spend a very long night at the casino and they keep track of the time when they are both ahead in their independent games they are not going to be too pleased. Unfortunately common sense prevails. There is a very large probability that they will be simultaneously happy only a small proportion of the time they are playing.

The figure shown earlier is the result of rather careful numerical simulation of Brownian motion. This could have been based on random Fourier series in the fashion of N. Wiener or random wavelet expansions in the fashion of P. Levy. We have chosen the later one. In particular, Fig 7 is an actual histogram which should be a good approximation to the unknown density for the distribution of μ.

Recall that Wiener's construction of Brownian motion is given by the following Fourier series with random coeffcients.

$$X(t,\omega) = \sum_{i=1}^{\infty} \sin\left(i+\frac{1}{2}\right)\pi t \; X_i(\omega)$$

$$X_i(\omega) \text{ independent } (0,1) \text{ Gaussians}$$

P. Levy's construction uses Haar's functions to give a random Fourier series that is meant to approximate white noise (the formal derivative of Brownian motion), and then he gets Brownian motion itself by integrating these Haar functions.

The explict expression for the distribution of μ is not known but it is in principle possible to compute the moments of its density. We have, for instance,

$$E(\mu(\omega)) = \frac{1}{4}$$

$$E(\mu^2(\omega)) = \frac{5}{32} - \frac{1}{8\pi^2} \cong 0.14358$$

$$E(\mu^3(\omega)) = \frac{53}{512} + \frac{7}{18\pi} - \frac{347}{288\pi^2} \cong .10522$$

The first result above is very natural, whereas the computation of each one of the other two is extremely laborious. We have never seen the values for higher order moments in the literature, but our simulations allow one to get some good approximations to them. These results are due to N. Bingham and R. Doney (1988), see [**BD**] while the numerical simulations shown earlier were carried out by one of us as part of a Master's Thesis at Berkeley (2002), see [**McG**].

Some conjectures to be alluded to later are given in work of T. Meyre and W. Werner (1995), see [**MW**].

The quality of this simulation has been validated by checking a few of the lower order moments of the *unknown* distribution. The results agree typically to order between 10−4 and 10−6. Another indication of the quality of this simulation comes from the case of one player. The plots we gave in Fig 3 and Fig 4 for the cumulative distribution and its density (which in this case are known analytically) are the result of the numerical simulations. They cannot be distinguished from the analytical results when plotted side by side.

In principle, all of these computations could be carried out directly if one had an explicit form of the distribution or its density.

This careful numerical study allows us to refine and correct some conjectures that have been made in the literature as to the asymptotic behaviour of the distribution close to the end points of the interval.

For example while in the case of one player one has

$$Pr(\omega, \tau(\omega) < x) \sim \frac{2}{\pi} x^{1/2} \qquad x \sim 0$$

in the case of two players we surmise

$$Pr(\omega, \mu(\omega) < x) \sim kx^{1/3} \qquad x \sim 0$$

2. Some historical comments

This is not the place to give a careful account of the historical roots for the relations between stochastic processes such as random walk or Brownian motion and the field of difference or differential equations. There are several books that document this very well, and we just mention [**La, IM, S**].

We just recall here the pioneering work of George Polya in connection with random walks in different dimensional integer lattices, the very influential paper by R. Courant, K. Friedrichs and H. Lewy on numerical methods for the solution of the heat equation, the paper of S. Kakutani relating the Dirichlet problem to Brownian motion and several contributions of J. von Neuman to what are called Monte Carlo methods.

3. Using the Feynman-Kac formula

All previous attempts to determine the explict expression for the distribution in the case of two players have failed. As far as we know these attempts have not tried to exploit any connection to PDE's, as we try to do now.

We want to describe some work in progress that could lead to an explicit expression for the distribution of the occupation time in any wedge anchored at the origin. We are interested in the case when the angle of the wedge is $\pi/2$

The problem consists of looking at a nice Poisson type equation in the plane and finding its value at the origin explicitly in terms of two parameters.

We start by recalling the way to relate the solution of certain PDEs to the evaluation of certain integrals in function space.

The historical motivation of this piece of mathematics done by Mark Kac, see [**K**], is the attempt by Richard Feynman [**Fey**] (in his thesis, 1948) to produce a Lagrangian formulation of quantum mechanics. You can consider the Schrödinger equation as the Hamiltonian version of the same story.

We will make no attempt to prove this formula here but we give an "after the fact" way of making it believable.

Recall how one solves the initial value problem

$$\begin{aligned} u_t &= u_x + V(x,t)u \qquad x \in R,\ t \geq 0 \\ u(x,0) &= f(x) \end{aligned}$$

The explicit formula is, as we know by the method of characteristics, given by

$$u(x,t) = e^{\int_0^t V(x+s,t-s)ds} f(x+t)$$

We want to interpret this formula by means of the picture in Fig 8. The interpretation is very appealing: you get the value of the solution at (x,t) by integrating the values of $V(x,t)$ along the characteristic starting at this point, taking the exponential of this integral and then multiplying this by the value of the initial data f at the point on the real line where the characteristic meets the x axis.

So, what do you do if you want to solve

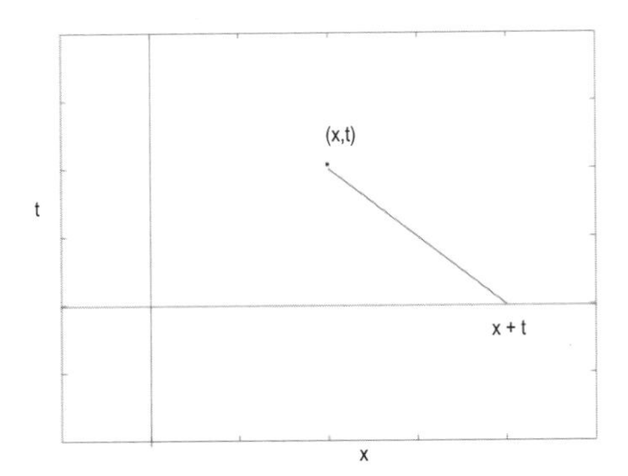

FIGURE 8. Labeling the states

$$\begin{aligned} u_t &= \frac{1}{2}u_{xx} + V(x,t)u \qquad x \in R,\ t \geq 0 \\ u(x,0) &= f(x) \end{aligned}$$

The answer is contained in Fig. 9. It simply says that you should use the previous recipe, but instead of using the one characteristic stemming from the point (x,t) you should pick a Brownian trajectory starting there and running towards the x axis. Along this random path you do as before and get a numerical value. Now start again with a different Brownian path. Repeat this over and over and take the average of the values you get. This solves your problem.

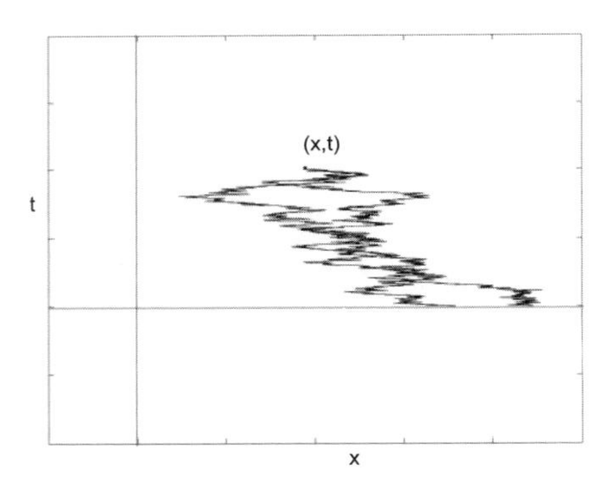

FIGURE 9. Labeling the states

The actual Feynman-Kac formula expresses the solution $u(x,t)$ by means of the formula

$$u(x,t) = E\left[e^{\int_0^t V(x+w(s),t-s)ds} f(x+w(t))\right].$$

The moral is simple: having replaced translation in one direction by a diffusion, one should replace the role of the one characteristic curve by an arbitrary Brownian path starting at (x, t) and running backwards, use the old formula and then average the result over all Brownian paths with respect to Wiener's measure.

Stochastic methods can also be used to solve problems that can be solved by others means. This is not the case of the Feynman-Kac formula where there is no alternative way to solve the problem, but here is a different example.

Define $X(t, y, \omega)$, $\tau(x, t)$ and $u(x, t)$ by means of the recipe

$$\begin{aligned} X(t, y, \omega) &= \sum_{i=1}^{n} e^{(4\xi_i^2 t - \xi_i y)} \sqrt{m_i} X_i(\omega) \\ & \quad X_i(\omega),\ 1 \leq i \leq n, \text{ independent } (0,1) \text{ Gaussians} \\ & \quad \xi_i, m_i \text{ positive} \\ \tau(x, t) &\equiv E_w \left(e^{-\frac{1}{2}\int_x^\infty X^2(t,y,\omega)dy} \right) \\ u(x, t) &= 2\frac{\partial^2}{\partial x^2} \log \tau(x, t) \end{aligned}$$

It is a remarkable result that this solves KdV. This result is due to S. Kotani, see [**Ko**].

4. Kac's proof of P. Levy's classical result

One of the first uses of the Feynman-Kac formula, in the hands of Mark Kac himself, consists in taking a double Laplace transform of the original problem so that one is led to consider the bounded solution of the ODE below (with both A and B positive and $B > A$)

$$\frac{1}{2}u_{xx} - V(x)u = -1 \qquad -\infty < x < \infty$$

$$V(x) = \begin{cases} B & \text{if } x \geq 0 \\ A & \text{if } x < 0 \end{cases}$$

and evaluating it at $x = 0$, to get its value to be

$$1/\sqrt{AB}$$

One then notices that this is already a double Laplace transform, namely

$$\frac{1}{\sqrt{AB}} = \int_0^\infty e^{-At} dt \int_0^t \left(\frac{1}{\pi} \frac{e^{-(B-A)s}}{\sqrt{(t-s)s}} \right) ds$$

This gives P. Levy's original result.

5. The two dimensional case

Here is the **PDE** that one needs to solve in the case of two players

$$\frac{1}{2}\nabla^2 u - V(x,y)u = -1 \qquad (x,y) \in R^2$$

$$V(x,y) = \begin{cases} B & \text{if } x, y \geq 0 \\ A & \text{otherwise} \end{cases}$$

Here A, B are arbitrary positive parameters. One only needs to evaluate the (bounded) solution at the origin, and then pray for some Laplace transform miracles.

It is very natural to embed this problem in the family of problems obtained when the opening of the wedge has magnitude γ. In the case of two players, Louis and Peter, we are dealing with a wedge in the plane of angle $\pi/2$. One could even dream of obtaining a *compatibility condition* for our function $u(0,0)$ as a function of all these different parameters, and maybe consider the corresponding nonlinear equation.

It is natural to assume that the value of u at the origin, with u the bounded solution of our problem, is given by a formula of the form

$$u(0,0) = \frac{1}{A^{\nu_1(\gamma)} B^{\nu_2(\gamma)}}$$

where the exponents are somehow related (maybe proportional to) the two angles in question, γ and its complement.

This is actually the case for $\gamma = \pi$ and the extreme case γ=0 or 2π.

Unfortunately, one can give a rigorous argument showing that this cannot be true for the case of interest, namely

$$\gamma = \pi/2.$$

If the value of the function $u(x,y)$ at $(0,0)$ were as suggested above one can see, by doing a double inverse Laplace transform as before, that the density of our random variable $\mu(w)$ should be given by a BETA distribution, i.e. the density would be

$$\frac{\Gamma(\alpha+\beta)}{\Gamma(\alpha)\Gamma(\beta)} t^{\alpha-1}(1-t)^{\beta-1} \qquad 0 < t < 1$$

$$\alpha, \beta > 0$$

This can be ruled out from the explicit computation of the first few moments reported earlier.

6. The cumulative distribution for several values of γ

In this section, we show a few more results of very careful numerical simulations undertaken in [**McG**].

For several values of the angle γ we display the cumulative distribution function of the corresponding random variable μ followed in the next plot by the histogram that gave rise to this distribution.

The values of γ for which we chose to dislay these plots are

$$\gamma = \pi/20, \pi/5, \pi/4, 11/20\pi, 3/4\pi.$$

The value $11/20\pi$ is chosen since it is close to the value where the distribution acquires an inflection point. This is reflected in the corresponding histogram.

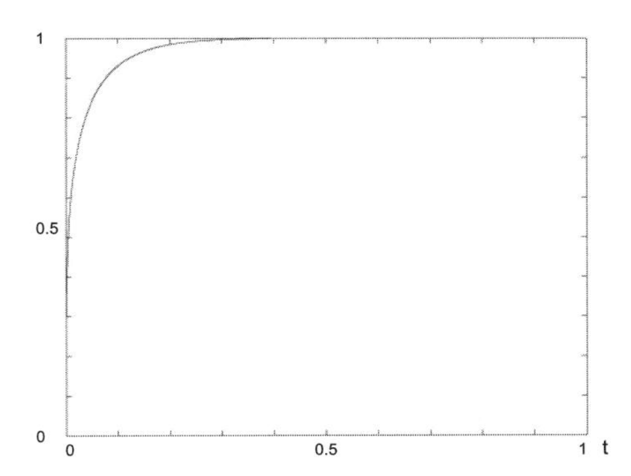

FIGURE 10. Distribution for $\gamma = \pi/20$

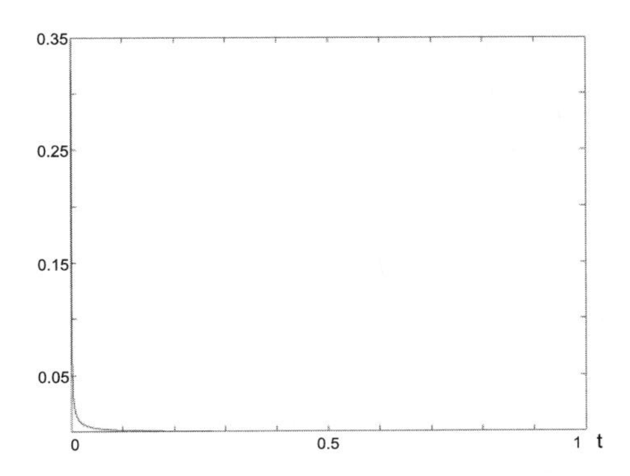

FIGURE 11. Histogram for $\gamma = \pi/20$

7. A related pair of integral equations

The purpose of this section is to indicate how the use of the so called Macdonald functions, sometimes known as modified Bessel functions, could allow one to solve the problem at hand.

Recall that we are trying to determine the value $u(0,0)$ of the bounded $\mathbb{C}^1$ solution of the equation

$$\frac{1}{2}\nabla^2 u - V(x,y)u = -1$$

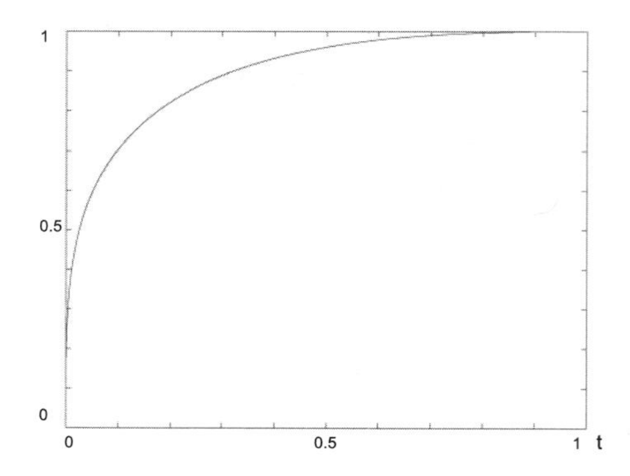

FIGURE 12. Distribution for $\gamma = \pi/5$

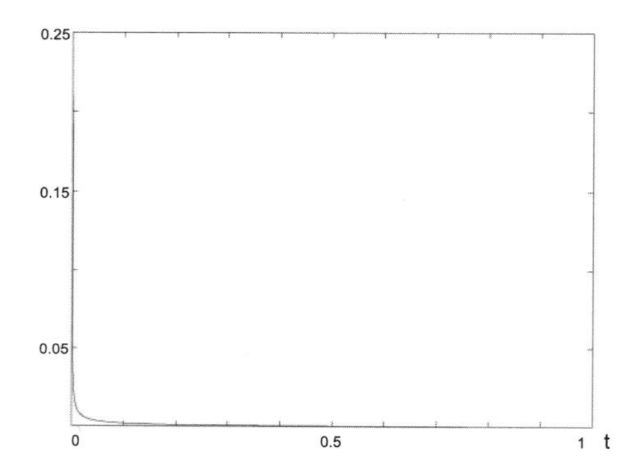

FIGURE 13. Histogram for $\gamma = \pi/5$

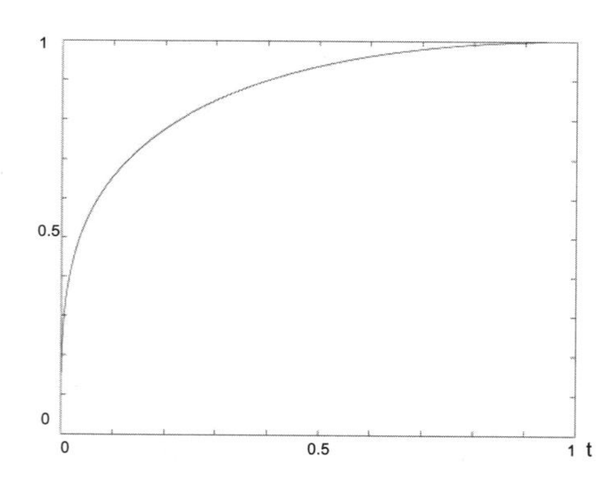

FIGURE 14. Distribution for $\gamma = \pi/4$

where

$$V(x,y) = \begin{cases} B & \text{in region I given by } |y| \leq x \\ A & \text{in region II, the complement of region I in the plane.} \end{cases}$$

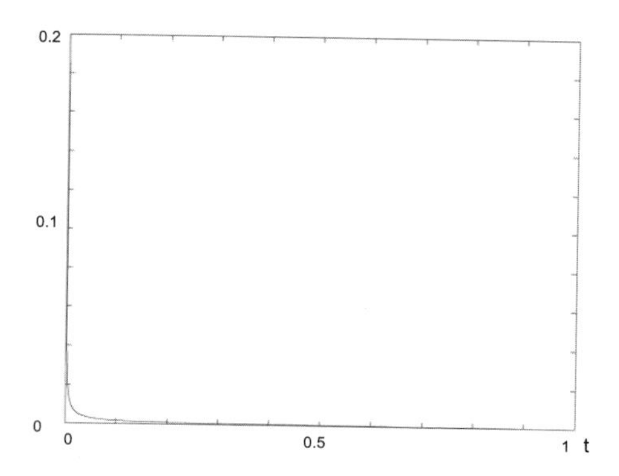

FIGURE 15. Histogram for $\gamma = \pi/4$

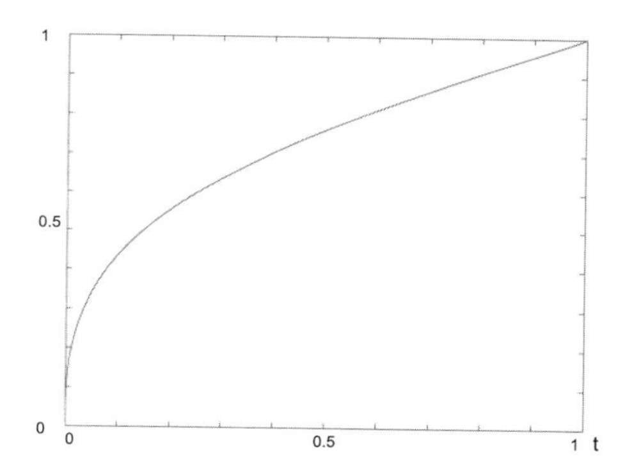

FIGURE 16. Distribution for $\gamma = 11/20\pi$

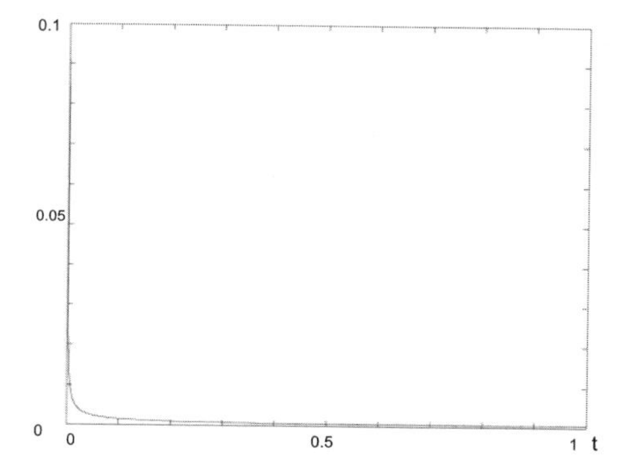

FIGURE 17. Histogram for $\gamma = 11/20\pi$

The smoothness conditions on $u(x, y)$ hold as one crosses the two lines that separate regions I and II. Notice that we have used the rotational invariance to replace the first quadrant by one obtained by a rotation of $\pi/4$.

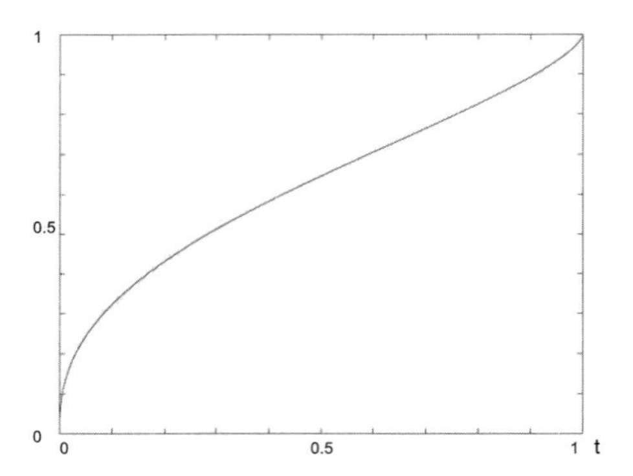

FIGURE 18. Distribution for $\gamma = 3/4\pi$

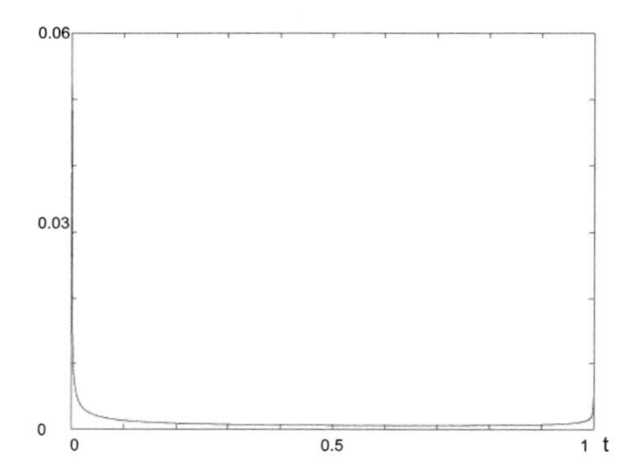

FIGURE 19. Histogram for $\gamma = 3/4\pi$

Here $A, B \geq 0$ and moreover $A \leq B$. One can see that

$$\frac{1}{B} \leq u(x, y) \leq \frac{1}{A}.$$

One can alse see that $\frac{1}{B}$ and $\frac{1}{A}$ are the radial limits as $r \to \infty$ inside of regions I and II, respectively.

We describe now an approach that could yield the value of $u(x, y)$ by solving a pair of integral equations. Bring in the so-called Macdonald function or modified Bessel function $K_{ik}(x)$ given by

$$K_{ik}(x) = \int_0^\infty e^{-x \cosh u} \cos(ku) du.$$

Here k is real and nonnegative. This is the kernel for the so-called Kontorovich–Lebedev transform.

We propose that in region I the function u should be given by

$$u(r, \theta) = \frac{1}{B} + \int_0^\infty \beta(k) K_{ik}(\sqrt{2B} r) \cosh k\theta dk$$

while in region II we should have

$$u(r,\theta) = \frac{1}{A} + \int_0^\infty \alpha(k) K_{ik}(\sqrt{2A}r) \cosh k\theta dk.$$

The functions $\alpha(k)$ and $\beta(k)$ are yet to be determined by trapping them in a pair of integral equations.

The continuity of u across the lines that bound regions I and II gives, for all $r > 0$

$$\frac{1}{B} + \int_0^\infty \beta(k) K_{ik}(\sqrt{2B}r) \cosh \frac{\pi k}{4} dk = \frac{1}{A} + \int_0^\infty \alpha(k) K_{ik}(\sqrt{2A}r) \cosh \frac{\pi k}{4} dk.$$

The continuity of the normal derivative of u across these same lines gives a second equation, namely

$$\int_0^\infty k\beta(k) K_{ik}(\sqrt{2B}r) \sinh \frac{\pi k}{4} dk = \int_0^\infty k\alpha(k) K_{ik}(\sqrt{2A}r) \sinh \frac{\pi k}{4} dk$$

for all $r > 0$.

It is clear that the solutions $\alpha(k), \beta(k)$ will depend on the free parameters A, B. If one were able to solve this pair of linear coupled equations for $\alpha(k), \beta(k)$ one could determine $u(0,0)$ as a function of A, B and then attempt a double inverse Laplace transform to yield the desired distribution.

References

[BD] Bingham, N. and Doney, R. , *On higher dimensional analogues of the arc-sine law*, J. Appl. Prob. 25, 1986.

[F] Feller, William, *An introduction to probability theory and its applications* Vol I, 3rd edition, J. Wiley, 1967.

[Fey] Feynman, R. , *Space-time approach to non-relativistic quatum mechanics*, Rev. Modern Phys. 20, 367–387, 1948.

[IM] Ito, K. and McKean, H. P., *Diffusion processes and their sample paths* Springer-Verlag, Berlin, 1965.

[K] Kac, M. , *On some connections between probability theory and differential and integral equations*, Proc. 2nd. berkeley Symp. on Prob. and Stat. 1951.

[Ko] Kotani, S. , *Probabilistic approach to reflectionless potentials*, Probabilistic methods in mathematical physics, Katata/Kyoto 1985 ed. by K. Ito and N. Ikeda, Academic Press, 1987, 219–250

[L] Levy, P., *Sur certain processus stochastiques homogenes*, Compositio Math. 7, 283–339, 1939.

[La] Lamperti, J. , *Probability* W.A. Benjamin, Inc. NY, NY, 1966.

[McG] McGrouther, Caroline, *Two-dimensional Brownian motion: a simulation study of the occupation time distribution for a cone*, Master Thesis,UC Berkeley, 2002.

[MW] Meyre, T. and Werner, W., *On the occupation times of cones by Brownian motion*, Prob. Theory Rel. Fileds, 101, 409–419, 1995.

[S] Stroock, D.. *Probability theory, an analytical view* , Cambridge University Press, 1993.

Department of Mathematics, University of California, Berkeley, CA 94720

School of Medicine, University of California, San Diego, CA 92093

Proceedings of Symposia in Applied Mathematics
Volume **65**, 2007

The semiclassical focusing nonlinear Schrödinger equation

Robert Buckingham, Alexander Tovbis, Stephanos Venakides,
and Xin Zhou

Dedicated to Peter D. Lax and Louis Nirenberg on their 80th birthdays

1. Overview

Our goal is to obtain the leading asymptotic behavior of the solution to the semiclassical focusing Nonlinear Schrödinger (NLS) equation,

$$i\varepsilon\partial_t q + \varepsilon^2\partial_x^2 q + |q|^2 q = 0, \qquad q(x,0) = A(x)e^{iS(x)/\varepsilon}, \tag{1}$$

as $\varepsilon \to 0$. The focusing NLS equation governs nonlinear transmission in optical fibers; its modulational instability (see below, also [**8**]), a source of problems in transmission, leads to a number of deep mathematical questions. A successful approach to this subject must address the following essential difficulties:

(a) The associated linear (Zakharov-Shabat [**19**]) eigenvalue problem (see below) is not selfadjoint; a lot of the analysis involves equations and inequalities along contours that are outside the real axis of the spectral variable;

(b) The semiclassical calculation of the scattering data (see below) is extremely delicate [**10**]; their determination for smooth nonanalytic initial data is still an almost completely open problem;

(c) The system of *modulation equations*, that govern the space-time large scale evolution of the wave parameters of quasiperiodic solutions has *complex characteristics* leading to modulational instability.

The existence of a particular initial amplitude $A(x)$ and phase $S(x)$ in (1) whose evolution overcomes the modulational instability and displays the ordered structure of modulated nonlinear waves was established by delicate numerical observations of Miller and Kamvissis [**13**]. They also observed wave breaking and trasnsition to more complex wave structures, phenomena that were later confirmed by numerical experiments by Ceniseros, Tian [**4**] , and Cai, Mc Laughlin (D.W.), Mc Laughlin (K. T-R) [**3**]. Analytic formulae describing such transitions were obtained by Kamvissis, McLaughlin and Miller [**9**] who studied a particular choice of pure soliton

The first author was supported by NSF Focused Research Group grant DMS-0354373.
The second author was supported by NSF grant DMS 0508779.
The third author was supported by NSF grant DMS 0207262.
The fourth author was supported by NSF grant DMS 0602344.

initial data. Analytic proof that such transitions do occur was obtained by Tovbis, Venakides, Zhou [**17**] who calculated the leading asymptotic solution to the initial value problem for a one-parameter family of initial data and showed that exactly one transition occurs for values of the parameter that corresponds to solitonless initial data. The same authors obtained formulae in terms of elementary functions for the breaking curve and for the wave parameters when time is large. Lyng and Miller [**12**] have proposed a mechanism for higher breaks in the pure soliton case; the mechanism is supported by a combination of numerical and analytical results.

The following paragraphs give a outline of the general phenomenology that has emerged following the above studies.

Separation of scales in space-time. A set of boundaries (curves) divides the x, t half-plane $t > 0$ into regions. When $t < t_0(x)$ (see figure 6), the solution has an amplitude/phase structure similar to the initial data; it is a plane wave (or 0-phase wave) with $O(\frac{1}{\varepsilon})$ wavenumber and frequency, both varying in space-time. The corresponding wavelength and waveperiod, that are of order $O(\varepsilon)$ consistent with the scaling of (1), constitute the small space-time scale in the solution. The fact that they vary smoothly in the large scale is referred to as *wave modulation*. The amplitude, determined in a simple calculation from the wavenumber and frequency (nonlinear dispersion relation) is also modulated as a result of the modulation of wavenumber and frequency. One may see directly from NLS that a plane wave solution is characterized by its wavenumber and frequency up to a constant factor of unit modulus.

When t becomes larger than the value $t_0(x)$, the solution breaks, in the sense that the slowly varying amplitude starts developing rapid oscillations as well. Two more wavenumber and frequency pairs are generated (bringing the total of real wave parameters to 6) and the solution is now characterized as a modulated 2-phase NLS wave. More breaks possibly occur as t increases further. In general, for initial data for which the character of the solution as a modulated wave is preserved by the evolution, the solution is a modulated n-phase NLS wave, characterized by $2n + 2$ real wave parameters that vary in space-time scale due to modulation. The oscillations generated after a 0-phase wave breaks are truely nonlinear, as opposed to the sinusoidal oscillations of a a 0-phase wave; hence, the attribute 0-phase for the plane wave.

The number of wave-phases n jumps across the region boundaries. These boundaries are somethimes referred to as *breaking curves*, *transitional curves* or *caustics* of the fully nonlinear problem. In the cartoon of figure 6, t increases from zero to a value at which the $n = 0$ initial phase *breaks* at a caustic point from which, as t increases further, a wave of higher n emerges and spreads in space. The waves thus created may break to form waves of an even higher phase and so on. Whether successive breaks occur for some types of initial data and what the breaking mechnisms are, are largely open research questions.

Breaking Curves (Nonlinear Caustics). The curves across which n jumps are the analog of the caustics of linear wave theory. For example, in the linearization of (1) in which the cubic term is omitted, the local wavenumber in the ε-scale $k(x, t) = S_x(x, t)$ develops a fold and becomes multi-valued in the geometric optics approximation $k_t + 2kk_x = 0$. The number of values k takes at (x, t) equals the number of wave-phases present at (x, t). In analogy to $k(x, t)$ of the linear case,

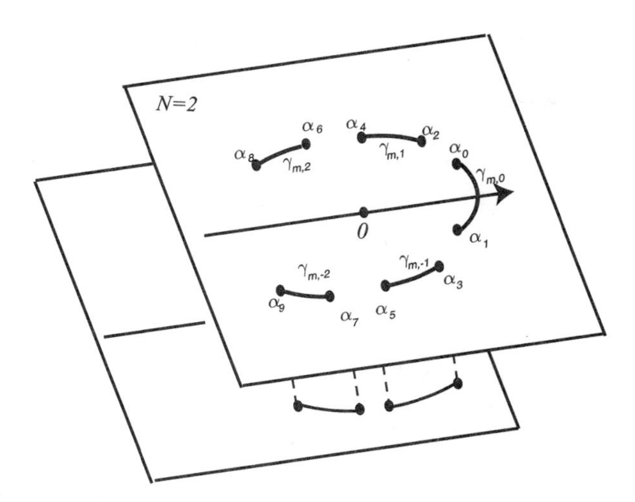

FIGURE 1. Riemann surface $\mathcal{R}(x,t)$; in this work, $\gamma_{m,i}$ is used below to denote the pair of arcs $\gamma_{m,i} \cup \gamma_{m,-i}$ of the figure.

there exists a complex *multivalued* function

$$\alpha(x,t) = (\alpha_0, \alpha_2, \alpha_4, \cdots, \alpha_{2n}), \qquad \Im a_{2j} > 0, \qquad n = n(x,t)$$

that describes the wave near (x,t) up to phase-shifts; its constitution and its time evolution differ greatly from the linear one. When α is independent of x and t, we have an exact (not modulated) quasiperiodic wave.

A mathematical description of the n-phase waves. Explicit expressions for n-phase wave solutions of integrable systems are given [**1**] in terms of the n-fold series of $\exp\{2\pi i \mathbf{z} \cdot \mathbf{m} + \pi i (B\mathbf{m}, \mathbf{m})\}$ summed over the multi-integer $\mathbf{m} = (m_1, \cdots, m_n)$. B is an $n \times n$ matrix with positive definite imaginary part that gives the series exponential quadratic convergence. The series, known as an n-phase Riemann theta function, has n complex arguments $\mathbf{z} = (z_1, \cdots z_n)$, each of which is of first degree in x and t and represents a phase of the wave; the series has a natural periodic structure. In the elliptic and hyperelliptic cases, which apply to NLS waves, the matrix B arises from periods of the elliptic or hyperelliptic Riemann surface (see figure 1) of the radical

$$R(z) = \left(\prod_{i=0}^{2n+1} (z - \alpha_i) \right)^{\frac{1}{2}}, \qquad \text{where} \quad \alpha_{2j+1} = \overline{\alpha}_{2j}; \tag{2}$$

meaning that the elements of B are linear combinations of integrals $\oint z^k/R(z),\ k = 0, 1 \cdots n-1$ along non-contractible closed contours on the Riemann surface of R [**1**]. The n-phase wave is, thus, parametrized by the $n+1$ complex quantities $\alpha_0,\ \alpha_2, \ldots, \alpha_{2n}$ that serve as the mathematically most convenient wave parameters; the wavenumbers and frequencies are expressed in terms of hyperelliptic integrals of the radical R and are thus functions of the α_{2j}s.

The semiclassical limit theory derives the emerging wave structure described above *without* any a priori assumption of such structure. The radical $R(z)$, and hence of the branchpoints $\alpha_j(x,t)$) that lie at the bottom of the wave structure, arise naturally in the asymptotic treatment of the Riemann-Hilbert problem (see below) to which solving (1) is reduced.

Outline of some results. The asymptotic behavior of the solution to the Riemann-Hilbert problem is the main focus of this work. We outline results in three domains. (I) The *semiclassical domain:* The NLS equation (1), (a) in the limit $\varepsilon \to 0$ with a one parameter family of initial data

$$A(x) = -\text{sech} x, \quad S'(x) = -\mu \tanh x, \tag{3}$$

[**17, 18**], (b) an extension of this calculation to a very broad class of initial data [**16**]. (II) Both semiclassical and long time regime (the order of limits is important) for these initial data. (III) The *long time domain with $\varepsilon = 1$ and shock initial data:* $A =$ constant $S'(x) = -\mu \text{sign} x$ in the limit $t \to \infty$. The approach utilizes the *steepest descent method for RHPs* introduced by Deift and Zhou [**6, 7**] and its extension through the g-function mechanism by Deift, Venakides, Zhou [**5**] that allows the recovery of fully nonlinear oscillations. The results include

1. Calculation of the scattering data for the semiclassical problem in terms of products of gamma functions [**15**]. When $\mu \geq 2$, the data are purely radiative; when $0 \leq \mu < 2$, there are $O(1/\varepsilon)$ eigenvalues spaced uniformly at $O(\varepsilon)$ spacing on the "positive" imaginary axis together with their complex conjugates.

2. Limit $\varepsilon \to 0$ [**17**], for the semiclassical problem: For *purely radiative* initial data, it is proven that there is only one breaking curve (caustic line) $t = t_0(x)$. Asymptotics of the curve are derived as $x \to 0$ and as $x \to \infty$. Leading asymptotic fromulae for the solution of (1) are derived both in the pre-break (0-phase)and post-break (2-phase) space-time regions. The results are global in time. In the *presence of solitons*, one can prove that the 2-phase regime cannot be sustained for all times; thus there should be some higher order regimes (return back to 0-phase is ruled out) beyond a certain curve $t = t_1(x) > t_0(x)$, or a *second break*; the above results are obtained up to times $t < (t_1(x)$ (the break mechanism at $t = t_1(x)$ is an ongoing research project). The derived asymptotic solution formulae are in terms of the radical $R(z)$; its branchpoints $\alpha_j(x,t)$ cannot be given by elementary functions, but the solvability of the system for these branchpoints is proven.

3. Limit $\varepsilon \to 0$, $t \to +\infty$ over a specified region of the t, ε plane [**18**], for (a): For purely radiative initial data, the branchpoints $\alpha_j(x,t)$ are pin-pointed asymptotically; their location and the asymptotic solution to NLS are expressed in terms of elementary functions.

4. Limit $\varepsilon \to 0$ [**16**], for a wide class of reflection coefficients (at $t = 0$): The prebreak solution and the breaking curve are calculated when the reflection coefficient satisfies certain conditions including analyticity in some sense. Gives insight to the way analyticity acts to neutralize the modulational instability.

5. The shock problem [**2**] in the long time limit: The asymptotic solution is obtained in the regions before the collision ($v_1 < |\frac{x}{t}|$, 0-phase), in the residual region ($|\frac{x}{t}| < v_2 < v_1$, 1-phase) and in the adjustment region ($v_2 < |\frac{x}{t}| < v_1$, 2-phase). A new mechanism for the generation of new phases is found, in which a complex conjugate pair of α_j emerges out of the real axis.

The Riemann-Hilbert approach to NLS. The NLS was solved by Zakharov and Shabat [**19**], who discovered a linearizing Lax pair [**11**] for it (see below for the solution process). The first operator of the pair is the Dirac operator with eigenvalue problem

$$\partial_x \Phi = -\frac{i}{\varepsilon}\begin{pmatrix} z & q(x,t,\varepsilon) \\ \bar{q}(x,t,\varepsilon) & -z \end{pmatrix}\Phi. \tag{4}$$

Here, $\Phi = \Phi(z;x,t,\varepsilon)$ is a 2×2 matrix valued wavefunction with spectral variable z; the (nonconstant) coefficient $q(x,t,\varepsilon)$ of this linear ODE system solves the NLS equation. We refer to (4) as the Zakharov-Shabat (ZS) eigenvalue problem. For a fixed value of the real parameter t, the NLS solution $q(x,t,\varepsilon)$ plays the role of the *scatterer* in this linear ODE system. The *scattering data* are obtained for those values of the spectral parameter z that produce bounded (ZS) solutions. For (real z), the *reflection coefficient* $r^0(z)$, *connects* the asymptotic behaviors of a wave solution of (ZS) as $x \to \pm\infty$, while the discrete spectrum, or eigenvalues z_j in complex conjugate pairs together with associated *norming constant* corresponding to bound states c_j characterize the square integrable solutions of (ZS). When $q(x,t,\varepsilon)$ satisfies NLS, the spectrum remains constant; the reflection coefficient and the norming constants evolve in t by a trivially solvable linear first order ODE. The reflection coefficient contributes *radiation* to the solution of NLS while the bound states contribute *solitons*.

Zakharov and Shabat [**19**] developed the inverse scattering procedure for deriving $q(x,t,\varepsilon)$ at any (x,t) given the scattering data. In the modern approach initiated by Shabat [**14**], the procedure is recast into a *Riemann-Hilbert problem* (RHP) for a 2×2 matrix on the complex plane of the spectral variable z. To be determined is the matrix

$$m(z) = \begin{pmatrix} m_{11} & m_{12} \\ m_{21} & m_{22} \end{pmatrix} = \Phi(z)\Phi_\infty^{-1}(z), \tag{5}$$

where $\Phi(z)$ is a fundamental matrix solution of (4) chosen by its asymptotic properties as $x \to \pm\infty$, and $\Phi_\infty(z)$ is the leading asymptotic behavior of $\Phi(z)$ as $z \to \infty$; clearly $m(z)$ satisfies the normalization condition

$$m(z) \to \text{Identity as } z \to \infty. \tag{6}$$

$m(z)$ is required to be analytic on the complex plane off an oriented contour Σ (analyticity). The contour consists of the real axis (points z corresponding to the scattering states) plus small circles about the bound states. Moreover, $m(z)$ satisfies a multiplicative jump condition

$$m_+(z) = m_-(z)V(z) \text{ when } z \in \Sigma. \tag{7}$$

where the subscripts $\pm$ indicate limits taken from the left/right of the contour. The jump condition (7) arises from the fact that matrices $\Phi_\pm$ are both fundamental solutions of the ODE system (4), thus, $\Phi_+ = \Phi_- J(z)$, where the matrix $J(z)$ is independent of x; the matrix V is related to the matrix J through the relation $V = \Phi_\infty J \Phi_\infty^{-1}$. The RHP consists of deriving $m(z)$ from its analyticity, from the knowledge of V in the jump condition (7) and from its behavior as $z \to \infty$. The RHP is linear, thus, the Lax pair [**11**] linearizes NLS.

The 2×2 matrix $V = V(z,x,t,\varepsilon)$, $z \in \Sigma$, is referred to as the *jump matrix* of the RHP. It is known if the scattering data are known because the choice of the fundamental matrix Φ is made only according to the asymptotic properties of Φ as $x \to \pm\infty$.

If $m(z)$ is obtained, the solution $q(x,t,\varepsilon)$ of (1) follows immediately. A convenient recovery formula for $q(x,t,\varepsilon)$ is derived from the Laurent expansion $m(z) =$

$I + \tilde{m}(x,t,\varepsilon)\frac{1}{z} + O\left(\frac{1}{z^2}\right)$ which gives

$$q(x,t,\varepsilon) = -2 \lim_{z\to\infty} z(m-I) = -2\tilde{m}_{12}(x,t,\varepsilon) \tag{8}$$

where the subscript indicates the 12 matrix entry.

The precise jump condition if the initial data decay fast enough as $|x| \to \infty$ is (see Appendix I)

$$m_+ = m_- \begin{pmatrix} 1+|r^2| & \overline{r} \\ r & 1 \end{pmatrix}, \quad r = r^{(0)}(z)e^{\frac{i}{\varepsilon}(2xz+4tz^2)}, \quad z \in \mathbb{R}\,, \tag{9}$$

where $r^{(0)}(z)$ is the reflection coefficient of the initial data. For positive parameter μ under study, solitons occur when $\mu < 2$; they manifest themselves as complex conjugate pairs of poles of the matrix $m(z)$. The effect of the jth pole on the Riemann-Hilbert problem, is an additional counter-clockwise closed contour (say, a circle) with jump matrix $\begin{pmatrix} 1 & 0 \\ \frac{c_j}{z-\zeta_j} & 1 \end{pmatrix}$ about the location of the pole.

2. Asymptotic Analysis: Decaying Initial Data as $|x| \to \infty$

The initial data and the scattering data. The study begins with initial data of the form (1) with

$$A(x) = -\text{sech} x; \qquad \frac{dS(x)}{dx} = -\mu \tanh x, \quad S(0) = 0, \quad \mu > 0 \tag{10}$$

The scattering data have been calculated exactly by Tovbis and Venakides [**15**] The reflection coefficient $r^{(0)}(z)$, defined on the real axis, is given by the formula

$$r^{(0)}(z) = \frac{b(z)}{a(z)} = -i\varepsilon 2^{-\frac{i\mu}{\varepsilon}} \frac{\Gamma(1-w+w_+ + w_-)\Gamma(w-w_+)\Gamma(w-w_-)}{\Gamma(w_+)\Gamma(w_-)\Gamma(w-w_+-w_-)}\,, \tag{11}$$

where

$$w_+ = -\frac{i}{\varepsilon}(T+\frac{\mu}{2}),\ w_- = \frac{i}{\varepsilon}(T-\frac{\mu}{2}),\ w = -z\frac{i}{\varepsilon} - \mu\frac{i}{2\varepsilon} + \frac{1}{2} \text{ and } T = \sqrt{\frac{\mu^2}{4} - 1}\,. \tag{12}$$

There are no bound states, hence no solitons, when $\mu \geq 2$. When $0 \leq \mu < 2$ there are bound states at $z_k = T - i\varepsilon(k - \frac{1}{2})$ with the corresponding norming constants

$$c_k^{(0)} = \frac{b(z_k)}{a'(z_k)} = \text{Res}_{\ z=z_k} r^{(0)}(z), \quad k = 1, 2, \dots < \frac{1}{2} + \frac{|T|}{\varepsilon}. \tag{13}$$

First jump matrix factoring and contour deformation. Jump matrix factoring is pivotal in the asymptotic solution of RHPs. When the jump matrix is a product V_1V_2, the left factor V_1 may split off to the right of the contour on a contour of its own provided it has an analytic continuation to the right. Similarly, the right factor $V_2(z)$ may split off to the left if it has an analytic continuation to the left. We factor the jump matrix on $z < \frac{\mu}{2}$,

$$V = \begin{pmatrix} 1+|r^2| & \overline{r} \\ r & 1 \end{pmatrix} = \begin{pmatrix} 1 & \overline{r} \\ 0 & 1 \end{pmatrix} \begin{pmatrix} 1 & 0 \\ r & 1 \end{pmatrix}. \tag{14}$$

An interesting cancelation occurs; the deforming contour anihilates the circular contours about each bound state as it interacts with them. Thus, after the deforming contour has swept through all poles, it remains the sole contour of the RHP. More precisely, the deforming RHP contour crosses poles by leaving a loop about each pole behind, in the same way as integration contours do. The jump on the loop and the jump on the RHP circular contour surrounding the pole cancel each other out.

Reversal of orientation and symmetry. We reverse the orientation of the contour in the upper half plane. The deformed contour of the RHP is shown in figure 3 panel 1. Now the jump matrix in the upper half plane becomes $V_{upper} = \begin{pmatrix} 1 & 0 \\ -r & 1 \end{pmatrix}$; it is the inverse of the second factor in (14), the inverse being taken to account for the change in orientation. The jump matrix in the lower half plane is the left factor $V_{lower} = \begin{pmatrix} 1 & \overline{r} \\ 0 & 1 \end{pmatrix}$. The jump matrices and the solution m to the RHP satisfy the symmetry relations

$$V_{lower}(\overline{z})V_{upper}(z)^* = \text{Identity}, \qquad m(z)m^*(\overline{z}) = \text{Identity}, \tag{15}$$

where superscript * stands for the matrix adjoint. All transformation of the problem will respect these symmetries; thus we only give jump matrices and relations in the upper half plane.

The Stirling approximation, pole condensation and redefinition of the initial data. We make a WKB-type approximation of the function $r(z)$ on the deformed contour; we replace the gamma functions by their Stirling approximation in (11). The effect of this is to *perturb the initial data.* For purely radiative initial data this is a higher order perturbation at $t = 0$. The same is true in the region $x > 0$ for initial data that contain solitons. For such data, recent work by Lyng and Miller [**12**] suggests that the solutions of the original and the perturbed problems eventually deviate. Assessing this rigorously and understanding how breaking mechanisms can differ in the two problems are subtle open questions.

Inserting the Stirling approximation, we obtain $r(z) = r^{(0)}(z)e^{(2ixz+4itz^2)/\varepsilon} = e^{(-2if_0(z)+2ixz+4itz^2)/\varepsilon}$ appearing in the jump matrix of the deformed contour from the equation

$$f_0(z) = (\frac{\mu}{2} - z)\left[\frac{i\pi}{2} + \ln(\frac{\mu}{2} - z)\right] + \frac{z+T}{2}\ln(z+T) + \frac{z-T}{2}\ln(z-T) \\ -T\tanh^{-1}\frac{T}{\frac{\mu}{2}} + \frac{\mu}{2}\ln 2, \tag{16}$$

where $f_0(z)$ is analytic in the upper complex half-plane (minus the imaginary segment $[0, T]$ when $\mu < 2$) and extends to the lower complex half-plane by Schwartz reflection, and where positive values have real logarithms. The function

$$f(z; x, t) = f_0(z) - xz - 2tz^2, \quad (\text{note } r(z; x, t) = e^{-\frac{2i}{\varepsilon}f(z)}) \tag{17}$$

encapsulates all the scattering information and constitutes the input to the RHP; it appears prominently in the subsequent analysis and forms the basis for generalizing the present asymptotic analysis to wider classes of initial data [**16**].

The function f acquires a logarithmic singularity at points $\pm T$ that are on the imaginary axis when solitons are present ($\mu < 2$); the slit between these points is taken as a branchcut (see fig. 4). The slit can be thought of as the continuum limit of the poles effected by the Stirling approximation; we refer to the approximation as *pole condensation.* It is seen below that the only possible breakdown of the solution following the first break is through the interaction of the RHP contour with the slit.

To avoid a technicality in this presentation, we also neglect the reflection coefficient when $z > \frac{\mu}{2}$ (proven in [**17**]).

Roadmap of the asymptotic solution of RHP. As with the steepest descent asymptotic evaluation of integrals, the issue is to *deform the contour* in a way that highlights the *main contribution* to the solution; we, then, say that the contour is at its *steepest descent position.*

Introduction of the g-function transformation : The transformation

$$m^{(1)} = m \begin{pmatrix} e^{\frac{2i}{\varepsilon}g(z)} & 0 \\ 0 & e^{-\frac{2i}{\varepsilon}g(z)} \end{pmatrix} \tag{18}$$

(due to upper-lower half plane symmetry, only expressions in the upper half plane are written) induces a corresponding transformation of the jump matrix $V \to V^{(1)}$

$$V^{(1)} = \begin{pmatrix} a & 0 \\ -b & a^{-1} \end{pmatrix}, \quad \begin{cases} a = e^{\frac{2i}{\varepsilon}(g_+ - g_-)} \\ b = e^{\frac{2i}{\varepsilon}(g_+ + g_- - f)} \end{cases} \tag{19}$$

In steps 2-8 below, we derive the phase function g by appropriately transforming the RHP while requiring that

(1) g is analytic off the RHP contour and Schwartz-reflection invariant,
(2) in the asymptotic limit $\varepsilon \to 0$,
- the jump matrix is piecewise independent of z (*constancy condition*);
- the jump matrix and its inverse are bounded uniformly in ε (*boundedness condition*).

These conditions lead to a solvable *model* RHP whose solution approximates the solution of the original RHP to leading order.

Factoring the jump matrix $V^{(1)}$: On a subset of the contour to be determined, typically a finite set of arcs called *main arcs*, we factor the jump matrix as described below. The rest of the contour, on which the jump matrix is not factored, consists of a number of *complementary arcs* plus two complex conjugate arcs to negative infinity . Figure 2 right panel lays out the geometry when there are five main arcs and four complementary arcs and introduces the notation $\gamma_{m,j}$ for the jth main arc and $\gamma_{c,j}$ for the jth complementary arc. Clearly, arcs $\gamma_{m,\pm j}$ are complex conjugate to each other; the same is true for arcs $\gamma_{c,\pm j}$. In general there $n+1$ main arcs and n complementary arcs; n is even due to symmetry with respect to the real axis and is referred to as the "genus" for a reason that will become clear shortly. All arcs inherit the orientation of the RHP contour.

The factorization of the jump matrix in the main arcs of the upper half plane is given by the formula

$$\begin{pmatrix} a & 0 \\ -b & a^{-1} \end{pmatrix} = \begin{pmatrix} 1 & -ab^{-1} \\ 0 & 1 \end{pmatrix} \begin{pmatrix} 0 & b^{-1} \\ -b & 0 \end{pmatrix} \begin{pmatrix} 1 & -a^{-1}b^{-1} \\ 0 & 1 \end{pmatrix}; \qquad \begin{cases} a = e^{\frac{2i}{\varepsilon}(g_+ - g_-)} \\ b = e^{\frac{2i}{\varepsilon}(g_+ + g_- - f)} \end{cases} \tag{20}$$

Clearly, for a product with three jump matrix factors $V_1 V_0 V_2$, if V_1 is analytic to the right and V_2 analytic on the left, the contour may deform to three contours, the initial one with jump V_0, plus one on the right with jump V_1 and one on the left with jump V_2 (see fig. 3, panel 2).

We enforce the constancy and boundedness conditions on the middle factor of (20) through the requirement that b have unit modulus and constant (independent of z) phase. We force the two other factors to tend to the identity matrix as

$\varepsilon \to 0$, when their contours deform; in order to achieve this, we require $ab^{-1} \to 0$ to the right of each arc (left factor) and $a^{-1}b^{-1} \to 0$ to the left of each arc (right factor). On the *complementary arcs* and the arc to negative infinity, we enforce the constancy and boundedness conditions in the unfactored matrix $V^{(1)}$, (left-hand side of (20)); we require a to have unit modulus and constant phase, and b to tend to the identity as $\varepsilon \to 0$.

Written in terms of g, the requirements in the upper complex half plane are:

(1) main arc,

$$g_+ + g_- - f = W = \text{real constant}; \quad \Im(2g - f) < 0 \text{ left and right of the contour}; \tag{21}$$

(2) complementary arc and arc to $-\infty$,

$$g_+ - g_- = \Omega = \text{real constant}; \quad \Im(2g - f) > 0 \tag{22}$$

Relations (21) imply that $\Im(2g_+ - f)$ is continuous and equal to zero on the main arcs,

$$\Im(2g_+ - f) = \Im(2g_- - f) = 0. \tag{23}$$

Relations (22) imply the *nonrigidity* of the complementary arcs; deforming the arcs while preserving the inequality automatically preserves the equality as well. In contrast, main arcs are *rigid* due to the equalities (23).

The conditions imposed on g in the upper complex half plane are mirrored by similar ones in the lower half plane through the upper/lower half plane symmetry for the jump matrix and from the Schwartz reflection invariance of the function g. The constant W takes generally different values on different main arcs; similarly Ω takes generally different values on different complementary arcs. By Schwartz reflection, the values of W on complex conjugate main arcs equal each other; the same is true for the values of Ω on complex conjugate complementary arcs. We normalize W to equal zero on the arc (α_1, α_0) and Ω to be zero on the arcs to negative infinity.

Derivation of formulae for g and introduction of the funciton h : Use of the radical $R(z)$ of (2) with the main arcs as branchcuts, shows that the jump of the function g/R across the full contour is known if the branchpoints $\alpha_0, \alpha_2, \cdots, \alpha_{2n}$ are known and if the real constants of equations (21) and (22) are known. Indeed,

$$\left(\frac{g}{R}\right)_+ - \left(\frac{g}{R}\right)_- = \begin{cases} \dfrac{g_+ + g_-}{R_+} = \dfrac{f + W}{R_+} & \text{on main arcs} \\ \dfrac{g_+ - g_-}{R} = \dfrac{\Omega}{R} & \text{on complementary arcs} \end{cases}. \tag{24}$$

Application of the Cauchy operator, yields the formula

$$g(z) = \frac{R(z)}{2\pi i}\left[\int_{\cup_j \gamma_{m,j}} \frac{f(\zeta)}{(\zeta - z)R(\zeta)_+} d\zeta \; + \; \sum_i \left(\int_{\gamma_{m,i}} \frac{W_i}{(\zeta - z)R(\zeta)_+} d\zeta \right.\right. \\ \left.\left. + \int_{\gamma_{c,i}} \frac{\Omega_i}{(\zeta - z)R(\zeta)}\right) d\zeta\right]. \tag{25}$$

Integration is over main and complementary arcs, as defined in the caption of Figure 2; $\cup_j$ stands for the union taken over the index j. Expressing the integrals over the

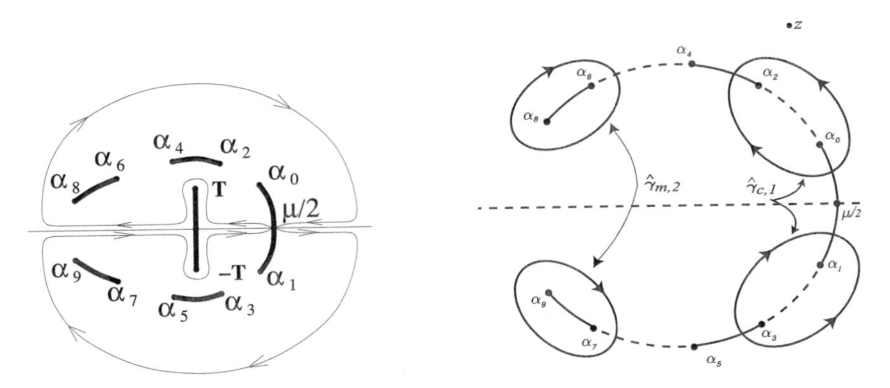

FIGURE 2. An example with five main arcs is shown. In formulae (25)-(28) the quantity $\gamma_{m,i}$ denotes the union of one main arc and its complex conjugate; the hatted quantiy $\hat{\gamma}_{m,i}$ denotes the union of two oriented loops about these two arcs as shown in the right panel ($i = 2$). The sum of all the loops $\hat{\gamma}_{m,i}$ equals the contour $\hat{\gamma}$ shown in the left panel. The quantiy $\gamma_{c,i}$ denotes the union of one complementary arc and its complex conjugate; the hatted quantiy $\hat{\gamma}_{c,i}$ denotes the union of four oriented arcs surrounding the complementary arcs $\gamma_{c,i}$ as shown in the right panel ($i = 1$).

arcs as integrals over the hatted contours defined in Figure 2, we obtain

$$g(z) = \frac{R(z)}{4\pi i}\left[\oint_{\hat{\gamma}} \frac{f(\zeta)}{(\zeta - z)R(\zeta)}d\zeta + \sum_i \left(\oint_{\hat{\gamma}_{m,i}} \frac{W_i}{(\zeta - z)R(\zeta)}d\zeta + \oint_{\hat{\gamma}_{c,i}} \frac{\Omega_i}{(\zeta - z)R(\zeta)}\right)d\zeta\right], \tag{26}$$

as long as the deforming paths of integration do not cut through z; thus, z is *outside* the contour $\hat{\gamma}$.

We now define the function

$$h(z) = 2g(z) - f(z) \tag{27}$$

when *z is inside the loop $\hat{\gamma}$* and still has not been cut by any of the other deforming contours. We obtain

$$h(z) = \frac{R(z)}{2\pi i}\left[\oint_{\hat{\gamma}} \frac{f(\zeta)}{(\zeta - z)R(\zeta)}d\zeta + \sum_i \left(\oint_{\hat{\gamma}_{m,i}} \frac{W_i}{(\zeta - z)R(\zeta)}d\zeta + \oint_{\hat{\gamma}_{c,i}} \frac{\Omega_i}{(\zeta - z)R(\zeta)}\right)d\zeta\right]. \tag{28}$$

The term $-f$ in the definition of h is simply the residue picked up as z cuts through the loop $\hat{\gamma}$.

If z approaches a main arc from either side, a residue is generated as z cuts through the corresponding contour $\gamma_{m,i}$; multiplied by the factor $\frac{R(z)}{2\pi i}$ outside the integral, the residue yields the contribution W_i to h. Similarly, if z approaches a complementary arc, the contribution to h from z cutting the corresponding contour

$\gamma_{c,i}$ is $+\Omega_i$ or $-\Omega_i$, depending on whether z approaches from the left or right of the contour. These observations lead directly to the relations $h_+ + h_- = 2W$ across main arcs and to $h_+ - h_- = 2\Omega$ across complementary arcs as requested by conditions (21) and (22); note that the bracket in (28) is analytic in a neighborhood of any α_j (we are assuming latter are distinct).

Behavior of $h(z)$ when z is near some $\alpha_j = \alpha$: From the above, it is clear that

$$h(z) \sim W + \pm\Omega + \nu_1(z-\alpha)^{\frac{1}{2}} + \nu_3(z-\alpha)^{\frac{3}{2}} + \cdots . \tag{29}$$

where $\pm$ refers to whether z is left or right of the contour.

The sign profile of the function $\Im h(z)$: The *sign profile* of the function $\Im h(z)$, plays a role that is central to the theory. Equation (23), written in terms of h, yields

$$\Im h(z) = 0, \qquad \text{when } z \text{ is on a main arc.} \tag{30}$$

The factorization of the jump matrix (20) on the main arcs, rewritten in terms of h, is

$$V^{(1)} = \begin{pmatrix} 1 & -e^{-\frac{2i}{\varepsilon}h_-} \\ 0 & 1 \end{pmatrix} \begin{pmatrix} 0 & e^{-\frac{2i}{\varepsilon}W} \\ -e^{-\frac{2i}{\varepsilon}W} & 0 \end{pmatrix} \begin{pmatrix} 1 & -e^{-\frac{2i}{\varepsilon}h_+} \\ 0 & 1 \end{pmatrix}. \tag{31}$$

In terms of h, the inequality conditions in (21) are that $\Im h < 0$ *on both sides of the main arc*; as $\varepsilon \to 0$, the left and right factors of (31) tend to the identity upon deformation.

Similarly, on a complementary arc, inequality (22) means $\Im h > 0$; the 21 entry of the jump matrix

$$V^{(1)} = \begin{pmatrix} e^{\frac{2i}{\varepsilon}\Omega} & 0 \\ -e^{\frac{2i}{\varepsilon}(h_+ + h_-)} & e^{-\frac{2i}{\varepsilon}\Omega} \end{pmatrix} \tag{32}$$

tends to zero (recall that $h_+ - h_-$ is real, hence $\Im h_+ = \Im h_- = \Im h$). Thus, asymptotically as $\varepsilon \to 0$, the jump matrix $V^{(1)}$ is piecewise constant on both the main and the complementary arcs,

$$\begin{pmatrix} 0 & e^{-\frac{2i}{\varepsilon}W} \\ -e^{-\frac{2i}{\varepsilon}W} & 0 \end{pmatrix} \text{ on main arcs,} \quad \begin{pmatrix} e^{\frac{2i}{\varepsilon}\Omega} & 0 \\ 0 & e^{-\frac{2i}{\varepsilon}\Omega} \end{pmatrix} \text{ on comp arcs.} \tag{33}$$

Figure 4 explains the mechanism of the first break in which a new main arc is generated; it shows the contour in the upper complex half plane and the sign profile of $\Im h$ during prebreak, at break and postbreak. Main arcs are shown thick, complementary arcs dashed. In both, arrows show the contour orientation.

Equations for the constants W_i and Ω_i: Formula (26) implies that $g(z)$ behaves like a polynomial with real coefficients of degree n as $z \to \infty$. The requirement that $g(z)$ be analytic at ∞ forces all the coefficients except the constant to be zero. These n real constraints are expressed as the n moment conditions

$$\oint_{\hat\gamma} \frac{\zeta^k f(\zeta)}{R(\zeta)} d\zeta + \sum_i \left(\oint_{\hat\gamma_{m,i}} \frac{W_i \zeta^k}{R(\zeta)} d\zeta + \oint_{\hat\gamma_{c,i}} \frac{\Omega_i \zeta^k}{R(\zeta)} d\zeta \right) = 0, \quad k = 0, 1, \cdots, n-1, \tag{34}$$

that we obtain by expanding $(\zeta - z)^{-1})$ in the integral in powers of z^{-1}. We solve these n equations explicitly and uniquely for the $n/2$ distinct values of W_i and the $n/2$ distinct values of Ω_i. Inserting the expressions for these values in (28) produces a formula for $h(z)$ solely in terms of the values of the branchpoints α_j.

Modulation equations: Necessarily in equation (29), $\nu_1 = 0$, otherwise the negative sign of $\Im h(z)$ left and right of the main arc from α would persist all around the branchpoint α; this would contradict the condition $\Im h(z) > 0$ on the complementary arc emanating from α. Thus,

$$h(z) \sim W_i \pm \Omega_i + \nu_3(z-\alpha)^{\frac{3}{2}} + \cdots \quad \text{as } z \to \alpha. \tag{35}$$

It follows that the bracket in equation (28), let us call it $B(z)$, vanishes at α, i.e.

$$B(z) = \oint_{\hat{\gamma}} \frac{f(\zeta)}{(\zeta-\alpha_{2j})R(\zeta)} d\zeta \; + \; \sum_i \left(\oint_{\hat{\gamma}_{m,i}} \frac{W_i}{(\zeta-\alpha_{2j})R(\zeta)} d\zeta + \oint_{\hat{\gamma}_{c,i}} \frac{\Omega_i}{(\zeta-\alpha_{2j})R(\zeta)} \right) d\zeta = 0, \tag{36}$$

where $j = 0, 1, \cdots, n$. The *modulation system of equations* (36) is a system of $n+1$ analytic conditions satisfied by the $n+1$ complex branchpoints α_{2j} and their complex conjugates α_{2j+1}; the system maybe used for the determination of the branchpoints for a given value of n. The determination of the α_js is a major step of our procedure that contains significant technical difficulties, especially on a caustic where the genus jumps. The analysis for the determination of the branchpoints [**17**] utilizes an alternative (equivalent) modulation system. *Alternative (equivalent) modulation system involving $h'(z)$:* A condition equivalent to $\nu_1 = 0$ is that g' (also h') are *bounded* in a neighborhood of the contour. This implies that we can calculate g' and h' by applying to them the procedure that we applied to the calculation of g and h; we obtain the formulae

$$g'(z) = \frac{R(z)}{4\pi i} \oint_{\hat{\gamma}} \frac{f'(\zeta)}{(\zeta - z)R(\zeta)} d\zeta, \quad z \text{ outside the contour } \hat{\gamma}, \tag{37}$$

and

$$h'(z) = \frac{R(z)}{2\pi i} \oint_{\hat{\gamma}} \frac{f'(\zeta)}{(\zeta - z)R(\zeta)} d\zeta, \quad z \text{ inside the contour } \hat{\gamma}. \tag{38}$$

Towards the determination of the α_i, we observe that the analyticity of $g(z)$ at ∞ implies $g'(z) \sim O(z^{-2})$ as $z \to \infty$. The latter is enforced by the $n+2$

$$\text{moment conditions } M_k : \oint_{\hat{\gamma}} \frac{\zeta^k f'(\zeta)}{R_+(\zeta)} d\zeta = 0, \qquad k = 0, 1, \cdots, n+1. \tag{39}$$

that we, again, obtain by expanding $(\zeta - z)^{-1})$ in powers of z^{-1}, in the integral formula for g'. To these moment conditions we adjoin the n *integral conditions* (called so because they were obtained in [**17**] by integrating the function h')

$$\int_{\alpha_{2k}}^{\alpha_{2k+2}} \Im h'(z) dz = 0, \qquad k = 1, \cdots, n. \tag{40}$$

We have a total of $2n+2$ real equations for the $n+1$ complex unknowns $\alpha_0, \alpha_2, \cdots, \alpha_{2n}$, called the system of moment and integral (MI) conditions . The solvability of this system, also referred to as the system of modulation equations, is examined in detail in [**17**]. *Determination of the genus $n = n(x,t)$:* The correct value of the genus $n = n(x,t)$ is obtained from the conditions

- the system of modulation equations yields the points $\alpha_0, \cdots \alpha_{2n}$,
- a contour exists that passes through these points and satisfies the obtained sign profile of the imaginary part of the function h..

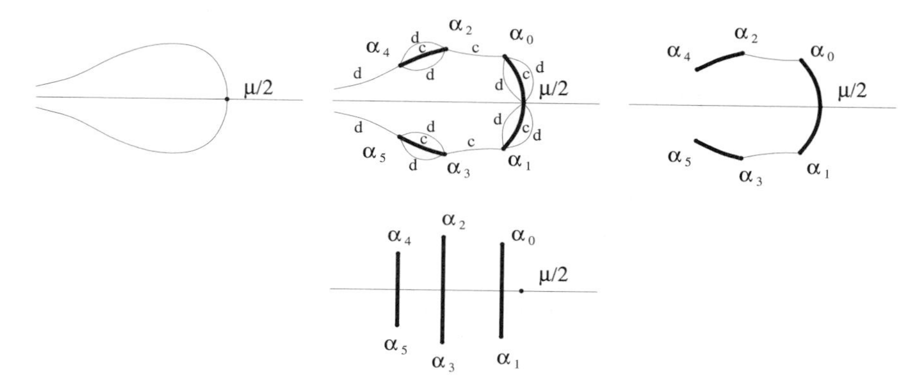

FIGURE 3. Pure radiation data: Succesive contour deformations; the real axis shown is not part of the contour. Main arcs are drawn bold; on arcs labeled d in the second panel, the jump matrix decays exponentially to the identity, while on arcs labeled c, the jump matrix is independnet of z. In the third and fourth panels, contours with jumps that decay exponentially to the identity as $\varepsilon \to 0$ have been omitted. A local parametrix is required near the endpoints (the α_js and $\frac{\mu}{2}$) of these contours at which there is no decay

Solution of the "model" RHP : the model RHP arises from neglecting the contours on which the jump matrix tends to the identity; the latter are labeled "d" in fig. 3. The solution to the model problem yields the leading behavior of the solution to NLS and is obtained by explicit formulae in terms of theta functions; this is ultimately attributable to the fact that a theta function associated with R satisfies a scalar RHP. The proof that, indeed, the solution to the model problem yields the leading behavior of the solution to NLS is based on the following two final steps.

Construction of parametrices: these are local solutions of the matrix RHP near "hot spots" (typically in the neighborhood of the α_j's and possibly of other special points, in our case $\frac{\mu}{2}$) where the convergence to the identity of the jump matrices neglected by the model problem is not uniform.

Error estimate: Estimation of the solution of the RHP that remains after the above parametrices and the solution of the model problem have been peeled off. This produces the error estimate in the calculation.

Uniqueness: Once an error estimate is established, the uniqueness of the procedure follows from the uniqueness of the solution of the initial value problem for the NLS equation. *The topology of the* $\Im h(z) = 0$ *contours and global* (x, t) *considerations*: For pure radiation initial data, $f(z)$ is analytic off the real axis. It is shown in **[17]** that n can equal at most 2, by examining all possible topologies. For data with solitons, new $\Im h(z) = 0$ contours may emerge from the interaction of the RHP contour with the slit $[0, T]$, making such an upper bound more elusive. The existence and mechanism of a second and higher breaks are subjects of current research.

Branchpoint degeneracies. Note that the formula (26) for g allows *degeneracies* in which main and complementary arcs collapse to points. Indeed:

- If the complementary interval $\gamma_{c,k}$ in the upper half plane collapses to a point, then $W_k = W_{k-1}$, the common value factors out of two terms in (26), the contours surrounding the two adjacent main arcs in the upper half-plane can be written as one contour surrounding both; no contour in the expression of $g(z)$ passes through the point of the collapsed interval and the collapsing produces no singularity in the formula;
- Similarly, if a main arc $\gamma_{m,k}$ collapses to a point, then $\Omega_k = \Omega_{k+1}$, the common value factors from two terms in (26) and the contours corresponding to the two adjacent complementary arcs can be replaced by a single one; no contour crosses to the lower sheet of the hyperelliptic Riemann surface $\mathcal{R}(x,t)$ at the point of the collapsed main arc and the collapse occurs without the appearance of a singularity in (26).

It is described below how this type of degeneracy occurs at the first break.

3. Analysis of the first break

For each genus n, we derived a system of nonlinear modulation equations (39) and (40)

$$F(\alpha, x, t) = F_n(\alpha, x, t) = 0 \tag{41}$$

for the $2n+2$ unknowns $\alpha = (\alpha_0, \alpha_1, \cdots, \alpha_{4N+1})$.

The Evolution and Degeneracy Theorems (see Appendix II) are our main tools to keep track of the time evolution of the Riemann surface $\mathcal{R}(x,t)$, associated with the solution. Leading order term $q_0(x,t,\varepsilon)$, reconstructed through $\mathcal{R}(x,t)$ in the genus zero and genus two regions, the breaking curve between these two regions, as well as error estimates are provided by Solution Theorem (Main Theorem, [**17**]).

Pre-break. The main result here is the construction of the function $h(z;x,t)$ ([**17**] section 4.2) with genus zero ($n=0$) when x,t are in a space-time region $0 \le t < t_0(x) < \infty$. System (41) has a single complex unknown, $\alpha_0(x,t) = \alpha = a + ib$, in the upper half plane. It reduces to a pair of real equations from which a and b are found. For all $t > 0$, there are two solution branches of these equations. Exactly one of these connects smoothly to the initial ($t=0$) values of $a = a(x,0) = \frac{\mu}{2}\tanh x$ and $b = b(x,0) = \operatorname{sech} x$. The connecting solution branch exists uniquely at all times for each $x \neq 0$; at $x = 0, t = \frac{1}{2(\mu+2)} = t_0(0)$ uniqueness is lost when the two branches yield the same α. In [**17**] sections 4.5 and 4.6, we show that for $0 \le t < t_0$ there is a zero level curve of $\Im h$ in the upper half plane connecting point $\mu/2$ to point $-\mu/2$, passing through the point α, and displaying the required sign profile $\Im h < 0$ on either side of a main arc, $\Im h > 0$ on a complementary arc (see figure 4 first panel). As shown in [**16**], in the pure radiational case $\mu \ge 2$, and even in the case of some wide class of pure radiational reflection coefficients described there (see also below), the existence of a solution to (41) at $t=0$ is equivalent to having "correct" signs of $\Im h(z)$ along the contour γ.

Breaking. Breaking occurs for a topological reason. For $x > 0$ fixed, and as time, increasing from zero, reaches the value $t = t_0(x)$, two branches of $\Im h = 0$ come in contact with each other at a point z_0 blocking the way of the arc from α_0 to $-\infty$ (figure 4 second). Necessarily $h'(z_0) = 0$,

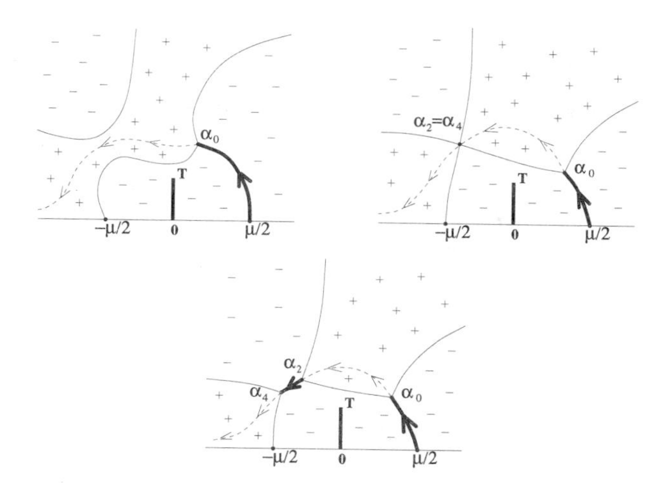

FIGURE 4. Prebreak, break and post break: Branches and sign profile of $\Im h$. Main arcs are bold, complementary arcs are dashed; generation of new main arc at breakpoint $\alpha_2 = \alpha_4$ is needed to keep the complementary arcs in regions of positive $\Im h$.

As t increases further, the four zero level curves of $\Im h(z; x, t) = 0$ emanating from z_0 as a result of the quadratic behavior of h at this point, interchange connections; a zero $\Im h$ level curve connection between points $\mu/2$ and $-\mu/2$, which is necessary for maintaining the "correct" signs of $\Im h$ along the contour γ, ceases to exist.

The breaking curve $t = t_0(x)$ is obtained in principle by eliminating z from the system of the three real equations $\Im h(z; x, t) = 0$, and $h'(z; x, t) = 0$. We have analytic formulae for the functions $h(z, x, t)$ and $h'(z, x, t)$, yet the above elimination of z cannot be performed explicitly. The existence of the breaking curve $t_0(x)$ is proved in [**17**] section 5.1; its asymptotic behaviors as $x \to \infty$ and $x \to 0^+$ are calculated explicitly in [**17**] section 5.3.

When the point (x, t) is on the breaking curve, the genus zero solution breaks down; the required sign structure on the part of the contour that connects α_0 to $-\frac{\mu}{2}$ is violated at the zero of h', i.e. at z_0. A degenerate genus two ($N = 1$) solution is obtained by identifying the point of contact z_0 of the two zero level curves of $\Im h$ with a double point $z_0 = \alpha_2 = \alpha_4$. As described in theorem 8.1 (Degeneracy Theorem, Appendix II), the function h, as a degenerate genus two solution of the scalar RHP, is identical to function h as a genus zero solution of the scalar RHP. This extends the region of our solution to $t \leq t_0(x)$.

In the case $x = 0$ and only in this case we have higher degeneracy at $t_0 = \frac{1}{2(\mu+2)}$ and $z_0 = i\sqrt{\mu + 2}$. Then the system (41) is satisfied with $\alpha_0 = \alpha_2 = \alpha_4 = z_0$ (triple point).

Post-break local calculation: $(t - t_0)$ small. As discussed above, we can solve (41) for genus two exactly on the breaking curve. We now show that we can solve the system for $n = 1$ in a vicinity of the breaking curve on either side of it (both pre and post break). We face the difficulty that the Jacobian $\frac{\partial F}{\partial \alpha}$ contains all factors of type $\alpha_k - \alpha_l$ and thus vanishes on the breaking curve, on which

$\alpha_2 = \alpha_4$. To overcome this, we make a change of the variables α ([**17**] section 6.2), a similar change of variable was made in [**9**]) so that the new Jacobian is different from zero in a neighborhood of the breaking curve. We then use the implicit function theorem to show the local solvability of (41) near the breaking curve for the three unknowns $\alpha_0, \alpha_2, \alpha_4$. The next step is to prove that the sign of $\Im h$ near the subcontour connecting α_2 and α_4, changes as we cross the breaking curve. This sign cannot satisfy the required sign profile of $\Im h$ in the pre-break region for, otherwise, conditions $(21) and (22)$ would be satisfied for $n = 0$ and $n = 1$; with the remainder of our analysis, *we would be able to construct two different asympotic behaviors for NLS*. Thus, the sign profile above the breaking curve (post-break region) and our solution, now of genus two, extends locally in a region above the breaking curve. The topology of zero level curves of $\Im h$ in the post-break region is shown in figure 4 third panel.

Post-break global calculation. In the pre-break region α_0 is found implicitly as solution to some transcendental equations. The leading asymptotic behavior of the solution of NLS in this region is given in terms of the α_0. For $\mu = 2$ and $\mu = 0$ expressions for α_0 are explicit.

By the Evolution Theorem (theorem 8.2), the existence of nondegenerate α for some x_0, t_0 implies the existence of nondegenerate α in a neighborhood of x_0, t_0 (if all α_{2k} are distinct then the Jacobian matrix $|\frac{\partial F}{\partial \alpha}| \neq 0$). Studying global solvability in the post-break region, i.e. solvability for all (or almost all) x, t in that region, includes two crucial elements: i) existence of a solution to (41) for at least one point x_0, t_0 with a nondegenerate α; ii) control over the degeneracy of α together with control over new breaks through collision of different branches of the zero level curve $\Im h = 0$ and through intersection of γ with singularities of $f(z)$.

Global solvability for solitonless initial data. We are able to establish the global post break solvability of (41) for genus two ($N = 1$) for the solitonless case $\mu \geq 2$ ([**17**] section 6.4) and to produce the leading asymptotic behavior of the solution of NLS in this region in terms of the α ([**17**] sections 7 and 8). Global post break solvability of (41) for genus two is achieved because we have control over the branches of the zero level curve $\Im h = 0$ (its branches can emanate only from the real axis and from infinity); we use this to show that the genus does not increse from two, and that the global solvability of (41) implies that the sign profile required for the function $\Im h$ can be satisfied everywhere above the breaking curve for $n = 2$. In the presence of solitons ($\mu < 2$), we can extend the solvability of (41) and calculate the leading order term $q_0(x, t, \varepsilon)$ of $q(x, t, \varepsilon)$ above the breaking curve as long as different branches of $\Im h = 0$ do not collide with each other or with the segment $[0, T]$, i.e. as long as the genus is equal to two. Global solvability of the g-function problem is still open.

Remarks on higher breaks. Higher breaks can occur in the presence of solitons when the RHP contour interacts with the slit. The mechanism of such interaction is still open. It is plausible that upon the contour touching the slit branchpoint T, a new branch of the contour will be generated. The roadmap for treating the problem on the entire contour, complemented by the new branch, will be the one described above, with the necessary adaptation due to the change of the topology. The proof of the solvability of the scheme in this case is still pending.

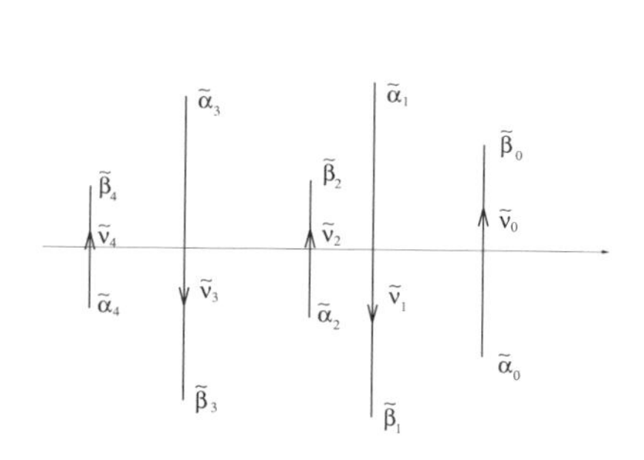

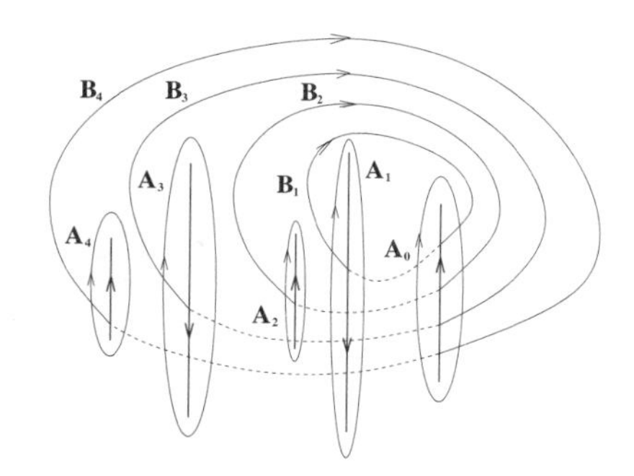

FIGURE 5. Example: The contour of a model problem with genus four; the contour, denoted by $\tilde{\Sigma}^{(mod)}$, consists of five vertical segments $\tilde{\nu}_k$ connecting the branchpoints α_{2k} and $\bar{\alpha}_{2k} = \alpha_{2k+1}$, $k = 0, 1, 2, 3, 4$; basic cycles $\boldsymbol{A}_j, \boldsymbol{B}_j$

4. Solution of the model problem and error estimation

Due to the Schwartz symmetry of the problem, the contour of the third panel of Figure 3 can be deformed to consist of vertical segments plus additional contours with exponentially decaying jump data that are neglected in the model problem (fourth panel of Figure 3) We solve such a model RHP that consists of $2N + 1$ vertical segments $\tilde{\nu}_k$ connecting the branchpoints α_{2k} and $\bar{\alpha}_{2k} = \alpha_{2k+1}$, $k = 0, 1, \cdots, 2N$ (see Figure 5). After removing the decaying portion of the jump and a normalization, we arrive at the piecewise constant model RHP with the following jump matrices

(42)

$$\tilde{V}^{(mod)} = \begin{cases} \begin{pmatrix} 0 & e^{-\frac{2i}{\varepsilon}(W_j - \Omega_{j+1} + \Omega_1)} \\ -e^{\frac{2i}{\varepsilon}(W_j - \Omega_{j+1} + \Omega_1)} & 0 \end{pmatrix} & \text{when } z \in \tilde{\nu}_{2j},\ j = 0, \cdots, N, \\ \begin{pmatrix} 0 & e^{-\frac{2i}{\varepsilon}(W_j - \Omega_j + \Omega_1)} \\ -e^{\frac{2i}{\varepsilon}(W_j - \Omega_j + \Omega_1)} & 0 \end{pmatrix} & \text{when } z \in \tilde{\nu}_{2j-1},\ j = 1, \cdots, N \end{cases}$$

To simplify the notation, we write the piece-wise constant matrix as

$$\tilde{V}^{(mod)} = \begin{pmatrix} 0 & e^{i\tilde{\Omega}} \\ -e^{-i\tilde{\Omega}} & 0 \end{pmatrix}, \tag{43}$$

where $\tilde{\Omega}$ attains the corresponding values on each vertical segment $\tilde{\nu}_j$, $j = 0, 1, \cdots, 2N$. We also denote by $\tilde{\alpha}_j$ and $\tilde{\beta}_j$ the beginning and the end points of the segment $\tilde{\nu}_j$ (so that $\tilde{\alpha}_0 = \alpha_1$, $\tilde{\beta}_0 = \alpha_0$, $\tilde{\alpha}_1 = \alpha_2, \cdots, \tilde{\beta}_{2N} = \alpha_{4N+1}$). As in Figure 5, we introduce the canonical homology basis $\boldsymbol{A}_j, \boldsymbol{B}_j$, $j = 1, \cdots, N$, of the hyperelliptic surface $\tilde{\mathcal{R}}(x,t)$, determined by the cuts $\tilde{\nu}_j$ of RHP $\tilde{P}^{(mod)}$. The dotted curves in Fig. 5 are passing through the second sheet. We introduce $u(z) = \int_{\tilde{\alpha}_0}^{z} \omega$, where $\omega = (\omega_1, \cdots, \omega_{2N})$ is the basis of holomorphic differentials dual to $\boldsymbol{A}$ cycles, i.e.

$$\int_{\boldsymbol{A}_j} \omega_k = \delta_{jk}\ , \tag{44}$$

$j, k = 1, \cdots, N$, and vector

$$\mathcal{M}(z,d) \equiv (\mathcal{M}_1, \mathcal{M}_2) = \left(\frac{\theta(u(z) - \frac{\hat{\Omega}}{2\pi} + d)}{\theta(u(z)+d)}, \frac{\theta(-u(z) - \frac{\hat{\Omega}}{2\pi} + d)}{\theta(-u(z)+d)} \right), \tag{45}$$

where $\hat{\Omega} = \left(\tilde{\Omega}_1, \cdots, \tilde{\Omega}_{2N}\right)^T$ and $d \in \mathbb{Z}^{2N}$ is a vector to be determined. Then, $\mathcal{M}$ satisfies

$$\mathcal{M}_+ = \mathcal{M}_- \begin{pmatrix} 0 & e^{i\tilde{\Omega}} \\ e^{-i\tilde{\Omega}} & 0 \end{pmatrix} \tag{46}$$

on $\tilde{\Sigma}^{(mod)}$ (see [20]).

Introduce

$$\lambda(z) = \left(\prod_{j=0}^{2N} \frac{z - \tilde{\beta}_j}{z - \tilde{\alpha}_j} \right)^{\frac{1}{4}} \tag{47}$$

with branch cuts along $\tilde{\nu}_j$. We choose the branch of $\lambda(z)$ such that $\lim_{z\to\infty} \lambda(z) = 1$ and that $\lambda_+ = i\lambda_-$ on $\tilde{\nu}_j$. One can verify directly that

$$\mathcal{L}_+ = \mathcal{L}_- \tilde{V}^{(mod)}, \tag{48}$$

where

$$\mathcal{L}(z) = \frac{1}{2} \begin{pmatrix} (\lambda(z) + \lambda^{-1}(z))\mathcal{M}_1(z,d) & -i(\lambda(z) - \lambda^{-1}(z))\mathcal{M}_2(z,d) \\ i(\lambda(z) - \lambda^{-1}(z))\mathcal{M}_1(z,-d) & (\lambda(z) + \lambda^{-1}(z))\mathcal{M}_2(z,-d) \end{pmatrix}. \tag{49}$$

If $\lambda(z) - \lambda^{-1}(z)$ has precisely $2N$ simple zeroes $z_1, \cdots, z_{2N}$, we choose

$$d = -\sum_{j=1}^{N} \int_{\tilde{\alpha}_j}^{X_2(z_j)} \omega_j, \tag{50}$$

where $X_2(z_j)$ is the preimage of z_j on the second sheet of the hyperelliptic surface to obtain a holomorphic solution $\mathcal{L}$ for the RHP.

Without the condition about zeros of $\lambda(z) - \lambda^{-1}(z)$, the model problem can still be solved with a more sophisticated expression for asymptotics as explained in the next section.

The successive deformations described in Figure 4 reduces the RHP to a model RHP on vertical slits plus "exponentially decaying" additional parts as $\epsilon \to 0$. However, these parts are exponentially decaying uniformly only away from the end points of the slits and the point $\mu/2$. The non-uniformness of this type was first dealt with in [7] in the study of asymptotics of PII equation, using parametrices near these nonuniform points. Near these points, parametrices are certain local solutions to the RHP, which may be used to obtain uniform estimates. In the particular work [16], the parametrices are constructed through Airy's functions, which are exactly what we need here at the end points of the slits. Thus it remains to construct the parametrix at $\frac{\mu}{2}$. In spite of the fact that we do not know how to construct such parametrix explicitly, we derive all required estimates based on the positivity of the jump matrix near $\mu/2$ using some general theory of RHP. In [17], such method shows that the error introduced by such parametrix is of order ϵ.

5. Main theorem for decaying initial data

The following theorem (Main Theorem, [**17**]) is an explicit statement of semi-classical limit for (1) that was discussed above.

THEOREM 5.1. *There exists a breaking curve $t = t_0(x)$, $x \in \mathbb{R}$, with the following properties:*

(1) *The genus of the solution $q_0(x,t,\varepsilon)$ is zero below the curve, i.e. in the region $0 \le t < t_0(x)$;*
(2) *In the solitonless (pure radiation) case $\mu \ge 2$ the solution in the entire region above the breaking curve has genus exactly two. In the case that includes solitons, i.e. $\mu < 2$, there exists some function $t_1(x)$, $x \in \mathbb{R}$, $t_0(x) < t_1(x) \le \infty$, such that the genus equals two in the region $t_0(x) < t < t_1(x)$, see Fig. 6;*
(3) *The breaking curve is an even function, smooth and monotonically increasing for $x > 0$ with the asymptotic behavior $t_0(x) \sim \frac{x}{2\mu}$ as $x \to +\infty$ and $t_0(x) = \frac{1}{2(\mu+2)} + 2\sqrt{\mu+2}\tan\frac{\pi}{5}x + o(x)$ as $x \to 0^+$;*
(4) *In the genus zero region $(0 \le t < t_0(x))$*

$$q_0(x,t,\varepsilon) = \Im\alpha_0(x,t)e^{-2\frac{i}{\varepsilon}\int_0^x \Re\alpha_0(s,t)ds}\ ; \tag{51}$$

(5) *In the genus two region $(t_0(x) < t < t_1(x))$*

$$q_0(x,t,\varepsilon) = \frac{2\theta(0)\theta(d_1)\theta(u(\infty)+\frac{\hat\Omega}{2\pi}+d_1)\theta(u(\infty)+\frac{\hat\Omega}{2\pi})}{\theta(u(\infty))\theta(u(\infty)+d_1)\theta(\frac{\hat\Omega}{2\pi}+d_1)\theta(\frac{\hat\Omega}{2\pi})}\left[\sum_{j=0}^{2}(-1)^j\Im\alpha_{2j}\right.$$
$$\left. - i\nabla\ln\frac{\theta(u(\infty)+\frac{\hat\Omega}{2\pi}+d_1)\theta(u(\infty))}{\theta(u(\infty)+d_1)\theta(u(\infty)+\frac{\hat\Omega}{2\pi})}\cdot\omega^0\right]e^{\frac{2}{\varepsilon\pi}a(x,t,\varepsilon)},\tag{52}$$
$$a(x,t,\varepsilon) = \int_{\gamma_m}\frac{f_0^{(0)}(\zeta)+\frac{\pi}{2}\varepsilon+x\zeta+2t\zeta^2+W}{R_+(\zeta)}\zeta^2d\zeta + \int_{\gamma_c}\frac{\Omega}{R(\zeta)}\zeta^2d\zeta + i\pi\Omega_1,$$

where: γ_m, γ_c denote union of all main and all complementary arcs respectively; the theta functions and the basic holomorphic differentials ω, dual to $\boldsymbol{\alpha}$-cycles $\boldsymbol{A}$, are associated with the hyperelliptic Riemann surface $\tilde{\mathcal{R}}(x,t)$, and the vector $\omega^0 \in \mathbb{C}^2$ is the leading coefficient of ω; $u(z) = \int_{\alpha_1}^z \omega$; $\hat\Omega_1 = -\frac{2}{\varepsilon}W_1$, $\hat\Omega_2 = -\frac{2}{\varepsilon}(W_1+\Omega_1)$ $d_1 = \frac{1}{2}(1,1)^T$; $f_0^{(0)}(z)$ is the leading order term of $\frac{i}{2\varepsilon}\ln r^{(0)}(z)$ as $\varepsilon \to 0$; $R(z) = \prod_{j=0}^{5}\sqrt{(z-\alpha_j)}$ and the branch $R_+(z) \to -z^3$ as $z \to \infty$; The real constant vectors W, Ω are determined through (34).

If, additionally, we asume that the function $\lambda(z) - \lambda^{-1}(z)$, where $\lambda(z)$ is expressed through the endpoints of the vertical cuts $\tilde\nu_j$ of $\tilde{\mathcal{R}}(x,t)$ as

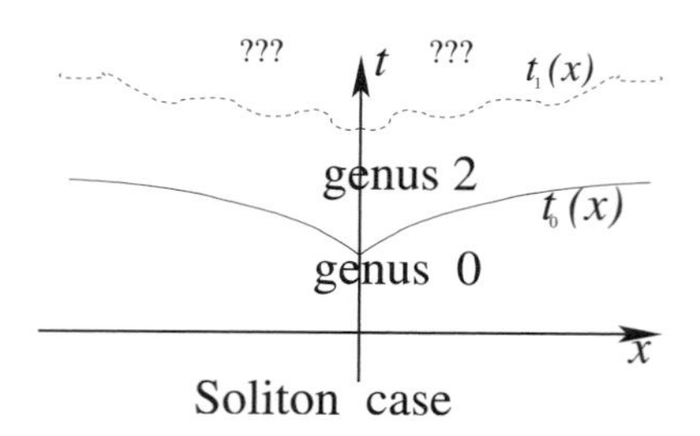

FIGURE 6. Regions in the (x,t) plane of different genus ; linear behavior as $x \to 0^+$, $x \to \infty$

$\lambda(z) = \left[\frac{(z-\alpha_0)(z-\alpha_3)(z-\alpha_4)}{(z-\alpha_1)(z-\alpha_2)(z-\alpha_5)}\right]^{\frac{1}{4}}$, *has 2 distinct zeroes* z_1, z_2, *then the expression for* $q_0(x,t,\varepsilon)$ *can be written as*

$$q_0(x,t,\varepsilon) = 2\frac{\theta(u(\infty) + \frac{\hat{\Omega}}{2\pi} - d)\theta(u(\infty)+d)}{\theta(u(\infty) - \frac{\hat{\Omega}}{2\pi} + d)\theta(u(\infty)-d)} \times$$

$$e^{\frac{2}{\varepsilon\pi}\left(\int_{\gamma_m} \frac{f_0^{(0)}(\zeta)+\frac{\pi}{2}\varepsilon+x\zeta+2t\zeta^2+W}{R_+(\zeta)}\zeta^2 d\zeta + \int_{\gamma_c}\frac{\Omega}{R(\zeta)}\zeta^2 d\zeta + i\pi\Omega_1\right)} \sum_{j=0}^{2}(-1)^j \Im\alpha_j \ . \tag{53}$$

Here $d = -\int_{\alpha_2}^{X_2(z_1)}\omega_1 - \int_{\alpha_5}^{X_2(z_2)}\omega_2$, *where* $X_2(z)$ *is the preimage of* z *on the second sheet of the hyperelliptic surface* $\tilde{\mathcal{R}}(x,t)$. *Similar to* (52) *expression for* $q_0(x,t,\varepsilon)$ *through theta functions, associated with the Riemann surface* $\mathcal{R}(x,t)$, *can be found in* (55) *below;*

(6) *The accuracy of the above leading term approximations is given by*

$$\mid q(x,t,\varepsilon) - q_0(x,t,\varepsilon) \mid = O(\varepsilon), \tag{54}$$

locally uniformly in x, t *away from the breaking curve.*

REMARK 5.2. *Another expression for* $q_0(x,t,\varepsilon)$ *in the genus two region is given by*

$$q_0(x,t,\varepsilon) = \frac{\theta(0)\theta(d_1)\theta(u(\infty)+\frac{\hat{W}}{2\pi}+d_1)\theta(u(\infty)+\frac{\hat{W}}{2\pi})}{\theta(u(\infty))\theta(u(\infty)+d_1)\theta(\frac{\hat{W}}{2\pi}+d_1)\theta(\frac{\hat{W}}{2\pi})}\left[\Im\sum_{j=1}^{2}(\alpha_{4j}-\alpha_{4j-2})\right.$$

$$\left. - i\nabla\ln\frac{\theta(u(\infty)+\frac{\hat{W}}{2\pi}+d_1)\theta(u(\infty))}{\theta(u(\infty)+d_1)\theta(u(\infty)+\frac{\hat{W}}{2\pi})}\cdot\omega^0\right] e^{\frac{2}{\varepsilon\pi}\int_{\gamma_m}\frac{f_0^{(0)}(\zeta)+\frac{\pi}{2}\varepsilon+x\zeta+2t\zeta^2}{R_+(\zeta)}P(\zeta)d\zeta}, \tag{55}$$

where: γ_m, γ_c *denote union of all main and all complementary arcs respectively; theta functions and the basic holomorphic differentials* ω, *dual to* $\boldsymbol{\alpha}$*-cycles* $\boldsymbol{\alpha}$, *are associated with the hyperelliptic Riemann surface* $\mathcal{R}(x,t)$, *and the vector* $\omega^0 \in \mathbb{C}^2$ *is the leading coefficient of* ω; $u(z) = \int_{\alpha_1}^{z}\omega$; $\hat{W} = -2\frac{i}{\varepsilon}\left(W + 2\int_{\gamma_c}\Omega\omega\right)$; $d_1 = \frac{1}{2}(1,1)^T$; $f_0^{(0)}(z)$ *is the leading order term of* $\frac{i}{2\varepsilon}\ln r^{(0)}(z)$ *as* $\varepsilon \to 0$; *the quadratic polynomial* $P(z) = z^2 + \sum_{j=\pm1}\int_{\gamma_{m,j}}\frac{\zeta^2 d\zeta}{R_+(\zeta)}\frac{\omega_j}{dz}$, *where* $R(z) = \prod_{j=0}^{5}\sqrt{(z-\alpha_j)}$ *and the branch* $R_+(z) \to -z^3$ *as* $z \to \infty$; *The real constants* W, Ω *are determined through* (34).

As it was mentioned in Subsection 3, control of branchpoints α in the genus two region is an important part of Theorem 5.1. In particular, we need to show that $\alpha_0, \alpha_2, \alpha_4$ do not collide insige the upper half-plane and cannot reach the boundary of the upper half-plane (in the latter case, they will collide with their complex conjugated branchpoints). Collisions inside the upper half-plane are ruled out through our control of signs of $\Im h_x(z)$ and $\Im h_t(z)$ ([**17**], Section 6.4), where the latter derivatives have explicit expressions ([**17**], Section 4.3)

$$h_x(z,x,t) = -R(z) \text{ and } h_t(z,x,t) = -2(z+a)R(z) \ . \tag{56}$$

To prove uniform boundedness of α in the genus zero region, as well that uniform separation from the real axis on compact subsets of this region, we use a collection of *a priori inequalities* ([**18**], Section 3) derived from the MI system (41)

Long time behavior of the semiclassical solution in the purely radiational case. Another interesting application of a priori inequalities, derived from the MI system (41), is the long time behavior of the leading semiclassical (small dispersion) term $q_0(x,t,\varepsilon)$ along the rays $x = \xi t$ of the x,t-plane. It is worth mentioning that the large time and small dispersion limits do not commute, so that we, in fact, calculate the large t limit of $q_0(x,t,\varepsilon)$. The central step in finding this limit is calculating of long time behavior of $\alpha(x,t)$, which are determined through (41) for a given x,t. Deforming the contour $\hat\gamma$ in the MI system (41) and using the fact that $f(z)$ has no singularities in the upper half-plane (solitonless case), the system (41) in the genus 2 region becomes

$$\begin{aligned}
&(M_0) \int_{|\zeta|\geq T} \frac{\text{sign}\zeta d\zeta}{|R(\zeta)|} = 0 \\
&(M_1) \int_{|\zeta|\geq T} \frac{\zeta\text{sign}\zeta d\zeta}{|R(\zeta)|} = 8t \\
&(M_2) \int_{|\zeta|\geq T} \frac{\zeta^2\text{sign}\zeta d\zeta}{|R(\zeta)|} = 2x + 8t\sum_{j=0}^{2} a_{2j} \\
&(M_3) \int_{|\zeta|\geq T} \frac{[\zeta^3\text{sign}\zeta - |R(\zeta)|]d\zeta}{|R(\zeta)|} = 2x\sum_{j=0}^{2} a_{2j} + 8tQ(\alpha) - \mu + 2T \\
&(I_1)\ \Im \int_{a_2}^{\alpha_2}\int_{|\zeta|\geq T} \frac{R(z)\text{sign}(\zeta)d\zeta dz}{(\zeta-z)|R(\zeta)|} = \pi\left(\frac{\mu}{2} - |\hat a_2|\right)\text{sign}(\frac{\mu}{2} - a_2) \\
&(I_2)\ \Im \int_{a_4}^{\alpha_4}\int_{|\zeta|\geq T} \frac{R(z)\text{sign}(\zeta)d\zeta dz}{(\zeta-z)|R(\zeta)|} = \pi\left(\frac{\mu}{2} - |\hat a_4|\right)\text{sign}(\frac{\mu}{2} - a_4) \ .
\end{aligned} \tag{57}$$

Here $T = \sqrt{\frac{\mu^2}{4} - 1}$, $\alpha_j = a_j + ib_j$, the quadratic form $Q(\alpha) = \frac{1}{2}\sum_{j<k}(a_{2j}+a_{2k})^2 - \frac{1}{2}\sum_{j=0}^{2} b_{2j}^2$ and $\hat a$ denotes $\max\{a,T\}$ when $a \geq 0$ and $\min\{a,-T\}$ when $a < 0$. The integrals in M_2, M_3 are principle value integrals. Here we used the fact that on the real axis $\Im f'(\zeta) = \frac{\pi}{2}\text{sign}\zeta$ if $|z| \geq T$ and $\Im f'(\zeta) = 0$ if $|z| < T$.

In fact, there is a heuristic way of calculating the long time limit behavior of α's that is based on the assumption that all $\alpha's$ are approaching the real axis as $t \to \infty$. It leads to a symplified system of MI conditions by replacing factors $\frac{\zeta - a_j}{|\zeta - \alpha_j|}$ and $\frac{1}{|\zeta - \alpha_j|}$, $j = 0,2,4$, in the integrals of (57) by $\text{sign}(\zeta - a_j)$ and $\delta(\zeta - a_j)$ respectively, where δ denotes the Dirac delta-function. However, a rigorous proof of the following theorem ([**18**], Section 4) requires the use of a priori inequalities.

THEOREM 5.3. *If $\mu \geq 2$ and $\xi \in [0, 2\mu)$, then the branch-points $\alpha_0, \alpha_2, \alpha_4$ converge to $\frac{\mu}{2}, -\frac{\xi}{4}, -\frac{\mu}{2}$ respectively as $t \to \infty$, $x = \xi t$.. The convergence of a_0, a_4 to $\frac{\mu}{2}, -\frac{\mu}{2}$ respectively is exponentially fast. Moreover,*

$$(58)\qquad \begin{aligned} &\ln b_4 = -4(\frac{\mu}{2} - \frac{\xi}{4})t + O(1), && \ln b_0 = -4(\frac{\mu}{2} + \frac{\xi}{4})t + O(1)\ , \\ &b_2 = \frac{\sqrt{\frac{\mu}{2} - \max\{\frac{\xi}{4}, T\}}}{\sqrt{2t}}\left(1 + O(t^{-\frac{1}{2}})\right), && a_2 = -\frac{\xi}{4} + \kappa(\frac{\xi}{4})\frac{\ln t}{8t} + O(\frac{1}{t})\ , \end{aligned}$$

where: $\kappa(s) = 0, \frac{1}{2}, 1$ if $s < T$, $s = T$ or $s > T$ respectively in the case $T > 0$; $\kappa(0) = 0$ and $\kappa(s) = 1$ if $s > 0$ in the case $T = 0$.

Using similar ideas, we can refine the long time asymptotics of the breaking curve (see Theorem 5.1) in the x, t plane and find the asymptotics of the corresponding double point $\alpha = \alpha_2 = \alpha_4$, [**18**], Section 5.

THEOREM 5.4. *The function $t = t_0(x)$, definding the breaking curve l, has asymptotics*

$$(59)\qquad t_0(x) = \frac{x}{2\mu} - \frac{1}{2\mu}\ln\frac{2\mu}{\mu + 2T} - \frac{T/\mu}{\mu + 2T} + O(\frac{1}{x}),$$

as $x \to \infty$. Moreover, along this curve

$$(60)\qquad \begin{aligned} b &= \frac{\pi}{8t}\left(1 - \frac{\mu^2/2 - 1}{4\mu t} + O(t^{-2})\right), \\ a &= -\frac{\mu}{2} - \frac{1}{4t}\left[\ln\frac{2\mu^2}{\mu + 2T} + \frac{2T}{\mu + 2T}\right] + O(t^{-2})\ . \end{aligned}$$

as $t \to \infty$, where $\alpha = a + ib$ is the double branch-point.

Our error estimate analysis ([**18**], Section 6) show that our results are valid for $t \to \infty$ as long as $\varepsilon = o\left(te^{-8t[\frac{\mu}{2} + \frac{\xi}{4}]}\right)$. It allows us ([**18**], Section 7) to calculate the leading order behavior of the solution $q(x, t, \varepsilon)$ along the rays $x = \xi t$ as $t \to \infty$.

THEOREM 5.5. *If $\mu \geq 2$ and $\xi \in [0, 2\mu)$, then the leading order behavior of the solution $q(x, t, \varepsilon)$ to (1),(3) as $\varepsilon \to 0$, $t \to \infty$ and $\varepsilon = o\left(b_0^2\sqrt{|\ln b_0|}\right)$ along the ray $x = \xi t$ is given by*

$$(61)\qquad \begin{aligned} q(x, t, \varepsilon) = -\sqrt{\tfrac{\frac{\mu}{2} - \max\{\frac{\xi}{4}, T\}}{2t}}\, e^{\frac{i}{\varepsilon}\left(\frac{t\xi^2}{4} + \ln t\left[\frac{\mu}{2} - \max\{\frac{\xi}{4}, T\}\right]\right)(1 + O(t^{-1}))}(1 + O(t^{-\frac{1}{2}})) \\ + O\left(\tfrac{\varepsilon}{b_0^2|\ln b_0|}\right), \end{aligned}$$

where $\ln b_0 = -4t\left(\frac{\mu}{2} + \frac{\xi}{4}\right) + O(1)$.

The long term behavior (based on Theorem 5.4) of q along the breaking curve $t_0(x)$ in the limit $t \to \infty$ is given by

$$(62)\qquad q_0(x, t, \varepsilon) = -e^{\left\{-\left(1 + \frac{i\mu}{2\varepsilon}\right)\left[3\mu t + \ln\frac{\mu}{2T + \mu} + \frac{2T}{2T + \mu}\right](1 + O(t^{-1}))\right\}}\ .$$

This result is formal in the sense that no error estimate for $q_0(x, t, \varepsilon)$ on the breaking curve has been provided. Making formal comparison of (61) and (62) for $\xi = 2\mu$, one can observe the abrupt change of phase near the breaking curve. Finally, ([**18**], Section 8) it was established that branches of zero level curve of $\Im h(z)$ in the upper half-plane approach the real axis and the vertical ray $\Re z = -\frac{\xi}{4}$ as $t \to \infty$ along $x = \xi t$.

General pure radiational case. It is widely accepted that solutions of the focusing NLS in the semiclassical limit are very unstable (modulation instability). However, working with the modified reflection coefficient $r(z) = r^{(0)}(z)e^{(2ixz+4itz^2)/\varepsilon} = e^{(-2if_0(z)+2ixz+4itz^2)/\varepsilon}$, where $f_0(z)$ is given by (16), in the solitonless case we observe that the MI conditions can be formulated in terms of $\Im f'(\zeta)$, $\zeta \in \mathbb{R}$, only. That indicates that, perhaps, our method of asymptotic solution of the solitonless inverse scattering problem is "robust" within a certain class of reflection coefficients. This idea was worked out in [**16**]. In particular, a wide class of reflection coefficients $r^{(0)}(z) = e^{-2if_0(z)/\varepsilon}$, $\zeta \in \mathbb{R}$, for which we can repeat the arguments outlined in Sections 2-3 was considered there. Without going into details, we can describe this class by the following requirements:

- $\Im f_0(\zeta) > 0$ on some interval $(\mu_-, \mu_+) \subset \mathbb{R}$ and negative elsewhere on $\mathbb{R}$;
- $\Im f_0(\zeta)$ is a piece-wise $C^1(\mathbb{R})$ function that has simple zeroes at μ_-, μ_+;
- (analyticity condition) there exists an analytic in $z > 0$ function $f_0(z)$, such that the function $w'(\zeta) = \lim_{z\to\zeta} \Im f_0'(z)$, $\zeta \in \mathbb{R}$, is defined on $\mathbb{R}$ and coincide with $\Im f_0'(\zeta)$ on (μ_-, μ_+);
- $\exists k \geq 0$ such that $w'(\zeta) - k = o(1)$ as $\zeta \to \infty$ and $w'(\zeta) - k \in L^1(\mathbb{R})$.

The latter two requirements seem to be essential for our method of asymptotic solution of the inverse scattering problem, whereas the requirements of only one positive interval (μ_-, μ_+) and of simple zeroes at $\mu_\pm$ of $\Im f_0(\zeta)$ seem to allow certain relaxation. In particular, the function

$$w(z) = \frac{1}{2}\left[w_0(z+T) + w_0(z-T)\right] + \frac{\pi}{2}\left(\frac{\mu}{2} - 1\right) \tag{63}$$

corresponds to $f_0(z)$ given by (16), where $T = \sqrt{\frac{\mu^2}{4} - 1}$. Another example studied in [**18**], Section 4 is

(64)

$$w(z) = w_c(z) = \begin{cases} \frac{\pi}{2}\left[\frac{1}{2}(1-|z+c|) + \frac{1}{2}(1-|z-c|) - c\right] & \text{if} z < 1-c \\ 0 & \text{if} 1-c \leq z \leq 1+c \\ -\frac{\pi}{2}(z-(1+c)) & \text{if} z > 1+c\ , \end{cases}$$

where $0 \leq c < 1$. Yet another example of $w(\zeta)$ that yields a compactly supported $q_0(x,t,\varepsilon)$ is under consideration now.

6. Long-time asymptotics of the shock problem

For the shock problem consider the NLS equation

$$i\partial_t q + \frac{1}{2}\partial_x^2 q + |q|^2 q = 0 \tag{65}$$

with initial condition

$$q(x,0) = Ae^{-i\mu|x|} \tag{66}$$

as $t \to 0$, where A and μ are real, positive constants. Equation (65) is the same as (1) with $\varepsilon = 1$ and an x scaling chosen to match results in [**2**]. If the initial condition was simply $Ae^{-i\mu x}$ (respectively, $Ae^{i\mu x}$), the solution would be a plane wave with negative (respectively, positive) velocity. Therefore, the initial condition (66) is thought of as two plane waves colliding at the origin at time zero. A novel feature is that, during evolution, two complex conjugate branchpoints $\alpha_{\pm i}$ emerge from the real axis, generating a novel contour geometry in the determination of the g-function; thus, in contrast to the initial data described in the previous sections, the genus n may take odd values. There are space-time regions of genus $n = 0$ (preshock), $n = 1$ (postshock, residual), $n = 2$ (postshock, transition).

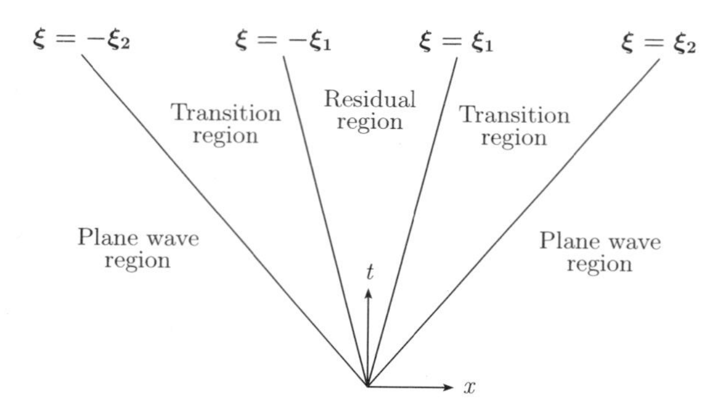

FIGURE 7. The three asymptotic solution regions for the shock problem.

The long-time solution has three qualitatively different behaviors depending on the value of $\xi = \frac{x}{t}$ (see figure 7). For $|\xi| > \xi_2 = \frac{\mu^2+A^2}{\mu}$, that is, outside the shock front, the leading-order solution is a phase-perturbed plane wave (see theorem 6.1A). The perturbation in the phase, which decays to zero as $|\xi| \to \infty$, shows the collision is felt far beyond the shock front. For $\xi_1 < |\xi| < \xi_2$, where ξ_1 depends only on A and μ and is defined below, the solution enters a transition region where the leading-order term is given in terms of Riemann theta functions with two nonlinear phases (see theorem 6.1B). For fixed x, the solution eventually settles down into a residual region. For $|\xi| < \xi_1$, the leading-order solution is given in terms of Riemann theta functions with one nonlinear phase (see theorem 6.1C).

The scattering data and RHP. The initial data have no point spectra. However, the reflection coefficient has two branch cuts that may be thought of as the continuum limit of infinitely many poles arising from the nondecaying nature of the initial data. The time evolution of the reflection coefficient is also affected by the nondecaying initial data. Define

$$\lambda_L(z) = \left(\left(z + \frac{\mu}{2}\right)^2 + A^2\right)^{1/2}, \quad \lambda_R(z) = \left(\left(z - \frac{\mu}{2}\right)^2 + A^2\right)^{1/2} \tag{67}$$

so $\lambda_L, \lambda_R \to z$ as $z \to \infty$, the branch cut for λ_L is a vertical line segment from $-\frac{\mu}{2} - iA$ to $-\frac{\mu}{2} + iA$, and the branch cut for λ_R is a vertical line segment from $\frac{\mu}{2} - iA$ to $\frac{\mu}{2} + iA$. At time $t = 0$ the reflection coefficient is

$$r^{(0)} = \frac{-\lambda_R - (z - \mu/2)}{A} \cdot \frac{(\lambda_R - \lambda_L + \mu)}{\lambda_R + \lambda_L - \mu} \tag{68}$$

with time evolution given by $r = r^{(0)} e^{ift}$, where

$$f(z, \xi) = 2\lambda_R \xi + 2\left(z + \frac{\mu}{2}\right)\lambda_R. \tag{69}$$

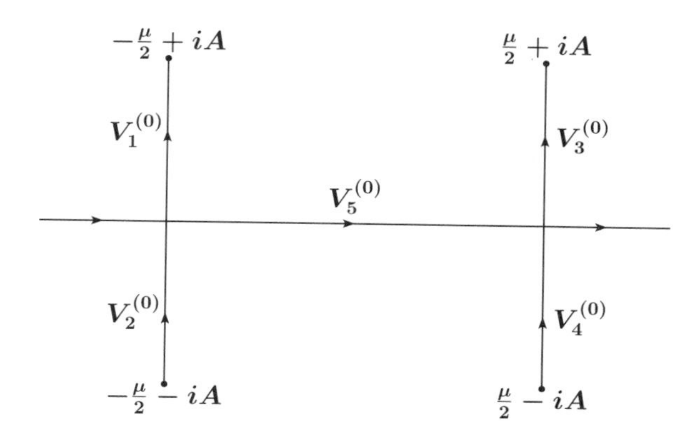

FIGURE 8. The RHP for m for the shock problem.

The contour for the RHP is the real axis and the λ_L and λ_R branch cuts, with jumps

$$V_1 = \begin{pmatrix} 1 & 0 \\ \frac{1}{D}(r_+^{(0)} - r_-^{(0)})e^{ift} & 1 \end{pmatrix}, \quad V_2 = \begin{pmatrix} 1 & \frac{1}{D}(-r_+^{(0)} + r_-^{(0)})e^{-ift} \\ 0 & 1 \end{pmatrix},$$

$$V_3 = \begin{pmatrix} -\Lambda_+ r_+^{(0)} e^{if_+ t} & D_-\Lambda_- \\ 0 & -\Lambda_- r_-^{(0)} e^{if_- t} \end{pmatrix}, \tag{70}$$

$$V_4 = \begin{pmatrix} -\Lambda_- r_-^{(0)} e^{-if_- t} & 0 \\ -D_-\Lambda_- & -\Lambda_+ r_+^{(0)} e^{-if_+ t} \end{pmatrix}, \quad V_5 = \begin{pmatrix} \frac{1+(r^{(0)})^2}{D} & r^{(0)}e^{-ift} \\ r^{(0)}e^{ift} & D \end{pmatrix}.$$

Here the functions Λ and D are defined by

$$\Lambda(z) = \frac{\lambda_R + z - \mu/2}{A} \quad \text{and} \quad D(z) = \frac{2\lambda_R^2 - 2(z-\mu/2)\lambda_R}{A^2}. \tag{71}$$

Asymptotic solution of the RHP in the plane-wave region. As with the semiclassical problem, the goal is to deform the contour and isolate the main contribution to the asymptotic solution.

The sign profile of the function $\Im f(z)$. Except for a bounded region near the λ_R branch cut, in the upper-half plane $\Im(f) < 0$ (respectively, $\Im(f) > 0$) to the left (respectively, right) of a contour which passes through the real axis at $z_0 = \frac{\mu-\xi-\sqrt{(\xi+\mu)^2-8A^2}}{4}$. See figure 9. The plane-wave region is defined to be when $\Im(f) > 0$ for all z on the λ_L branch cut. As ξ decreases, z_0 moves to the right. When $\xi = \xi_2$, z_0 hits the λ_L branch cut, and for $\xi < \xi_2$ part or all of the λ_L branch cut is in the region where $\Im(f) < 0$.

Factoring the jump matrix V. The jump matrix V_5 on the real axis has the two alternate factorizations

$$V_5 = V_2^{(1)} V_6^{(1)} V_1^{(1)} \quad \text{and} \quad \mathrm{V}_5 = \mathrm{V}_4^{(1)} \mathrm{V}_3^{(1)}. \tag{72}$$

The first factorization is used for $z < z_0$ because $V_1^{(1)}$ (respectively, $V_2^{(1)}$) decays to the identity when $\Im(f) < 0$ (respectively, $\Im(f) > 0$), and the second is used for $z > z_0$ because $V_3^{(1)}$ (respectively, $V_4^{(1)}$) decays to the identity when $\Im(f) > 0$ (respectively, $\Im(f) < 0$).

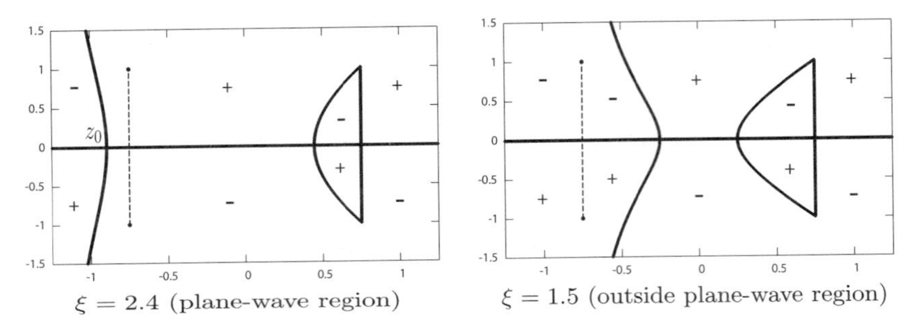

$\xi = 2.4$ (plane-wave region) $\qquad$ $\xi = 1.5$ (outside plane-wave region)

FIGURE 9. Sign structure of $\Im(f)$ for $A = 1, \mu = 1.5$ inside and outside the plane-wave region.

On the λ_L branch cut in the upper-half plane, the jump matrix has the factorization

$$V_1 = (V_{3-}^{(1)})^{-1} V_{3+}^{(1)}.$$

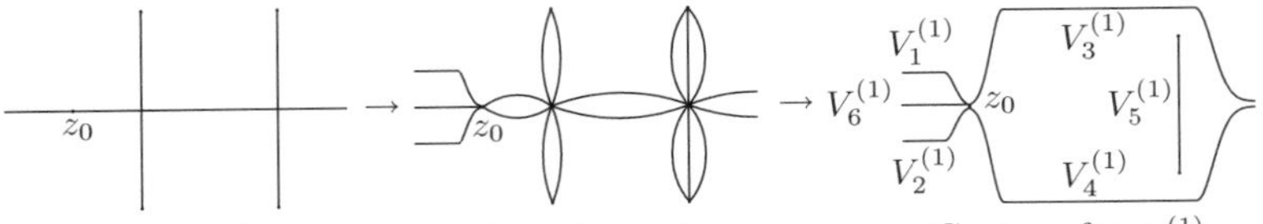

z_0 $\rightarrow$ z_0 $\rightarrow$ $V_6^{(1)}$ $V_1^{(1)}$ $V_2^{(1)}$ z_0 $V_3^{(1)}$ $V_5^{(1)}$ $V_4^{(1)}$

Contour for m $\qquad$ Initial deformation $\qquad$ Contour for $m^{(1)}$

FIGURE 10. The transformations of the contour in the plane-wave region.

On the λ_R branch cut in the upper-half plane, the jump matrix can be factored as

$$V_3 = (V_{3-}^{(1)})^{-1} V_5^{(1)} V_{3+}^{(1)}.$$

After reversing the orientation of the contours with jump $(V_3^{(1)})^{-1}$, there are now seven contours with jump $V_3^{(1)}$ that can be deformed into a single contour with a jump that decays uniformly in time except near $z = z_0$. The procedure for the contours in the lower-half plane with jump $V_4^{(1)}$ is analagous. Here the jump matrices $V_i^{(1)}$ are

$$V_1^{(1)} = \begin{pmatrix} D^{-1/2} & D^{1/2}P^{-1}e^{-ift} \\ 0 & D^{1/2} \end{pmatrix}, \quad V_2^{(1)} = \begin{pmatrix} D^{-1/2} & 0 \\ D^{1/2}P^{-1}e^{ift} & D^{1/2} \end{pmatrix},$$

$$V_3^{(1)} = \begin{pmatrix} D^{-1/2} & 0 \\ \frac{r^{(0)}}{D^{1/2}}e^{ift} & D^{1/2} \end{pmatrix}, \quad V_4^{(1)} = \begin{pmatrix} D^{-1/2} & \frac{r^{(0)}}{D^{1/2}}e^{-ift} \\ 0 & D^{1/2} \end{pmatrix}, \tag{73}$$

$$V_5^{(1)} = \begin{pmatrix} 0 & 1 \\ -1 & 0 \end{pmatrix}, \quad V_6^{(1)} = \begin{pmatrix} 1 + (r^{(0)})^2 & 0 \\ 0 & \frac{1}{1+(r^{(0)})^2} \end{pmatrix},$$

with

$$P(z) = \frac{1 + (r^{(0)})^2}{r^{(0)}}. \tag{74}$$

The g-function transformation. The g-function removes the nondecaying jump $V_6^{(1)}$ on $(-\infty, z_0)$. This modifies the constant jump matrix $V_5^{(1)}$, so the g-function also restores a constant jump on the λ_R branch cut. This is accomplished by the transformation

$$m^{(2)} = m^{(1)} \begin{pmatrix} e^{ig(z)} & 0 \\ 0 & e^{-ig(z)} \end{pmatrix} \tag{75}$$

where

$$g(z) = \ln(i\delta) + \frac{\lambda_R}{2\pi^2 i} \int_{\mu/2-iA}^{\mu/2+iA} \frac{1}{(\eta - z)\lambda_{R-}(\eta)} \int_{-\infty}^{z_0} \frac{\ln(1 + (r^{(0)})^2)}{\zeta - z} d\zeta \tag{76}$$

and

$$\delta(z) = \exp\left(\frac{1}{2\pi i} \int_{-\infty}^{z_0} \frac{\ln(1 + (r^{(0)}(\zeta))^2)}{\zeta - z} d\zeta\right). \tag{77}$$

Solution of the model RHP. Ignoring all the jump matrices which decay in time leads to a model problem with a constant jump $V^{(\mathrm{mod})} = V_5^{(1)}$ on a single main arc, the λ_R branch cut. This problem is solved explicitly by

$$m^{(\mathrm{mod})} = \frac{1}{2} \begin{pmatrix} e^{ig(\infty)} & 0 \\ 0 & e^{-ig(\infty)} \end{pmatrix} \begin{pmatrix} L + L^{-1} & -iL + iL^{-1} \\ iL - iL^{-1} & L + L^{-1} \end{pmatrix}, \tag{78}$$

where

$$L(z) = \left(\frac{z - \mu/2 - iA}{z - \mu/2 + iA}\right)^{1/4} \quad \text{and} \quad g(\infty) = \lim_{z\to\infty} g(z). \tag{79}$$

In the reconstruction of the solution $q(x,t)$ to the NLS equation, the underlying plane wave is encoded in the jump on the λ_R branch cut, and the phase perturbation is encoded in the g-function that removes the jump on $(-\infty, z_0)$.

Error estimate. The leading-order correction term $O(t^{-1/2})$ comes from the nonuniform decay of the jump matrices near $z = z_0$.

Analysis in the residual region. For $\xi < \xi_2$, the factorizations and deformations used in the plane-wave region break down around the λ_L branch cut and an alternate factorization of V is necessary.

Factoring the jump matrix V. On the λ_L branch cut in the upper half-plane, we use the alternate factorization

$$V_1 = \begin{pmatrix} D^{-1/2} & D^{1/2}P_-^{-1}e^{-ift} \\ 0 & D^{1/2} \end{pmatrix}^{-1} \begin{pmatrix} 0 & P_-^{-1}e^{-ift} \\ -P_-e^{-ift} & \end{pmatrix} \begin{pmatrix} D^{-1/2} & D^{1/2}P_+^{-1}e^{-ift} \\ 0 & D^{1/2} \end{pmatrix} \tag{80}$$

There is an analogous factorization in the lower half-plane.

The g-function. With the new factorization, it is possible to deform contours so the jumps off the λ_L and λ_R branch cuts decay in the residual region. However, the resulting jump matrices on the λ_L branch cut are exponentially growing in time. This is dealt with by the g-function

$$G(z) = \frac{-\lambda_L \lambda_R}{2\pi i z} \int_{-\mu/2-iA}^{-\mu/2+iA} \frac{f(\zeta) - \Omega_L}{(\zeta - z)\lambda_{L-}(\zeta)\lambda_R(\zeta)} d\zeta. \tag{81}$$

The constant Ω_L is found by the moment conditions guaranteeing

$$G(\infty) = \lim_{z\to\infty} G(z) \tag{82}$$

is finite. The introduction of G changes the time dependence of the jumps, and the controlling phase function is no longer $f(z)$ but

$$h(z) = f(z) + 2G(z), \tag{83}$$

Now z_0 is chosen according to the sign of $\Im(h)$ instead of $\Im(f)$. From the sign structure of $\Im(h)$ (see figure 11), the jump matrices decay off the λ_L and λ_R branch cuts.

The model RHP. Define $\mathcal{G}$ on the λ_L and λ_R branch cuts by

$$\mathcal{G}(z) = \begin{cases} \omega_L - i\ln(\delta^2/P_-) & \lambda_L \text{ cut in the UHP} \\ \omega_L - i\ln(\delta^2 P_-) & \lambda_L \text{ cut in the LHP} \\ -i\ln(\delta^2) & \lambda_R \text{ cut} \end{cases}. \tag{84}$$

Then the g-function

$$g(z) = -\frac{\lambda_L\lambda_R}{2\pi i}\int_{\Sigma^\omega} \frac{\mathcal{G}(\zeta)}{(\zeta - z)\lambda_{L-}(\zeta)\lambda_{R-}(\zeta)}\mathrm{d}\zeta, \tag{85}$$

with integation along the appropriate branch cuts, creates a constant jump on the two main arcs, the λ_L and λ_R branch cuts. The constant ω_L is determined by requiring

$$g(\infty) = \lim_{z\to\infty} g(z) \tag{86}$$

is finite. The constant jumps on the λ_L and λ_R branch cuts define the model problem. The leading-order solution in the residual region is therefore given in terms of theta functions defined on a Riemann surface of genus one.

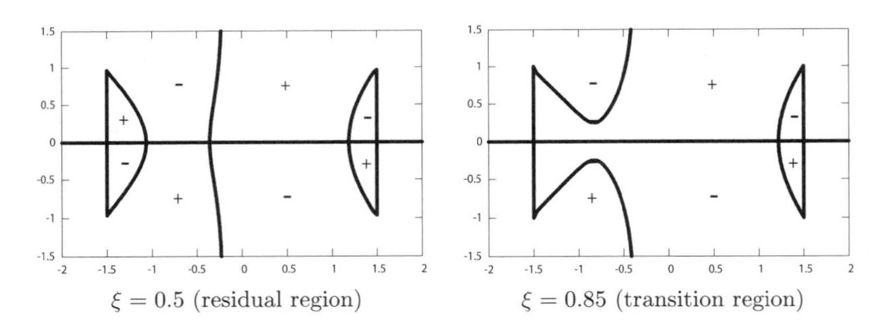

FIGURE 11. Sign structure of $\Im(h)$ for $A = 1, \mu = 3$ in the residual and transition regions.

Analysis in the transition region. The factorizations used in the residual region rely on the existence of a single contour in the upper-half plane from $-\infty$ to z_0 along which $\Im(h) < 0$. However, for $\xi > \xi_1$, the curve $\Im(h) = 0$ lifts off the real axis (see figure 11), and both the residual and plane-wave region deformations fail. At the border between the transition and residual regions, $h(z)$ exhibits cubic behavior. Therefore ξ_1 is found (along with the value of z_0 at the border) by the system

$$h'(z_0) = 0, \quad h''(z_0) = 0. \tag{87}$$

New factorization of the jump matrix. In the transition region it is necessary to use the factorization

$$V_1^{(1)} = \begin{pmatrix} 1 & 0 \\ Pe^{ift} & 1 \end{pmatrix} \begin{pmatrix} 0 & P^{-1}e^{-ift} \\ -Pe^{ift} & 0 \end{pmatrix} \begin{pmatrix} 1 & 0 \\ Pe^{ift} & 1 \end{pmatrix} \tag{88}$$

along part of the contour near z_0 in the upper half-plane, with an analogous deformation in the lower half-plane. This new deformation will be used on the branch cut of

$$\lambda_C(z) = ((z-\alpha)(z-\overline{\alpha}))^{1/2}. \tag{89}$$

The branch cut is chosen to pass through z_0, with z_0 and α given by (92).

The g-function. The g-function

$$\begin{aligned} G(z) \quad &= \quad \frac{\lambda_L\lambda_C\lambda_R}{2\pi i}\left[\int_{-\mu/2-iA}^{-\mu/2+iA} \frac{f(\zeta)-\Omega_L}{(\zeta-z)\lambda_L(\zeta)\lambda_C(\zeta)\lambda_R(\zeta)}d\zeta \right. \\ &+ \left. \int_{\overline{\alpha}}^{\alpha} \frac{f(\zeta)-\Omega_C}{(\zeta-z)\lambda_L(\zeta)\lambda_C(\zeta)\lambda_R(\zeta)}d\zeta\right] \end{aligned} \tag{90}$$

removes the growth in time in the jump matrices on the three branch cuts. Define

$$G(\infty) = \lim_{z\to\infty} G(z). \tag{91}$$

Now the three real quantities Ω_L, Ω_C, and z_0 and the complex quantity α are found simultaneously by the system

$$\begin{aligned} &\cdot\, (z-\alpha)^{1/2}h'(z)|_{z=\alpha} = 0 \text{ (two real equations)} \\ &\cdot\, \Im\left(\frac{1}{\nu}h(z_0+i\nu)\right)\Bigg|_{\nu=0} = 0 \text{ (one real equation)} \\ &\cdot \text{ The moment conditions requiring } G(\infty) \text{ to be finite (two real equations)} \end{aligned} \tag{92}$$

The first equation ensures three branches of $\Im(h)$ emanate from α, and the second ensures $\Im(h) = 0$ along a curve passing through z_0 (in addition to the real axis).

The model RHP. Define $\mathcal{G}$ on the three branch cuts by

$$\mathcal{G}(z) = \begin{cases} \omega_L - i\ln(\delta^2/P_-) & \lambda_L \text{ cut in the UHP} \\ \omega_L - i\ln(\delta^2 P_-) & \lambda_L \text{ cut in the LHP} \\ \omega_C - i\ln(\delta^2/P) & \lambda_C \text{ cut in the UHP} \\ \omega_C - i\ln(\delta^2 P) & \lambda_C \text{ cut in the LHP} \\ -i\ln(\delta^2) & \lambda_R \text{ cut} \end{cases}. \tag{93}$$

Then the g-function

$$g(z) = -\frac{\lambda_L\lambda_C\lambda_R}{2\pi i}\int_{\Sigma^\omega} \frac{\mathcal{G}(\zeta)}{(\zeta-z)\lambda_{L-}(\zeta)\lambda_{C-}(\zeta)\lambda_{R-}(\zeta)}\mathrm{d}\zeta, \tag{94}$$

with integration along the appropriate branch cuts, creates a constant jump on these three branch cuts. The constants ω_L and ω_C are obtained from the moment conditions that ensure

$$g(\infty) = \lim_{z\to\infty} g(z) \tag{95}$$

is finite. These three constant jumps define the model problem, and the leading-order solution in the transition region is given by theta functions defined on a Riemann surface of genus two.

The length of the third main arc increases from zero as ξ increases from ξ_1 (see figure 12). This breaking mechanism between the two regions where a complex conjugate pair of branch points emerges from the real axis has not been observed in previous problems. As ξ increases, the new branch cut eventually coincides with the λ_L branch cut at $\xi = \xi_2$ and the two jumps cancel each other out, which returns to the plane-wave region.

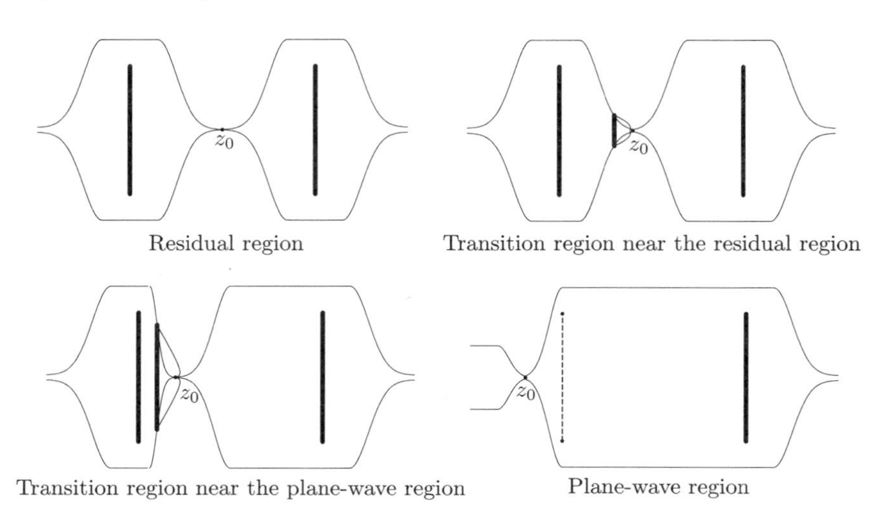

FIGURE 12. The final transformed RHP for varying ξ. The jump on the thin contours decay in time; the jump on the thick contours contribute to the leading-order solution. The dashed line in the last picture is not part of the RHP but indicates the λ_L cut.

Results for the shock case. Because of the symmetry of the problem, results are given for nonnegative x.

THEOREM 6.1. *For $x \geq 0$, the leading order behavior of the solution $q(x,t)$ to (65)-(66) as $t \to \infty$ along the ray $x = \xi t$ is given by*

A. The plane-wave region. *For $|\xi| \geq \xi_2$,*

$$q(x,t) = Ae^{-i[(\mu|\xi|+\mu^2/2-A^2)t-2g(\infty)]} + O\left(t^{-1/2}\right), \tag{96}$$

where $\xi_2 = \frac{\mu^2+A^2}{\mu}$ and $g(\infty)$ defined by (79).

B. The residual region. *For $|\xi| \leq \xi_1$,*

$$q(x,t) = 2A\frac{\theta(\frac{\Omega_L}{2\pi}t + \frac{\omega_L}{2\pi} - u(\infty) + d)\theta(u(\infty)+d)}{\theta(\frac{\Omega_L}{2\pi}t + \frac{\omega_L}{2\pi} + u(\infty) + d)\theta(-u(\infty)+d)}e^{iH(t)} + O(t^{-1/2}). \tag{97}$$

Here: ξ_1 is defined by the system (87); $H(t) = [2G(\infty) - \mu|\xi| - \mu^2/2 + A^2]t + 2g(\infty)$; $G(\infty)$ and $g(\infty)$ are defined by (82) and (86); Ω_L and ω_L are found by the moment conditions ensuring $G(\infty)$ and $g(\infty)$ are finite; the theta function is associated to the Riemann surface $\lambda_L\lambda_R$ of genus one with canonical α-homology cycle A_1; $u(\infty) = \lim_{z\to\infty}\int_{\mu/2-iA}^{z}\omega_1$, where ω_1

is the basic holomorphic differential dual to A_1; $d = -\int_{-\mu/2-iA}^{X_2(z_1)} \omega_1$, *where* $X_2(z)$ *is the preimage of* z *on the second sheet of the Riemann surface,* z_1 *is the unique zero of* $\gamma - \gamma^{-1}$, *and* $\gamma = \left(\frac{(z-(\mu/2+iA))(z-(-\mu/2+iA))}{(z-(\mu/2-iA))(z-(-\mu/2-iA))}\right)^{1/4}$.

C. The transition region. *For* $\xi_1 < |\xi| < \xi_2$,

(98)
$$q(x,t) = (2A + \Im(\alpha))\frac{\theta(\frac{\boldsymbol{\Omega}}{2\pi}t + \frac{\boldsymbol{\omega}}{2\pi} - \boldsymbol{u}(\infty) + \boldsymbol{d})\theta(\boldsymbol{u}(\infty) + \boldsymbol{d})}{\theta(\frac{\boldsymbol{\Omega}}{2\pi}t + \frac{\boldsymbol{\omega}}{2\pi} + \boldsymbol{u}(\infty) + \boldsymbol{d})\theta(-\boldsymbol{u}(\infty) + \boldsymbol{d})}e^{iH(t)} + O(t^{-1/2}).$$

Here: $H(t) = [2G(\infty) - \mu|\xi| - \mu^2/2 + A^2]t + 2g(\infty)$; $G(\infty)$ *and* $g(\infty)$ *are defined by* (91) *and* (95); $\boldsymbol{\omega} = (\omega_L, \omega_C)$ *is found by the moment conditions ensuring* $g(\infty)$ *is finite;* $\boldsymbol{\Omega} = (\Omega_L, \Omega_C)$ *and the branch point* α *are defined by the system of equations* (92); *the theta function is associated to the Riemann surface* $\lambda_L\lambda_C\lambda_R$ *of genus two with canonical homology* α*-cycles* $\{A_1, A_2\}$; $\boldsymbol{u}(\infty) = \lim_{z\to\infty}\left(\int_{\mu/2-iA}^{z}\omega_1, \int_{\mu/2-iA}^{z}\omega_2\right)^{\mathsf{T}}$, *where* ω_i *is the basic holomorphic differentials dual to* A_i; $\boldsymbol{d} = -\int_{\overline{\alpha}}^{X_2(z_1)}\begin{pmatrix}\omega_1\\ \omega_2\end{pmatrix} - \int_{-\mu/2-iA}^{X_2(z_2)}\begin{pmatrix}\omega_1\\ \omega_2\end{pmatrix}$, $X_2(z)$ *is the preimage of* z *on the second sheet,* z_1 *and* z_2 *are the two zeros of* $\gamma - \gamma^{-1}$, *and* $\gamma = \left(\frac{(z-(\mu/2+iA))(z-\alpha)(z-(-\mu/2+iA))}{(z-(\mu/2-iA))(z-\overline{\alpha})(z-(-\mu/2-iA))}\right)^{1/4}$.

7. Appendix I: Inverse scattering approach to (1) through a Riemann-Hilbert Problem (RHP)

We begin with the eigenvalue problem for the Dirac operator, the first of the two linear operators of the Lax pair [**11**] that linearizes NLS. Referred to as the Zakharov-Shabat eigenvalue problem, it is a linear first order ODE system for the When z is real and $q \equiv 0$, each vector solution of the system is a linear combination of the two plane waves $\begin{pmatrix}1\\0\end{pmatrix}e^{-\frac{i}{\varepsilon}zx}$ and $\begin{pmatrix}0\\1\end{pmatrix}e^{\frac{i}{\varepsilon}zx}$. If q is nonzero but decays as $|x| \to \infty$, this is true only asymptotically. The solution satisfying $\begin{pmatrix}1\\0\end{pmatrix}e^{-\frac{i}{\varepsilon}zx}$ as $x \to -\infty$, has behavior $a(z)\begin{pmatrix}1\\0\end{pmatrix}e^{-\frac{i}{\varepsilon}zx} + b(z)\begin{pmatrix}0\\1\end{pmatrix}e^{\frac{i}{\varepsilon}zx}$ as $x \to +\infty$, displaying a reflection coefficient $r(z) = b/a$ and a transmision coefficient $1/a$. A wronskian argument shows that $|a^2| + |b^2| = 1$ (the modulus of the transmission coefficient equals $\sqrt{|r|^2+1}$). Moreover, one proves that $a(z)$ extends to an analytic function in the upper complex half plane with $a(z) \to 1$ as $z \to \infty$. This allows deriving a formula for the transmission coefficient given the reflection coefficient as well as soliton information (solitons correspond to zeros of $a(z)$). By the Lax pair theory [**11**], when $q(x,t,e)$ evolves in t as an NLS solution, the reflection coefficient r is calculated to evolve by $r(z,t,\varepsilon) = e^{\frac{i}{\varepsilon}z^2 t}r(0,t,\varepsilon)$. In the solitonless case treated in this work, the *(direct) scattering transform* is the map $q(x) \mapsto r(z)$, while the *inverse scattering transform* is given by the inverse map $r(z) \mapsto q(x)$. A characterization of functional families for which the map is $1-1$ and bounded has been given by Zhou [**21**].

The modern approach to inverse scattering is through a Riemann-Hilbert problem (RHP) and was initiated by Shabat [**14**]. In general a 2×2 matrix RHP seeks a 2×2 matrix $m = m(z)$ that is analytic in the closed complex plane minus a given oriented contour on which m satisfies a given jump relation $m_+ = m_- V$ (subscripts $+$ and $-$ indicate the left and right limits of m along the contour respectively). A normalization of m at some point z, in our case, $m(z) \to I =$ Identity matrix as $z \to \infty$, is also given. The relevance of this to the first order ODE system (4), is that if a fundamental solution $\Phi(z,x)$ is chosen to jump across a contour in th ecomplex plane, its limiting values $\Phi_\pm(z,x)$ on the contour satisfy a jump condition $\Phi_+(z,x) = \Phi_-(z,x)J(z)$, which expresses the fact that each columns of fundamental matrix $\Phi_+(z,x)$ depends linearly on the columns of fundamental matrix Φ_-; clearly, the jump matrix J is independent of x. For real z's, the asymptotic behavior of the two columns of $\Phi(z,x)$ is chosen to be $a(z)^{-1} \begin{pmatrix} 1 \\ 0 \end{pmatrix} e^{-\frac{i}{\varepsilon}zx}$ as $x \to -\infty$ for the first column and $\begin{pmatrix} 0 \\ 1 \end{pmatrix} e^{\frac{i}{\varepsilon}zx}$ as $x \to \infty$ for the second column ($a(z)^{-1}$ is the transmission coefficent, see (4)). The asymptotic behavior at the other end of the real line is a linear combination of the two plane waves with coefficents expressed in terms of the reflection coefficient. This choice allows $\Phi(z,x)$ to extend into the upper complex half plane as a meromorphic function whose poles correspond to solitons; in our present solitonless case it is analytic. The symmetry relation $\Phi(z,x)\Phi(\overline{z})^* = I$ is implemented to define Φ in the lower complex half plane, producing a jump on the real axis with limit values $\Phi_\pm$ both of which are fundamental matrices of (4) on the real axis. Clearly, the jump matrix J is expressed in terms of the reflection coefficient. Finally, a normalization on Φ multiplies its first column by $e^{izx/\varepsilon}$ and its second column by $e^{-izx/\varepsilon}$; the ensuing matrix $m(z)$ (a) is analytic in $\mathbb{C}/\mathbb{R}$, (b) is normalized so that $m(z) \to I$, where I is the identity matrix as $z \to \infty$, (c) satisfies the jump relation

$$m_+ = m_- \begin{pmatrix} 1 + |r^2| & \overline{r} \\ r & 1 \end{pmatrix}, \quad r = r^{(0)}(z) e^{\frac{i}{\varepsilon}(2xz + 4tz^2)}, \quad z \in \mathbb{R}\,, \tag{99}$$

where $r^{(0)}(z)$ is the reflection coefficient of the initial data.

For positive parameter μ under study, solitons occur when $\mu < 2$; they manifest themselves in the analysis as poles of the matrix $m(z)$. The effect of the jth pole on the Riemann-Hilbert problem, is an additional counter-clockwise closed contour (say, a circle) with jump matrix $\begin{pmatrix} 1 & 0 \\ \frac{c_j}{z - \zeta_j} & 1 \end{pmatrix}$ about the location of the pole. The soliton poles are distributed on the imaginary axis over the imaginary interval $(-T, T)$ with $O(\varepsilon)$ spacing, where $T = \pm i\sqrt{1 - \frac{\mu^2}{4}}$. We refer to this interval of tightly packed poles as the *slit*.

A convenient recovery formula for $q(x,t,\varepsilon)$ is derived from the expansion $m = I + \dfrac{\tilde{m}(x,t,\varepsilon)}{z} + O\left(\frac{1}{z^2}\right)$ which gives

$$q(x,t,\varepsilon) = -2 \lim_{z \to \infty} z(m - I) = -2\tilde{m}_{12}(x) \tag{100}$$

where the subscript indicates the 12 matrix entry.

Obtaining $m(z)$ from conditions (a), (b), (c) is the *Riemann-Hilbert problem* (RHP) that solves the inverse scattering problem.

8. Appendix II: Important tools

THEOREM 8.1. Degeneracy Theorem *A) Suppose*

$$\frac{h'(z_0)}{R(z_0)} = 0 \quad \left(\text{by symmetry also: } \quad \frac{h'(\bar{z}_0)}{R(\bar{z}_0)} = 0\right) \tag{101}$$

for some point $z_0 \in \gamma$. *Then:*

1. *Replacing* $R(z)$ *in* (38) *with* $\tilde{R} = R(z)(z-z_0)(z-\bar{z}_0)$ *(the multiplicities of* z_0 *and* $\bar{z}_0$ *are thus increased by two) does not change the functions* $h'(z)$ *and* $h(z)$, *i.e.* $h'(z;\tilde{R}) = h'(z;R)$ *and* $h(z;\tilde{R}) = h(z;R)$;
2. *If the original* α *satisfy the MI conditions with genus* n, *then the new* α *corresponding to* $\tilde{R}$, *also satisfy the MI conditions with genus* $n+2$.

B) Conversely, if a degenerate $\alpha = (\alpha_0, \alpha_2, \cdots \alpha_{2n})$ *with* $\alpha_{2k} = \alpha_{2k+2} = z_0$ *satisfy the MI conditions with genus* n, *then the* α *that is obtained by removing the degenerate pair and its complex conjugate satisfies the MI conditions for genus* $n-2$. *Furthermore, after the removal,* $h'/R = 0$ *at the site* z_0 *of the removed pair.*

THEOREM 8.2. Evolution Theorem
Let $\alpha = (\alpha_0, \alpha_2, \alpha_4, \cdots \alpha_{2n})$ *with distinct* α_{2k} *be a solution of* (41) *with genus* n *at some point* (x_0, t_0). *Then:*

- *The solution* $\alpha(x,t)$ *can be continued uniquely with the same genus into a neighborhood of* (x_0,t_0) *and* $\alpha(x,t)$ *is a smooth function of* x *and* t;
- W *and* Ω *are smooth functions of* x *and* t;
- *if the function* $h(z) = h(z;\alpha(x,t), W(x,t), \Omega(x,t))$ *satisfies conditions* $\Im h = 0$ *on the main arcs,* $\Im h < 0$ *on both sides of any main arc and* $\Im h > 0$ *on the complementary arcs and on the arcs to infinity at* $(x,t) = (x_0,t_0)$ *then it also satisfies these conditions in a neighborhood of* (x_0,t_0).

THEOREM 8.3. Jacobian *For genus* $n = 2N$, *the Jacobian* $|\frac{\partial F}{\partial \alpha}|$ *is given by*

$$|\frac{\partial F}{\partial \alpha}| = \prod_{j=0}^{2N} |\frac{h'(\alpha_{2j})}{2R(\alpha_{2j})}|^2 \prod_{j<l}(\alpha_l - \alpha_j) \int_{\hat{\gamma}_{m,1}} \int_{\hat{\gamma}_{c,1}} \cdots \int_{\hat{\gamma}_{m,N}} \int_{\hat{\gamma}_{c,N}} \prod_{j<l}^{2N}(z_j - z_l) \prod_{k=1}^{2N} \frac{dz_k}{R(z_k)}, \tag{102}$$

where $h'(z)$ *is defined by* (38) *and the latter integral in* (102) *is equal to*

$$\det \begin{pmatrix} \int_{\hat{\gamma}_{m,1}} \frac{dz_1}{R(z_1)} & \int_{\hat{\gamma}_{c,1}} \frac{dz_2}{R(z_2)} & \int_{\hat{\gamma}_{m,2}} \frac{dz_3}{R(z_3)} & \cdots & \int_{\hat{\gamma}_{c,N}} \frac{dz_{2N}}{R(z_{2N})} \\ \int_{\hat{\gamma}_{m,1}} \frac{z_1 dz_1}{R(z_1)} & \int_{\hat{\gamma}_{c,1}} \frac{z_2 dz_2}{R(z_2)} & \int_{\hat{\gamma}_{m,2}} \frac{z_3 dz_3}{R(z_3)} & \cdots & \int_{\hat{\gamma}_{c,N}} \frac{z_{2N} dz_{2N}}{R(z_{2N})} \\ \vdots & \vdots & \vdots & \vdots & \vdots \\ \int_{\hat{\gamma}_{m,1}} \frac{z_1^{2N-1} dz_1}{R(z_1)} & \int_{\hat{\gamma}_{c,1}} \frac{z_2^{2N-1} dz_2}{R(z_2)} & \int_{\hat{\gamma}_{m,2}} \frac{z_3^{2N-1} dz_3}{R(z_3)} & \cdots & \int_{\hat{\gamma}_{c,N}} \frac{z_{2N}^{2N-1} dz_{2N}}{R(z_{2N})} \end{pmatrix}. \tag{103}$$

References

[1] E. D. Belokolos, A.I. Bobenko, V.Z Enol'Skii, and A.R. Its. Algebro-geometric approach to nonlinear integrable equations. Springer, 1994.

[2] R. Buckingham and S. Venakides. Long-time asymptotics of the nonlinear Schrödinger equation shock problem. Comm. Pure Appl. Math., 2007.

[3] D. Cai, D.W. McLaughlin, and K.T-R. McLaughlin. The nonlinear Schrödinger equation as both a pde and a dynamical system. Handbook of dynamical systems, North-Holland, Amsterdam, 2:599–675, 2002.

[4] H.D. Ceniceros and F.-R. Tian. A numerical study of the semi-classical limit of the focusing nonlinear Schrödinger equation. Phys. Lett. A, 306(1):25–34, 2002.

[5] P. Deift, S. Venakides, and X. Zhou. New results in small dispersion KdV by an extension of the steepest descent method for Riemann-Hilbert problems. Internat. Math. Res. Notices, 6:286–299, 1997.

[6] P. Deift and X. Zhou. A steepest descent method for oscillatory Riemann-Hilbert problems. asymptotics for the mKdV equation. Ann. of Math., 137:295–370, 1993.

[7] P. Deift and X. Zhou. Asymptotics for the Painlevé ii equation. Comm. Pure and Appl. Math., 48:277–337, 1995.

[8] M.G. Forest and J.E. Lee. Geometry and modulation theory for the periodic nonlinear Schrödinger equation. C. Dafermos, et. al. (Eds.), Oscillation Theory, Computation, and Methods of Compensated Compactness, Vol. 2, IMA, Springer, 1986.

[9] S. Kamvissis, K. T.-R. McLaughlin, and P.D. Miller. Semiclassical soliton ensembles for the focusing nonlinear Schrödinger equation. Annals of Mathematics Studies 154, Princeton Unversity Press, 2003.

[10] M. Klaus and J. K. Shaw. Purely imaginary eigenvalues of Zakharov-Shabat systems. Phys. Rev. E(3)., 65(3):036607, 5pp, 2002.

[11] P. D. Lax. Integrals of nonlinear equations of evolution and solitary waves. Comm. Pure Appl. Math., 21:467–490, 1968.

[12] G. Lyng and P. D. Miller. The n-soliton of the focusing nonlinear Schrödinger equation for n large. Comm. Pure Appl. Math., 2007.

[13] P. D. Miller and S. Kamvissis. On the semiclassical limit of the focusing nonlinear Schrödinger equation. Phys. Lett. A, 247(1-2):75 –86, 1998.

[14] A. B. Shabat. One-dimensional perturbations of a differential operator and the inverse scattering problem. Problems in Mechanics and mathematical physics, Nauka, Moscow, 1976.

[15] A. Tovbis and S. Venakides. The eigenvalue problem for the focusing nonlinear Schrödinger equation: new solvable cases. Phys. D, 146(1-4):150–164, 2000.

[16] A. Tovbis, S. Venakides, and X. Zhou. Semiclassical focusing nonlinear Schrödinger equation i: Inverse scattering map and its evolution for radiative initial data. Preprint.

[17] A. Tovbis, S. Venakides, and X. Zhou. On semiclassical (zero dispersion limit) solutions of the focusing nonlinear Schrödinger equation. Comm. Pure Appl. Math, 57(7):877–985, 2004.

[18] A. Tovbis, S. Venakides, and X. Zhou. On the long-time limit of semiclassical (zero dispersion limit) solutions of the focusing nonlinear Schrödinger equation: Pure radiation case. Comm. Pure and Appl. Math., 59(10):1379–1432, 2006.

[19] V. E. Zakharov and A. B. Shabat. Exact theory of two-dimensional self-focusing and one-dimensional self-modulation of waves in nonlinear media. Soviet Physics JETP, 34(1):62–69, 1972.

[20] X. Zhou. Riemann-Hilbert problems and integrable systems. Preprint.

[21] X. Zhou. The L2-Sobolev space bijectivity of the scattering and inverse scattering transforms. Comm. Pure Appl. Math., 51:697–731, 1998.

University of Michigan, Ann Arbor, MI 48109
E-mail address: robbiejb@umich.edu

Department of Mathematics, University of Central Florida, Orlando, FL 32816
E-mail address: atovbis@pegasus.cc.ucf.edu

Department of Mathematics, Duke University, Durham, NC 27708
E-mail address: ven@math.duke.edu

Department of Mathematics, Duke University, Durham, NC 27708
E-mail address: zhou@math.duke.edu

Proceedings of Symposia in Applied Mathematics
Volume **65**, 2007

An extension to a classical theorem of Liouville and applications

YanYan Li

Dedicated to Louis Nirenberg and Peter Lax with admiration and friendship

A classical theorem of Liouville says that a positive harmonic function in $\mathbb{R}^n$ must be a constant. The Laplace operator Δ is invariant under rigid motion: For any rigid motion $T : \mathbb{R}^n \to \mathbb{R}^n$, and for any function $u : \mathbb{R}^n \to \mathbb{R}$,

$$\Delta(u \circ T) = (\Delta u) \circ T.$$

Recall that T is called a rigid motion if $Tx \equiv Ox + b$ for some orthogonal $n \times n$ matrix O and some vector $b \in \mathbb{R}^n$.

Instead of rigid motions, let us look at Möbius transformations of $\mathbb{R}^n \cup \{\infty\}$, and we look at operators invariant under Möbius transformations. A map φ : $\mathbb{R}^n \cup \{\infty\} \to \mathbb{R}^n \cup \{\infty\}$ is called a Möbius transformation, if it is a composition of a finitely many of the following three types of transformations:

$$\begin{aligned} \text{A translation}: &\quad x \to x + \bar{x}, \text{ where } \bar{x} \text{ is a given point in } \mathbb{R}^n, \\ \text{A dilation}: &\quad x \to ax, \text{ where } a \text{ is a positive number}, \\ \text{A Kelvin transformation}: &\quad x \to \tfrac{x}{|x|^2}. \end{aligned}$$

For a function u on $\mathbb{R}^n$, let

$$u_\varphi := |J_\varphi|^{\frac{n-2}{2n}} (u \circ \varphi)$$

where J_φ denotes the Jacobian of φ.

Let $H(s, p, M)$ be a smooth function in its variables, where $s > 0$, $p \in \mathbb{R}^n$ and $M \in \mathcal{S}^{n \times n}$, the set of $n \times n$ real symmetric matrices. We say that the second order fully nonlinear operator $H(u, \nabla u, \nabla^2 u)$ is conformally invariant if

$$H(u_\varphi, \nabla u_\varphi, \nabla^2 u_\varphi) \equiv H(u, \nabla u, \nabla^2 u) \circ \varphi$$

holds for all positive smooth functions u and all Möbius transformations φ.

It was proved in [**11**] that a conformally invariant operator $H(u, \nabla u, \nabla^2 u)$ must be of the form

$$H(u, \nabla u, \nabla^2 u) \equiv f(\lambda(A^u))$$

Partially supported by an NSF grant.

where

$$A^u := w\nabla^2 w - \frac{|\nabla w|^2}{2} I, \qquad \text{with } w = u^{-\frac{2}{n-2}},$$

$\lambda(A^u)$ denotes the eigenvalues of A^u, I is the $n \times n$ identity matrix, $f(\lambda)$ is some symmetric function in $\lambda = (\lambda_1, \cdots, \lambda_n)$.

Let φ be a Möbius transformation, then for some $n \times n$ orthogonal matrix functions $O(x)$ (i.e. $O(x)O(x)^t = I$), depending on φ,

$$A^{u_\varphi}(x) \equiv O(x)A^u(\varphi(x))O^t(x).$$

Thus $f(\lambda(A^u))$ is a conformally invariant operator for all symmetric functions f.

Taking $f(\lambda) = \sigma_1(\lambda) := \lambda_1 + \cdots + \lambda_n$, we have a simple expression:

$$\sigma_1(\lambda(A^u)) \equiv -\frac{2}{n-2} u^{-\frac{n+2}{n-2}} \Delta u.$$

In general, $f(\lambda(A^u))$ is a fully nonlinear operator, and is rather complex even for $f(\lambda) = \sigma_k(\lambda) := \sum_{1 \le i_1 < \cdots < i_k \le n} \lambda_{i_1} \cdots \lambda_{i_k}$, the $k-$th elementary symmetric function. The expression for σ_2 is still quite pleasant:

$$\sigma_2(\lambda(A^u)) \equiv \frac{1}{2}\left(\sigma_1(A^u)^2 - (A^u)^t A^u\right),$$

where $(A^u)^t$ denotes the transpose matrix of A^u.

For a Riemannian metric g,

$$A_g := \frac{1}{n-2}\left(Ric_g - \frac{R_g}{2(n-1)} g\right)$$

is called Schouten tensor. Let $g_{Eucl.} = \delta_{ij} dx^i dx^j$ be the Euclidean metric, and let, for a positive function u, $g := u^{\frac{4}{n-2}} g_{Eucl}$ be conformal to the Euclidean metric, then A^u is associated with the Schouten tensor A_g as follows:

$$A_g = u^{\frac{4}{n-2}} A^u_{ij} dx^i dx^j.$$

The so called σ_k-Yamabe problem was studied in [**19**] and [**20**], which involves the Schouten tensor A_g. Equations involving σ_2 and A_g, with applications in topology and geometry, were studied in [**2**].

Let

(1) $\qquad \Gamma \subset \mathbb{R}^n$ be open convex symmetric cone with vertex at the origin

satisfying

$$\Gamma_n := \{\lambda \mid \lambda_i > 0, \forall\, i\} \subset \Gamma \subset \{\lambda \mid \sum_{i=1}^n \lambda_i > 0\} =: \Gamma_1. \tag{2}$$

Naturally, Γ being symmetric means that $(\lambda_1, \cdots, \lambda_n) \in \Gamma$ implies $(\lambda_{i_1}, \cdots, \lambda_{i_n}) \in \Gamma$ for any permutation $(\lambda_{i_1}, \cdots, \lambda_{i_n})$ of $(\lambda_1, \cdots, \lambda_n)$.

THEOREM 1. *([**15**], [**16**]) Let u be a positive function in $C^{0,1}_{loc}(\mathbb{R}^n)$ satisfying*

$$\lambda(A^u) \in \partial\Gamma \quad \text{in } \mathbb{R}^n \text{ in the viscosity sense.} \tag{3}$$

Then $u \equiv Constant$.

A positive continuous function u is said to satisfy (3) when the following holds: if $x_0 \in \mathbb{R}^n$, $\psi \in C^2$, $(u-\psi)(x_0) = 0$, $u - \psi \leq 0$ near x_0, then $\lambda(A^{\psi}(x_0)) \in \mathbb{R}^n \setminus \Gamma$; if $\psi \in C^2$, $(u-\psi)(x_0) = 0$, $u - \psi \geq 0$ near x_0, then $\lambda(A^{\psi}(x_0)) \in \overline{\Gamma}$.

It is not difficult to see that if u is a positive $C^{1,1}$ function in $\mathbb{R}^n$ satisfying $\lambda(A^u) \in \partial\Gamma$ a.e. in $\mathbb{R}^n$, then u satisfies (3).

Fully nonlinear second order elliptic equations with $\lambda(\nabla^2 u)$ in such Γ was first studied by Caffarelli, Nirenberg and Spruck in [**1**]. Equation (3) is a fully nonlinear degenerate elliptic equation. The proof of Theorem 1 makes use of the method of moving spheres, a variant of the method of moving planes as in Gidas, Ni and Nirenberg [**5**].

For $u \in C^2$, $\lambda(A^u) \in \partial\Gamma_1$ means $\Delta u = 0$, and Theorem 1 says that a positive harmonic function in $\mathbb{R}^n$ is constant — a classical theorem of Liouville.

Let

$$\Gamma_k := \{\lambda \in \mathbb{R}^n \mid \sigma_1(\lambda) > 0, \cdots, \sigma_k(\lambda) > 0\},$$

then $\Gamma = \Gamma_k$ satisfies (1) and (2)..

Such Liouville theorem was proved, and used in deriving global apriori C^0 and C^1 estimates for solutions, in [**3**] for $u \in C^{1,1}_{loc}$, $\Gamma = \Gamma_2$ and $n = 4$; in [**10**] for $u \in C^{1,1}_{loc}$, $\Gamma = \Gamma_2$ and $n = 3$; in ([14, v1]) for $u \in C^{1,1}_{loc}$, Γ satisfying (1), (2) and $n \geq 3$. Our proofs of the Liouville theorems in [**14**] and [**15**] are entirely different from arguments in [**3**] and [**10**]. In particular, the proof of Theorem 1 in [**15**] uses only the following properties of the set of $C^{0,1}_{loc}(\mathbb{R}^n)$ viscosity solutions of $\lambda(A^u) \in \partial\Gamma$: It is a subset of the set of positive continuous superharmonic functions in $\mathbb{R}^n$, it is conformally invariant, it is invariant under positive constant scalar multiplication, and some weak comparison principle holds for functions in it.

The motivation of our study of such Liouville properties of entire solutions of $\lambda(A^u) \in \partial\Gamma$ is to answer the following questions concerning local gradient estimates of solutions to general second order conformally invariant fully nonlinear elliptic equations.

Assume

$$f \in C^1(\Gamma) \cap C^0(\overline{\Gamma}) \text{ is a symmetric function,} \tag{4}$$

$$f \text{ is homogeneous of degree 1,} \tag{5}$$

$$f > 0, f_{\lambda_i} := \frac{\partial f}{\partial \lambda_i} > 0 \text{ in } \Gamma, \ f|_{\partial\Gamma} = 0, \tag{6}$$

$$\sum_{i=1}^n f_{\lambda_i} \geq \delta, \quad \text{in } \Gamma \text{ for some } \delta > 0. \tag{7}$$

Note that $(f, \Gamma) = (\sigma_k^{\frac{1}{k}}, \Gamma_k)$ satisfies (4)-(7).

QUESTION 1. *Let $n \geq 3$, (f, Γ) satisfy (1), (2), (4)-(7). For constants $0 < a < \infty$ and $0 < h \leq \bar{h} \leq 1$, let u be a C^3 function in B_2, a ball of radius 2 in $\mathbb{R}^n$, satisfying*

$$f(\lambda(A^u)) = h, \ 0 < u \leq a, \ \lambda(A^u) \in \Gamma, \quad \text{in } B_2. \tag{8}$$

Is it true that

$$|\nabla \log u| \leq C \quad \text{in } B_1 \tag{9}$$

for some constant C depending only on a, $\bar{h}$ and (f, Γ)?

A more general question on Riemannian manifolds is

QUESTION 2. *Let g be a smooth Riemannian metric on $B_2 \subset \mathbb{R}^n$, $n \geq 3$, (f,Γ) satisfy (1), (2), (4)-(7). For a positive number a and a positive function $h \in C^1(B_2)$, let $u \in C^3(B_2)$ satisfy, with $\tilde{g} := u^{\frac{4}{n-2}} g$,*

$$f(\lambda(A_{\tilde{g}})) = h, \; 0 < u \leq a, \; \lambda(A_{\tilde{g}}) \in \Gamma, \quad in\ B_2,$$

where $\lambda(A_{\tilde{g}})$ denotes the eigenvalues of $A_{\tilde{g}}$ with respect to $\tilde{g}$. Is it true that

$$\|\nabla \log u\|_g \leq C \quad in\ B_1$$

for some constant C depending only on a, g, $\|h\|_{C^1(B_2)}$ and (f,Γ)?

Theorem 1 is a crucial ingredient in our proof of the following optimal local gradient estimate.

THEOREM 2. *([**15**], [**16**]) The answer to Question 2 is "Yes".*

Such local gradient estimates have been studied by a number of authors. See [**7**] for $(f,\Gamma) = (\sigma_k^{\frac{1}{k}}, \Gamma_k)$ as well as the efforts of achieving further generality in [**11**], [**8**], [**6**] and [**17**]. If f is assumed in addition to be concave in Γ which includes $(\sigma_k^{\frac{1}{k}}, \Gamma_k)$, such local gradient estimate, depending on $\|h\|_{C^2(B_2)}$ instead of $\|h\|_{C^1(B_2)}$, is a consequence of the Liouville theorem for $C^{1,1}_{loc}$ solutions in ([14, v1]) and the estimates in [**11**], see [**15**], [**16**] and [**18**]; a completely different proof of this is given in [**4**]. The concavity of f is not assumed in Theorem 2.

In the rest of this note, we assume Theorem 1 and outline the arguments, in three steps, which give an affirmative answer to Question 1. For details, as well as for the proofs of Theorem 1 and Theorem 2, see [**15**].

A subtlety of (9) is that the constant C does not depend on the lower bound of u. Indeed we have

Step 1. ([**12**]) If we further assume in (8) that $u \geq b$ in B_2 for some positive constant b, then (9) holds for some constant C depending only on $a, b, \bar{h}$ and (f,Γ).

This result was extended to manifolds with boundary under prescribed mean curvature boundary conditions in [**9**], see theorem 1.3 there.

Next, we first estimate semi-Hölder norm of $\log u$ instead of the gradient of $\log u$. Namely, we have

Step 2. For $0 < \alpha < 1$, $[\log u]_{\alpha, B_1} \leq C(\alpha)$.

For $|x| < 3/2$ and $0 < \delta < 1/8$, let

$$[v]_{\alpha,\delta}(x) := \sup_{|y-x|<\delta} \frac{|v(y) - v(x)|}{|y-x|^\alpha},$$

$$\delta(v, x; \alpha) := \begin{cases} \infty & \text{if } [v]_{\alpha,1}(x) < 1, \\ \mu \text{ where } \mu \in (0,1], \; \mu^\alpha [v]_{\alpha,\mu}(x) = 1 & \text{if } [v]_{\alpha,1}(x) \geq 1. \end{cases}$$

We establish Step 2 by contradiction. Suppose the contrary, then, for some positive constants a and $h_i \leq 1$, there exists a sequence of solutions $\{u_i\}$ to $f(\lambda(A^{u_i})) = h_i$, $\lambda(A^{u_i}) \in \Gamma$ in B_2 such that $\inf_{x \in B_1} \delta(\log u_i, x; \alpha) \to 0$. For simplicity, we assume

$$\delta(\log u_i, 0) := \delta(\log u_i, 0; \alpha) = \inf_{x \in B_1} \delta(\log u_i, x; \alpha) \to 0.$$

Rescale u_i:

$$v_i(y) := \frac{1}{u_i(0)} u_i(\epsilon_i y), \qquad |y| \le \frac{1}{\epsilon_i},$$

where

$$\epsilon_i := \delta(\log u_i, 0).$$

Then,

$$[\log v_i]_{\alpha,1}(0) = 1, \tag{10}$$

and, for all $1 < \beta < \infty$ there exists some positive constant $C(\beta)$ such that

$$[\log v_i]_{\alpha,1}(x) \le C(\beta) \qquad \forall\ |x| \le \beta.$$

Since $\log v_i(0) = 0$, we have, for some positive constant $C(\beta)$,

$$\frac{1}{C(\beta)} \le v_i(x) \le C(\beta) \qquad \forall\ |x| \le \beta. \tag{11}$$

Since u_i satisfies

$$f(\lambda(A^{u_i})) = h_i,\ 0 < u_i < a,\ \lambda(A^{u_i}) \in \Gamma,$$

v_i satisfies

$$f(\lambda(A^{v_i})) = \epsilon_i^2 u_i(0)^{\frac{4}{n-2}} h_i,\ \lambda(A^{v_i}) \in \Gamma. \tag{12}$$

Thus, in view of (11) and Step 1,

$$|\nabla v_i(x)| \le C(\beta) \qquad \forall\ |x| \le \beta.$$

Passing to a subsequence, we have, for any $\alpha < \gamma < 1$,

$$v_i \to v \qquad \text{in } C^{\gamma}_{loc}(\mathbb{R}^n) \tag{13}$$

for some positive function v in $C^{0,1}_{loc}(\mathbb{R}^n)$.

It follows from (10) and (13) that

$$[\log v]_{\alpha,1}(0) = 1.$$

In particular, v is not a constant.

On the other hand, it is easy to see from (12) that v is a $C^{0,1}_{loc}(R^n)$ viscosity solution of $\lambda(A^v) \in \partial\Gamma$. By Theorem 1, v is a constant. A contradiction. Step 2 is established.

Step 3. $|\nabla \log u| \le C$.

Step 2 implies the Harnack inequality:

$$\sup_{B_{\frac{3}{2}}} u \le C \inf_{B_{\frac{3}{2}}} u.$$

Let

$$w := \frac{u}{u(0)}.$$

Then w satisfies

$$f(\lambda(A^w)) = u(0)^{\frac{4}{n-2}} h, \quad \frac{1}{C} \le w \le C, \qquad \text{in } B_{\frac{3}{2}}.$$

By Step 1,

$$|\nabla \log u| = |\nabla \log w| \le C \qquad \text{in } B_1.$$

Step 3 is established.

References

[1] L. Caffarelli, L. Nirenberg and J. Spruck, The Dirichlet problem for nonlinear second-order elliptic equations, III: Functions of the eigenvalues of the Hessian. Acta Math. 155 (1985), 261-301.

[2] S.Y. A. Chang, M. Gursky and P. Yang, An equation of Monge-Ampere type in conformal geometry, and four-manifolds of positive Ricci curvature, Ann. of Math. 155 (2002), 709-787.

[3] S.Y.A. Chang, M. Gursky and P. Yang, A prior estimate for a class of nonlinear equations on 4-manifolds, Journal D'Analyse Journal Mathematique 87 (2002), 151-186.

[4] S. Chen, Local estimates for some fully nonlinear elliptic equations, Int. Math. Res. Not. 2005, 3403-3425.

[5] B. Gidas, W.M. Ni and L. Nirenberg, Symmetry and related properties via the maximum principle, Comm. Math. Phys. 68 (1979), 209-243.

[6] P. Guan, C.S. Lin and G. Wang, Local gradient estimates for for conformal quotient equations, preprint.

[7] P. Guan and G. Wang, Local estimates for a class of fully nonlinear equations arising from conformal geometry, Int. Math. Res. Not. 2003, no. 26, 1413-1432.

[8] P. Guan and G. Wang, Geometric inequalities on locally conformally flat manifolds, Duke Math. J. 124 (2004), 177-212.

[9] Q. Jin, A. Li and Y.Y. Li, Estimates and existence results for a fully nonlinear Yamabe problem on manifolds with boundary, Calculus of Variations and PDE's 28 (2007), 509-543.

[10] A. Li, Liouville type theorem for some degenerate conformally invariant fully nonlinear equation, private communication, 2003.

[11] A. Li and Y.Y. Li, On some conformally invariant fully nonlinear equations, Comm. Pure Appl. Math. 56 (2003), 1416-1464.

[12] A. Li and Y.Y. Li, Unpublished note, 2004.

[13] A. Li and Y.Y. Li, On some conformally invariant fully nonlinear equations, Part II: Liouville, Harnack and Yamabe, Acta Math. 195 (2005), 117-154.

[14] Y.Y. Li, Degenerate conformally invariant fully nonlinear elliptic equations, arXiv:math.AP/0504598 v1 29 Apr 2005; v2 24 May 2005; final version, to appear in Arch. Rational Mech. Anal.

[15] Y.Y. Li, Local gradient estimates of solutions to some conformally invariant fully nonlinear equations, arXiv:math.AP/0605559.

[16] Y.Y. Li, Local gradient estimates of solutions to some conformally invariant fully nonlinear equations, C. R. Acad. Sci. Paris, Ser. I 343 (2006), 249-252.

[17] W. Sheng, N.S. Trudinger and X.J. Wang, The Yamabe problem for higher order curvatures, arXiv:math.DG/0505463 v1 23 May 2005.

[18] X.J. Wang, Apriori estimates and existence for a class of fully nonlinear elliptic equations in conformal geometry, Chin. Ann. Math. Ser. B 27 (2006), 169-178.

[19] J. Viaclovsky, Conformal geometry, contact geometry, and the calculus of variations, Duke Math. J. 101 (2000), 283-316.

[20] J. Viaclovsky, Estimates and existence results for some fully nonlinear elliptic equations on Riemannian manifolds, Comm. Anal. Geom. 10 (2002), 815-846.

Department of Mathematics, Rutgers University, 110 Frelinghuysen Road, Piscataway, NJ 08854, USA

Proceedings of Symposia in Applied Mathematics
Volume **65**, 2007

From Green to Lax via Fourier

A.S. Fokas

1. Introduction

Almost forty years ago, an ingenious new method was discovered [1] for the solution of the Cauchy problem of the Korteweg-de Vries (KdV) equation. This equation had been derived earlier [2], [3], as a model for small amplitude, long wave inviscid, irrotational water waves. The new method, which was later called the *inverse scattering transform* (IST) was based on the mysterious fact that the KdV equation is equivalent to two linear eigenvalue equations called a *Lax pair* in honor of Peter Lax, who first understood [4] that the IST method was the consequence of this remarkable property. The KdV equation belong to a large class of nonlinear equations which are called *integrable*. Although there exist several types of integrable equations, which include PDEs, ODEs, singular-integrodifferential equations, difference equations and cellular automata, the existence of an associated Lax pair provides a common feature of all these equations.

The IST method had *no* linear analogue, nevertheless it was noted in [5] that this method has certain conceptual similarities with the classical Fourier transform method. Furthermore, for small initial conditions, $q(x,0) = \varepsilon u_0(x)$, ε small, the solution $q(x,t)$ obtained via the IST can be approximated by the Fourier transform solution of the Cauchy problem of the associated linearized equation.

For example, in the case of the nonlinear Schrödinger equation (NLS), the linearized equation is

$$iq_t + q_{xx} = 0. \tag{1.1}$$

The Fourier transform solution for the Cauchy problem on the line for equation (1.1) is a consequence of the reverent approach of *separation of variables*. Indeed, letting $q(x,t) = X(x;k)T(t;k)$, equation (1.1) yields

$$\frac{d^2X}{dx^2} - k^2X = 0, \quad \frac{dT}{dt} + ik^2T = 0, \quad k \in \mathbb{C}. \tag{1.2}$$

The assumption $q = XT$ is highly restrictive, thus equations (1.2) are *not* equivalent with equation (1.1). However, these equations are still useful, since it is possible to construct $q(x,t)$ via superposition. This can be achieved in a systematic and rigorous way using spectral theory. For example, the spectral analysis of the first

This work was partially supported by EPSRC and the European Commission.

of equations (1.2) with $x \in \mathbb{R}$ and with decaying conditions as $|x| \to \infty$, yields the Fourier transform pair.

Lax Pairs for Linear PDEs. Equations (1.2) are based on the fact that equation (1.1) is linear, thus such equations do *not* exist for the nonlinear Schrödinger equation. Does there exist an alternative formulation that can be nonlinearized? The answer is affirmative; this formulation is based on the notion of Lax pairs for linear PDEs introduced in [6]. For example, equation (1.1) is *equivalent* with the following pair of linear eigenvalue equations for the scalar function $\mu(x,t,k)$,

$$\mu_x - ik\mu = q, \quad \mu_t + ik^2\mu = iq_x - kq, \quad k \in \mathbb{C}. \tag{1.3}$$

Indeed, equations (1.3), which are *two* equations for the *single* function μ, are compatible iff q solves equation (1.1): Equations (1.3) can be rewritten in the form

$$\frac{\partial}{\partial x}\mu e^{-ikx+ik^2t} = qe^{-ikx+ik^2t}, \quad \frac{\partial}{\partial t}\mu e^{-ikx+ik^2t} = (iq_x - kq)e^{-ikx+ik^2t}. \tag{1.4}$$

Thus equations (1.3) are compatible iff q satisfies

$$\frac{\partial}{\partial t}\left[qe^{-ikx+ik^2t}\right] - \frac{\partial}{\partial x}\left[(iq_x - kq)e^{-ikx+ik^2t}\right] = 0. \tag{1.5}$$

It is straightforward to verify that this equation is equivalent with equation (1.1). Equations (1.3) constitute a Lax pair for the linear equation (1.1).

Equations (1.4) and (1.5) indicate that in order to construct a Lax pair for a linear PDE in two-dimensions, it is sufficient to rewrite this PDE as a *one-parameter family of PDEs, each of which is in a divergence form.* If a scalar function q satisfies a linear PDE in two dimensions, it is straightforward to write this PDE in divergence form by considering the given PDE together with the PDE satisfied by the formal adjoint, which will be denoted by $\tilde{q}$. For a large class of PDEs, and in particular for PDEs with constant coefficients, it is possible to construct a one-parameter family of solutions for $\tilde{q}$; this often can be achieved using separation of variables.

For example, if q satisfies equation (1.1), $\tilde{q}$ satisfies a similar equation with i replaced by $-i$. The equations for q and $\tilde{q}$ imply

$$i(q\tilde{q})_t - (\tilde{q}q_x - q\tilde{q}_x)_x = 0.$$

The particular choice of $\tilde{q} = \exp[-ikx + ik^2t]$, yields equation (1.5).

From Green's Identities to Lax Pairs. The fundamental Green's identities involve a formulation in terms of a divergence form. In this sense, George Green came the closest to discovering Lax pairs for linear PDEs. For example, let $q(x,y)$ satisfy the Helmholtz equation

$$q_{xx} + q_{yy} + q = 0. \tag{1.6}$$

In this case, $\tilde{q}$ also satisfies equation (1.6). The equations for q and $\tilde{q}$ imply

$$(\tilde{q}q_x - \tilde{q}_xq)_x - (q\tilde{q}_y - q_y\tilde{q})_y = 0. \tag{1.7}$$

Separation of variables yields $\tilde{q} = \exp[ik_1x + ik_2y]$, where $k_1^2 + k_2^2 = 1$, i.e. $k_1 = \cos\alpha$, $k_2 = \sin\alpha$. Letting $\exp[i\alpha] = k$, equations (1.7) yield the following Lax pair for the Helmholtz equation,

$$\mu_y + \frac{1}{2}\left(k - \frac{1}{k}\right)\mu = q_x - \frac{i}{2}\left(k + \frac{1}{k}\right)q,$$

$$\mu_x + \frac{i}{2}\left(k + \frac{1}{k}\right)\mu = -q_y + \frac{1}{2}\left(k - \frac{1}{k}\right)q, \quad k \in \mathbb{C}. \tag{1.8}$$

Lax Pairs and the Construction of Integrable Nonlinear PDEs. The Lax pair formulation of equation (1.1) was used in [6] mainly for pedagogical reasons. Namely, it was shown in [6] that if one solves the Cauchy problem on the infinite line of equation (1.1) by utilizing equations (1.3), then one can solve the corresponding problem for the NLS equation by following identical conceptual steps utilizing the associated Lax pair of the NLS. However, later developments indicate that the Lax pair formulation of linear PDEs is an important concept in its own right. Some of the implications of the Lax pair formulation will be discussed in this subsection, as well as in the following two subsections.

For the solution of the Cauchy problem, a decisive role is played by the first of equations (1.3). Indeed, it is the spectral analysis of this equation that yields a Riemann-Hilbert (RH) formulation expressing $\mu(x,t,k)$ in terms of the Fourier transform of $q(x,t)$ denoted by $\hat{q}(k,t)$. The time evolution of $\hat{q}(k,t)$ can be determined from the second of equations (1.3) (or from equation (1.1) itself). This RH problem takes the following form for the sectionally analytic scalar function $\mu(x,t,k)$,

$$\mu^+(x,t,k) - \mu^-(x,t,k) = e^{ikx-ik^2t}\hat{q}_0(k), \quad k \in \mathbb{R}, \tag{1.9a}$$

$$\mu = O\left(\frac{1}{k}\right), \quad k \to \infty, \tag{1.9b}$$

where $\hat{q}_0(k)$ denotes the Fourier transform of the initial condition $q(x,0)$. Equations (1.9) can be rewritten in the following matrix form for the sectionally analytic 2×2 matrix $M(x,t,k)$,

$$M^+(x,t,k) = M^-(x,t,k)\begin{pmatrix} 1 & e^{ikx-ik^2t}\hat{q}_0(k) \\ 0 & 1 \end{pmatrix}, \quad k \in \mathbb{R} \tag{1.10a}$$

$$M(x,t,k) = I_2 + O\left(\frac{1}{k}\right), \quad k \to \infty, \tag{1.10b}$$

where $I_2 = diag(1,1)$.

We will now show that it is possible, by deforming equation (1.10a), to construct a genuine matrix RH problem. In this respect we must eliminate the restriction of the upper diagonality of the jump matrix appearing in the rhs of equation (1.10a), in such a way that the unimodular nature of this matrix is preserved. This yields

$$M^+(x,t,k) = M^-(x,t,k)\begin{pmatrix} 1 & e^{ikx-ik^2t}\rho_1(k) \\ e^{-ikx+ik^2t}\rho_2(k) & 1+\rho_1(k)\rho_2(k) \end{pmatrix}, \quad k \in \mathbb{R}, \tag{1.11}$$

where ρ_1 and ρ_2 are some functions of k.

Starting from the RH problem defined by equations (1.11) and (1.10b), and using a powerful method introduced by Zakharov and Shabat [7]-[9], the so-called *dressing method*, it is possible to construct *algorithmically* the following Lax pair,

$$M_x + ik[\sigma_3, \mu] = QM, \tag{1.12a}$$

$$M_t + 2ik^2[\sigma_3, \mu] = (2kQ - iQ_x\sigma_3 - iQ^2\sigma_3)M, \tag{1.12b}$$

where [,] denotes the usual matrix commutator and

$$\sigma_3 = \begin{pmatrix} 1 & 0 \\ 0 & -1 \end{pmatrix}, \quad Q = \begin{pmatrix} 0 & q(x,t) \\ r(x,t) & 0 \end{pmatrix}. \tag{1.13}$$

The compatibility condition of equations (1.12) yields the following pair of nonlinear evolution PDEs for $q(x,t)$ and $r(x,t)$,

$$\begin{aligned} &iq_t + q_{xx} - 2rq^2 = 0 \\ &-ir_t + r_{xx} - 2r^2 q = 0. \end{aligned} \tag{1.14}$$

The reduction $r = \sigma\bar{q}$, $\sigma = \pm 1$, yields the celebrated NLS equation,

$$iq_t + q_{xx} - 2\sigma|q|^2 q = 0, \quad \sigma = \pm 1. \tag{1.15}$$

In summary: Starting from the *linearized* PDE (1.1), it is straightforward to construct the associated Lax pair (1.3). The analysis of equations (1.3) yields the RH problem (1.9), which is equivalent with the RH problem (1.10). The deformation of this RH problem yields the RH problem defined by equations (1.11) and (1.10b), which through the dressing method yields the Lax pair (1.12) and hence the *nonlinear* PDEs (1.14).

Spectral Analysis of Single Eigenvalue Equations and the Inversion of Integrals. If the time-dependence of $q(x,t)$ is ignored in the first of equations (1.3), then this equation can be considered as an ODE in x for the scalar function $\mu(x,k)$. The spectral analysis of this ODE yields a RH problem similar to the one formulated by equations (1.9), but without the term $\exp[-ik^2t]$. The solution of this scalar RH problem yields the classical inverse transform formula,

$$q(x) = \frac{1}{2\pi}\int_{-\infty}^{\infty} e^{ikx}\hat{q}(k)dk, \quad x \in \mathbb{R}. \tag{1.15a}$$

In other words, by using the above RH formulation, it is possible to invert the integral

$$\hat{q}(k) = \int_{-\infty}^{\infty} e^{-ikx}q(x)dx, \quad k \in \mathbb{R}. \tag{1.15b}$$

Thus, this approach provides *an algorithmic method for inverting certain integrals.* Novel applications of this method include the inversion of the so-called attenuated Radon transform [10], [11], which is a fundamental transform in medical imaging, as well as the inversion of the integrals that characterize the Dirichlet to Neumann map for evolution equations formulated in domains involving a moving boundary [12], [13].

Simultaneous Spectral Analysis of Lax Pairs and Integral Representations of Boundary Value Problems. The development of an effective method for the analysis of initial-boundary value problems for integrable nonlinear evolution PDEs remained an open problem for more than thirty years. This was apparently the consequence of the fact that this analysis requires going beyond the basic concept of separation of variables. We recall that the Cauchy problem of equation (1.1) is solved by the x-Fourier transform, i.e. a transform based on *one* of the two independent variables (x,t). This transform can be derived through the spectral analysis of either equation (1.2a) or equation (1.3a). Similarly, the solution of the Cauchy problem of the NLS (1.15) is based on a "nonlinear x-Fourier transform". This transform is constructed through the spectral analysis of equation

(1.12a), which can be considered as an ODE in x. Thus, the concept of the separation of variables, is the basic concept required for the solution of the Cauchy problem of both linear and integrable nonlinear PDEs. However, the solution of initial-boundary value problems requires the *synthesis* as opposed the separation of variables. A Lax pair is precisely the proper mathematical tool for implementing this new concept: The synthesis of separation of variables is implemented through the *simultaneous* spectral analysis of the Lax pair [14].

For example, let $q(x,t)$ satisfy the following linear PDE on the half-line,

$$q_t + w(-i\partial_x)q = 0, \quad 0 < x < \infty, \quad 0 < t < T, \tag{1.16}$$

where T is a positive constant, and $w(k)$ is a polynomial of degree n such that Re $w(k) \geq 0$ for k real. The simultaneous spectral analysis of the associated Lax pair yields the following integral representation [15], [16],

$$q(x,t) = \frac{1}{2\pi}\int_{-\infty}^{\infty} e^{ikx-w(k)t}\hat{q}_0(k)dk - \frac{1}{2\pi}\int_{\partial D_+} e^{ikx-w(k)t}\tilde{g}(k)dk, \tag{1.17a}$$

where ∂D_+ is the boundary of the domain D_+,

$$D_+ = \{k \in \mathbb{C} : \text{Re } w(k) < 0, \text{Im } k > 0\}, \tag{1.17b}$$

oriented so that D_+ is to the left of the increasing direction, $\hat{q}_0(k)$ is the x-Fourier transform of the initial condition $q(x,0)$,

$$\hat{q}_0(k) = \int_0^{\infty} e^{-ikx}q(x,0)dx, \quad \text{Im } k \leq 0, \tag{1.17c}$$

and $\tilde{g}(k)$ involves the t-transform of the boundary values $\{\partial_x^j q(0,t)\}_0^{n-1}$,

$$\tilde{g}(k) = \sum_{j=0}^{n-1} c_j(k)g_j(w(k)), \tilde{g}_j(k) = \int_0^T e^{ks}\partial_x^j q(0,s)ds, k \in \mathbb{C}, j = 0, 1, \cdots, n-1; \tag{1.17d}$$

the constants $c_j(k)$ can be determined explicitly in terms of $w(k)$.

The integral representation (1.17a) can *not* be used directly for the solution of a given initial-boundary value problem, because not *all* boundary values can be prescribed as boundary conditions. Indeed, a well posed problem requires the initial condition as well as N boundary conditions, where $N = n/2$ if n is even and N equals either $(n+1)/2$ or $(n-1)/2$ if N is odd (depending on the sign of the coefficient of k^n in $w(k)$). Thus in order for equation (1.7a) to be used effectively, the unknown boundary values must be determined in terms of the given initial and boundary conditions. This problem, which will be referred to as the *generalized Dirichlet to Neumann map*, will be discussed in the next subsection.

In the same way that the Cauchy problem of equation (1.1) can be solved without using the spectral analysis of equation (1.3a), it is also possible to derive equations (1.7) *without* using the associated Lax pair formulation. However, for certain linear PDEs in complicated domains, such as the Helmholtz equation in the interior of a convex polygon, the simultaneous spectral analysis provides the most efficient way of constructing an integral representation for the solution [17]. Furthermore, this approach is apparently the only effective method for analyzing initial-boundary value problems for nonlinear integrable evolution PDEs [18]-[26].

For the latter problems, the integral representation for $q(x,t)$ now involves the entries of a 2×2-matrix valued function $M(x,t,k)$. This function satisfies a 2×2-matrix RH problem, whose jump matrix has explicit exponential (x,t) dependence and also contains certain functions of k called spectral functions. These functions depend on the initial condition and on *all* boundary values. Thus, again it is necessary to characterize the generalized Dirichlet to Neumann map.

The Global Relation and the Dirichlet to Neumann Map. It has been shown by the author that this map can be characterized by analyzing a certain equation coupling all the boundary values, which the author has called the *global relation.*

For example, for equation (1.16) the associated global relation is

$$\hat{q}_0(k) - \tilde{g}(k) = e^{w(k)T}\int_0^\infty e^{-ikx}q(x,T)dx, \quad \text{Im } k \le 0. \tag{1.18}$$

The function $\tilde{g}(k)$ involves $n-N$ unknown boundary values, thus equation (1.18) is a *single* equation for $n-N$ unknown functions. Furthermore, equation (1.18) depends on the unknown function $q(x,T)$. However, the functions $\tilde{g}_j(w(k))$ defined by the second of equations (1.17d), depend on k only through $w(k)$, thus these functions remain *invariant* by those transformations in the complex k-plane which leave $w(k)$ invariant. Utilizing these transformations, it is actually possible to characterize the generalized Dirichlet to Neumann map in two different ways.

(i) *Determination of the t-transforms of the boundary values.* The integral representation (1.17a) contains the functions $\{\tilde{g}_j\}_0^{n-1}$. Thus it is more convenient to determine these functions instead of the boundary values $\{\partial_x^j q(0,t)\}_0^{n-1}$.

For example, for equation (1.1) the global relation (1.18) becomes

$$\hat{q}_0(k) - i\int_0^T e^{ik^2 s}q_x(0,s)ds + k\int_0^T e^{ik^2 s}q(0,s)ds = e^{ik^2 T}\hat{q}_T(k), \quad \text{Im } k \le 0, \tag{1.19}$$

where $\hat{q}_T(k)$ denotes the Fourier transform of $q(x,T)$, i.e. the integral appearing in the rhs of equation (1.18). For the Dirichlet problem, $q(0,t)$ is given, thus the only unknown in (1.17a) is $\tilde{g}_1(w(k))$, where $\tilde{g}_1$ is the first integral in the lhs of equation (1.19). Since $\tilde{g}_1$ is defined for $k\in\partial D_+$, i.e., for Im $k>0$, we replace in equation (1.19) k by $-k$ and solve the resulting equation for $\tilde{g}_1$,

$$i\tilde{g}_1 = -k\tilde{g}_0 + \hat{q}_0(-k) - e^{ik^2T}\hat{q}_T(-k), \quad \text{Im } k \ge 0. \tag{1.20}$$

Thus, $\tilde{g}_1$ is determined to within *an equivalent class of functions*, namely equation (1.20) involves $\hat{q}_T(-k)$, which is a function analytic for Im $k>0$ and of $O(1/k)$ as $k\to\infty$. It turns out that this lack of uniqueness does *not* affect the solution $q(x,t)$, since the contribution of $\exp[ik^2T]\hat{q}_T(-k)$ vanishes [15].

(ii) *Determination of the boundary values themselves.* Using the global relation, it is also possible to determine the boundary values themselves. However, whereas (i) above requires only the *algebraic* manipulation of the global relation (and of the equations obtained from the global relation by using certain invariant transformations), the determination of the boundary values themselves requires the inversion of certain integrals.

For example, for the Dirichlet problem of equation (1.1), the determination of $q_x(0,t)$ requires the inversion of the integral $\tilde{g}_1$ which is defined by the rhs of equation (1.20). This inversion yields, see [33],

$$q_x(0,t) = -\frac{1}{\sqrt{\pi}} e^{-\frac{i\pi}{4}} \left[\frac{1}{\sqrt{t}} \int_0^\infty e^{\frac{ix^2}{4t}} \dot{q}_0(x) dx + \int_0^t \frac{\dot{q}(0,s)}{\sqrt{t-s}} ds \right], \quad 0 < t < T, \tag{1.21}$$

where $\dot{q}_0(x)$ and $\dot{q}(0,s)$ denote the derivatives of $q_0(x)$ and $q(0,s)$.

There exist particular initial-boundary value problems for certain integrable nonlinear evolution PDEs for which it is possible to use the characterization (i) mentioned earlier. These problems, which can be solved with the same level of efficiency as the Cauchy problem, are referred to as *linearizable*. An example of such a problem is the KdVII (the KdV for the case that q_{xxx} and q_t have opposite signs) formulated on the half-line with $q(0,t) = \chi$ and $q_{xx}(0,t) = \chi + 3\chi^2$, where χ is a real constant.

General initial-boundary value problems for integrable nonlinear PDEs, require the characterization (ii) mentioned earlier. In this respect the main difficulty stems from the fact that the global relation now contains certain eigenfunctions of the associated Lax pair evaluated at $x = 0$, and these eigenfunctions denoted by $(\Phi_1(t,k), \Phi_2(t,k))$ depend on $q_x(0,t)$ and $q(0,t)$. Hence, the global relation is now a highly *nonlinear* equation coupling $q(0,t)$ and $q_x(0,t)$. In spite of this difficulty, it is still possible to solve the global relation in closed form [27]-[29]. For example, for the Dirichlet problem of the NLS on the half-line, equation (1.20) is now replaced by an equation which expresses $q_x(0,t)$ in terms of $q(x,0)$, $q(0,t)$ and $\{M_1(t,s), M_2(t,s)\}$, where these functions characterize the Gel'fand-Levitan-Marchenko representation of $\{\Phi_1(t,k), \Phi_2(t,k)\}$.

Organization of the Paper. The derivation of the RH problem (1.9) and the implementation of the dressing method for the derivation of the Lax pair (1.12), is presented in section 2. An algorithm for inventing certain types of integrals is presented in section 3 and is illustrated using the inversion of two particular examples. The derivation of integral representations for linear PDEs via the simultaneous spectral analysis of the associated Lax pairs is discussed in section 4 and is illustrated using equation (1.16) and the modified Helmholtz equation as examples. The integral representation for the NLS is derived in section 5. The nonlinear analogue of equation (1.21) for the NLS equation is derived in section 6. Linearizable boundary conditions are discussed in section 7. The emergence of a unified approach for studying boundary value problems is discussed in section 8.

Notation and Assumptions. For simplicity we assume throughout this paper that the initial condition $q(x,0)$ belongs to the Schwartz space,

$$q(x,0) = q_0(x) \in S(\mathbb{R}^+), \tag{1.22}$$

and that the boundary conditions have sufficient smoothness and are compatible with $q_0(x)$ for $x = t = 0$. It is of course possible to obtain similar results for less restrictive spaces.

2. From Linear to Integrable Nonlinear PDEs

Equation (1.16) can be rewritten in the form

$$\left(e^{-ikx+w(k)t} q\right)_t - \left(e^{-ikx+w(k)t} g\right)_x = 0, \quad k \in \mathbb{C}, \tag{2.1}$$

where the function g is the following combination of q and its spatial derivatives,

$$g = i \left. \frac{w(k) - w(l)}{k - l} \right|_{l=-i\partial_x} q(x,t). \tag{2.2}$$

This equation implies that g can be written in the form

$$g = \sum_{j=0}^{n-1} c_j(k) \partial_x^j q(x,t), \tag{2.3}$$

and it also provides an algorithm for determining $\{c_j(k)\}_0^{n-1}$.

Equation (2.1) yields the following Lax pair formulation of equation (1.16),

$$\mu_x - ik\mu = q, \quad \mu_t + w(k)\mu = g. \tag{2.4}$$

For example, substituting $q = \exp[ikx - w(k)t]$ in equation (1.1) it follows that in this case $w(k) = ik^2$ Hence, equation (2.2) becomes

$$g = -(k - i\partial_x)q = iq_x - kq,$$

and equations (2.4) yield equations (1.3).

We will use the Lax pair formulation (2.4) in order to solve the Cauchy problem on the infinite line, i.e. equation (1.16) with $x \in \mathbb{R}$, $t > 0$ and $q(x,0) = q_0(x) \in S(\mathbb{R})$. In the process we will also derive the RH problem (1.19).

We first *assume* that the solution of the Cauchy problem exists and has sufficient smoothness and decay. As it was mentioned earlier, the solution of the Cauchy problem is based on the spectral analysis of one of the two equations forming the Lax pair, namely equation (2.4a). We consider this eigenvalue equation as an ODE in x defining the unknown function $\mu(x,t,k)$ in terms of $q(x,t)$, where $t > 0$ is fixed, and k is an arbitrary *complex* parameter. By integrating equation (2.4a) with respect to x from either $-\infty$ or $+\infty$, it follows that

$$\mu = \left[\begin{array}{ll} \mu^+, & \text{Im } k \geq 0 \\ \\ \mu^-, & \text{Im } k \leq 0 \end{array} \right. \tag{2.5}$$

where

$$\mu^+(x,t,k) = \int_{-\infty}^{x} e^{ik(x-\xi)} q(\xi,t) d\xi, \quad \text{Im } k \geq 0,$$

$$\mu^-(x,t,k) = -\int_{x}^{\infty} e^{ik(x-\xi)} q(\xi,t) d\xi, \quad \text{Im } k \leq 0. \tag{2.6}$$

Equations (2.6) imply that $\mu = 0(1/k)$, $k \to \infty$. Thus, given $q(x,t)$, equations (2.5) and (2.6) define a function $\mu(x,t,k)$ which is sectionally analytic in the entire complex k-plane including infinity. This fact implies that there exists an *alternative* representation of μ, which instead of $q(x,t)$ involves $\hat{q}(k,t)$. Indeed, subtracting equations (2.6) we find

$$\mu^+(x,t,k) - \mu^-(x,t,k) = e^{ikx} \hat{q}(k,t), \quad k \in \mathbb{R}, \tag{2.7}$$

where $\hat{q}(k,t)$ is the Fourier transform of $q(x,t)$.

In order to find the time evolution of $\hat{q}(k,t)$ we note that equation (2.6a) implies

$$\hat{q}(k,t) = \lim_{x\to\infty} e^{-ikx} \mu^+(x,t,k). \tag{2.8}$$

Substituting this expression in equation (2.4b) and using the fact that $\{\partial_x^j q(x,t)\}_0^{n-1}$ vanish as $x \to \infty$, we find $\hat{q}_t + w(k)\hat{q} = 0$. Hence, equation (2.7) becomes

$$\mu^+(x,t,k) - \mu^-(x,t,k) = e^{ikx-w(k)t}\hat{q}_0(k), \quad k \in \mathbb{R}, \tag{2.9}$$

where $\hat{q}_0(k)$ is the Fourier transform of $q_0(x)$,

$$\hat{q}_0(k) = \int_{-\infty}^{\infty} e^{-ikx} q_0(x)dx, \quad k \in \mathbb{R}. \tag{2.10}$$

Equation (2.9) expresses the "jump condition" of the sectionally analytic function $\mu(x,t,k)$. Hence, μ satisfies an elementary scalar RH problem whose unique solution is given by

$$\mu(x,t,k) = \frac{1}{2\pi i}\int_{-\infty}^{\infty} \frac{e^{ilx-w(l)t}}{l-k}\hat{q}_0(l)dl, \quad k \in \mathbb{C}, \quad \text{Im } k \neq 0. \tag{2.11}$$

Equations { (2.5), (2.6) } and (2.11) express $\mu(x,t,k)$ in terms of $q(x,t)$ and of $\hat{q}_0(k)$ respectively. Substituting equation (2.11) into equation (2.4a) we find

$$q(x,t) = \frac{1}{2\pi}\int_{-\infty}^{\infty} e^{ilx-w(l)t}\hat{q}_0(l)dl. \tag{2.12}$$

Although we have constructed $q(x,t)$ under the assumption of existence, we can rigorously justify a posteriori this result *without* this assumption. Indeed, the above construction motivates the *definitions* of both the direct and the inverse maps, i.e. equations (2.10) and (2.12). Equation (2.12) shows that q solves equation (1.16). In order to verify that $q(x,0) = q_0(x)$, we must derive the inverse Fourier transform of $q_0(x)$ in terms of $\hat{q}_0(k)$. This can be achieved by repeating the spectral analysis performed earlier, where equation (2.4a) is now evaluated at $t = 0$. In this case instead of $q(x,t)$ we have the *known* function $q_0(x)$, thus every step of the spectral analysis can be rigorously justified.

The Dressing Method. The jump condition (2.9) with $w(k) = ik^2$ becomes equation (1.9a). The deformation of this equation yields equation (1.11a). It is more convenient to replace k by $-2k$ in this equation, and then equation (1.11) becomes

$$M^+(x,t,k) = M^-(x,t,k)e^{-i(kx+2k^2t)\sigma_3}S(k)e^{i(kx+2k^2t)\sigma_3}, \quad k \in \mathbb{C}, \tag{2.13}$$

where $S(k)$ is an arbitrary 2×2 unimodular matrix with $(S)_{11} = 1$, and σ_3 denotes the third Pauli matrix defined by the first of equations (1.13).

The main idea of the dressing method is to construct two linear operators L and N such that: (i) LM and NM satisfy the same jump condition as M. (ii) LM and NM are of $O(1/k)$ as $k \to \infty$. Then, under the *assumption* that the RH problem defined by equations (1.10b) and (2.13) has a unique solution, it follows that

$$LM = 0, \quad NM = 0. \tag{2.14}$$

These equations constitute the Lax pair associated with the above RH problem.

In order to construct L we introduce the operator $\hat{\sigma}_3$ defined by

$$\hat{\sigma}_3 M = [\sigma_3, M]. \tag{2.15}$$

Equation (2.13) can be rewritten in the form

$$M^+ = M^- e^{-i(kx+2k^2t)\hat{\sigma}_3}S(k). \tag{2.16}$$

This equation immediately implies that M satisfies the equation

$$\{(\partial_x + ik\hat{\sigma}_3) M^+\} = \{(\partial_x + ik\hat{\sigma}_3) M^-\} e^{-i(kx+2k^2t)\hat{\sigma}_3} S(k), \tag{2.17}$$

as well as a similar equation with the operator $\partial_x + ik\hat{\sigma}_3$ replaced by the operator $\partial_t + 2ik^2\hat{\sigma}_3$.

Since M satisfies the boundary condition (1.10b), it follows that $(\partial_x + ik\hat{\sigma}_3)M$ is of $O(1)$. Thus in order to construct an operator L such that LM is of $O(1/k)$ we must subtract $Q(x,t)M$ (note that QM satisfies the same jump condition as M). Thus, we let

$$LM \doteqdot M_x + ik[\sigma_3, M] - QM. \tag{2.18}$$

Substituting the asymptotic expansion

$$M(x,t,k) = I_2 + \frac{M_1(x,t)}{k} + \frac{M_2(x,t)}{k^2} + O\left(\frac{1}{k^3}\right), \tag{2.19}$$

into equation (2.18) we find

$$LM = \{i[\sigma_3, M_1] - Q\} + O\left(\frac{1}{k}\right).$$

Thus, if Q is defined by the equation

$$Q(x,t) = i[\sigma_3, M_1(x,t)], \tag{2.20}$$

then LM satisfies the following homogeneous RH problem,

$$(LM)^+ = (LM)^- e^{-i(kx+2k^2t)\sigma_3} S(k) e^{i(kx+2k^2t)\sigma_3},$$

$$LM = O\left(\frac{1}{k}\right), \quad k \to \infty.$$

Hence, $LM = 0$, i.e. M and Q satisfy equation (1.12a).

The operator $(\partial_t + 2ik^2\hat{\sigma}_3)M$ is of $O(k)$, thus we define N by

$$NM \doteqdot M_t + 2ik^2[\sigma_3, M] - kA(x,t)M - B(x,t)M. \tag{2.21}$$

Substituting the asymptotic expansion (2.19) into equation (2.21) we find

$$NM = k\{2i[\sigma_3, M_2] - A\} + \{2i[\sigma_3, M_2] - AM_1 - B\} + O\left(\frac{1}{k}\right).$$

Thus, we define A and B by the equations

$$A = 2i[\sigma_3, M_1], \tag{2.22}$$

$$B = 2i[\sigma_3, M_2] - AM_1. \tag{2.23}$$

Comparing equations (2.20) and (2.22) it follows that

$$A = 2Q. \tag{2.24}$$

Then equation (2.23) becomes

$$B = -2\left(QM_1 - i[\sigma_3, M_2]\right). \tag{2.25}$$

The $O(1/k)$ term in the asymptotic expansion of equation $LM = 0$ yields

$$M_{1_x} + i[\sigma_3, M_2] = QM_1. \tag{2.26}$$

Comparing this equation with equation (2.25) it follows that $B = -2M_{1_x}$, i.e.

$$B = -2\partial_x\left[M_1^{(0)} + M_1^{(D)}\right], \tag{2.27}$$

where the superscripts refer to the off-diagonal and the diagonal parts of the matrix M_1. Equation(2.20) implies that

$$M_1^{(0)} = -\frac{i}{2}Q\sigma_3. \tag{2.28}$$

The diagonal part of equation (2.26) yields

$$M_{1_x}^{(D)} = QM_1^{(0)} = -\frac{i}{2}Q^2\sigma_3. \tag{2.29}$$

Hence, equations (2.27)-(2.29) imply

$$B = i(Q\sigma_3 + Q^2\sigma_3). \tag{2.30}$$

Equation $NM = 0$, where N is defined by (2.21) and A, B are defined by (2.24), (2.30), is equation (1.12b). The case of $r = \sigma\bar{q}$ corresponds to the celebrated NLS equation [30].

3. A New Method for Inverting Certain Integrals

Let $\mu(x, y, k)$ satisfy the eigenvalue equation $L\mu = q(x, y)$, where L is a linear differential operator in ∂_x and ∂_y depending on $k \in \mathbb{C}$. The spectral analysis of this equation consists of the construction of two maps: In the construction of the *direct map*, the equation $L\mu = q$ is solved for μ in terms of q for *all complex k*. In the construction of the *inverse map*, μ is expressed in terms of an appropriate spectral function of q by formulating either a RH or a $\bar{\partial}$ problem in the complex k-plane.

Suppose that the spectral function of q involves an integral transform of q. Then the above formalism provides a method for inverting this integral transform.

The above general formalism will be illustrated with the aid of two concrete examples.

3.1. The Radon and the Attenuated Radon Transforms. Let the line L make an angle θ with the positive x_1-axis. A point (x_1, x_2) on this line can be specified by the variables (τ, ρ), where ρ is the distance from the origin and τ is a parameter along the line, see Figure 3.1.

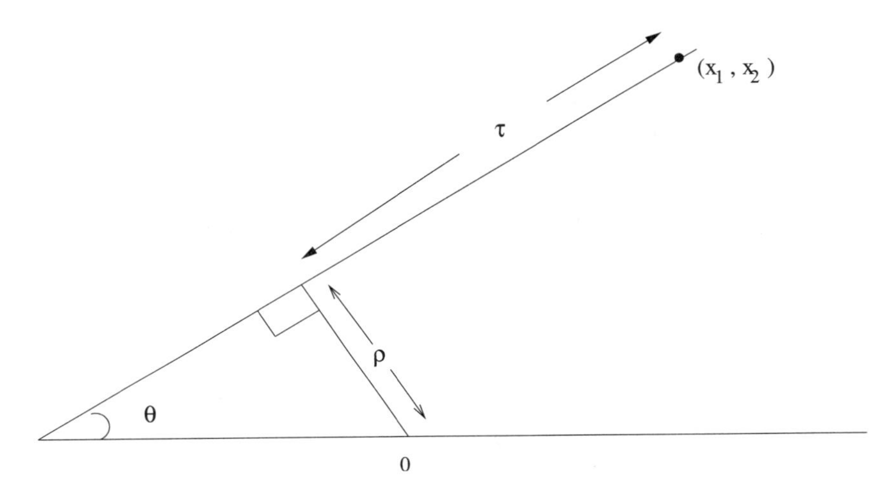

Figure 3.1

A unit vector $\underset{\sim}{k}$ along L is given by $(\cos\theta, \sin\theta)$, thus

$$(x_1, x_2) = \tau(\cos\theta, \sin\theta) + \rho(-\sin\theta, \cos\theta).$$

Hence

$$\begin{aligned} x_1 &= \tau\cos\theta - \rho\sin\theta, \quad & \rho &= x_2\cos\theta - x_1\sin\theta, \\ x_2 &= \tau\sin\theta + \rho\cos\theta, \quad & \tau &= x_2\sin\theta + x_1\cos\theta. \end{aligned} \tag{3.1}$$

The Radon transform of a function $f(x_1, x_2) \in S(\mathbb{R}^2)$, which will be noted by $\hat{f}(\rho, \theta)$, is defined as the integral of f along L, i.e.,

$$\hat{f}(\rho, \theta) = \int_{-\infty}^{\infty} f(\tau\cos\theta - \rho\sin\theta, \tau\sin\theta + \rho\cos\theta)d\tau, \quad \rho \in \mathbb{R}, \quad 0 < \theta < \pi. \tag{3.2}$$

The attenuated Radon transform of a function $g(x_1, x_2) \in S(\mathbb{R}^2)$ with attenuation $f(x_1, x_2) \in S(\mathbb{R}^2)$, which will be denoted by $\hat{g}_f(\rho, \theta)$, is defined by

$$\hat{g}_f(\rho, \theta) = \int_{-\infty}^{\infty} \Big(e^{-\int_\tau^\infty f(s\cos\theta - \rho\sin\theta, s\sin\theta + \rho\cos\theta)ds} \times$$

$$g(\tau\cos\theta - \rho\sin\theta, \tau\sin\theta + \rho\cos\theta)d\tau \Big), \quad \rho \in \mathbb{R}, \quad 0 < \theta < \pi. \tag{3.3}$$

The Radon transform can be inverted in a simple way using the Fourier transform. Also, it can be inverted via the spectral analysis of the following eigenvalue equation for the scalar function $\mu(x_1, x_2, k)$,

$$\frac{1}{2}(k + \frac{1}{k})\frac{\partial\mu}{\partial x_1} + \frac{1}{2i}(k - \frac{1}{k})\frac{\partial\mu}{\partial x_2} = f(x_1, x_2), \quad -\infty < x_1, x_2 < \infty, \quad k \in \mathbb{C}. \tag{3.4}$$

Indeed, using this equation it was shown in [31] that

$$f(x_1, x_2) = -\frac{i}{4\pi^2}\left(\partial_{x_1} - i\partial_{x_2}\right)\int_0^{2\pi} e^{i\theta} J(x_1, x_2, \theta)d\theta, \tag{3.5}$$

where J is the Hilbert transform with respect to ρ of $\hat{f}(\rho, \theta)$ evaluated at $\rho = x_2\cos\theta - x_1\sin\theta$, i.e.,

$$J(x_1, x_2, \theta) = \fint_{-\infty}^{\infty} \frac{\hat{f}(\rho', \theta)d\rho'}{\rho' - (x_2\cos\theta - x_1\sin\theta)}. \tag{3.6}$$

It was shown by Novikov in [10] that the spectral analysis of the eigenvalue equation

$$\frac{1}{2}(k + \frac{1}{k})\frac{\partial\mu}{\partial x_1} + \frac{1}{2i}(k - \frac{1}{k})\frac{\partial\mu}{\partial x_2} + f(x_1, x_2)\mu = g(x_1, x_2), \tag{3.7}$$

$$-\infty < x_1, x_2 < \infty, \quad k \in \mathbb{C},$$

yields the following formula for the inverse attenuated Radon transform: $g(x_1, x_2)$ is given by the rhs of equation (3.5), where

$$J(x_1, x_2, \theta) = -e^{\int_\tau^\infty f(s\cos\theta - \rho\sin\theta, s\sin\theta + \rho\cos\theta)ds} \times$$

$$\left(e^{P^- \hat{f}(\rho,\theta)} P^- e^{-P^- \hat{f}(\rho,\theta)} + e^{-P^+ \hat{f}(\rho,\theta)} P^+ e^{P^+ \hat{f}(\rho,\theta)} \right) \hat{g}_f(\rho, \theta); \tag{3.8}$$

in this equation, ρ and θ are given in terms of x_1 and x_2 by the second set of equations (3.1) and $P^\mp$ denote the usual projectors in the variable ρ, i.e.

$$(P^\mp f)(\rho) = \mp\frac{f}{2} + \frac{1}{2i\pi}Hf. \tag{3.9}$$

It was shown in [11] that equation (3.7) can be reduced to equation (3.4), hence the inversion formula for the attenuated Radon transform is the direct consequence of the spectral analysis of the equation (3.4) performed in [31].

In order to construct the direct map associated with equation (3.4) we change variables from (x_1, x_2) to $(z, \bar{z})$, where

$$z = \frac{1}{2i}(k - \frac{1}{k})x_1 - \frac{1}{2}(k + \frac{1}{k})x_2. \tag{3.10}$$

Then equation (3.4) becomes

$$\nu(k)\frac{\partial \mu}{\partial \bar{z}} = f, \quad \nu(k) := \frac{1}{2i}\left(\frac{1}{|k|^2} - |k|^2\right). \tag{3.11}$$

We supplement equation (3.11) with the boundary condition

$$\mu = O\left(\frac{1}{z}\right), \quad z \to \infty. \tag{3.12}$$

Using the inverse ∂-bar formula, it follows that the unique solution of the first of equations (3.11) satisfying the boundary condition (3.12), is given by

$$\mu(x_1, x_2, k) = \frac{1}{2\pi i} sgn\left(\frac{1}{|k|^2} - |k|^2\right) \int\int_{\mathbb{R}^2} \frac{f(x_1', x_2')dx_1'dx_2'}{z' - z}, \quad |k| \neq 1. \tag{3.13}$$

This equation provides the direct map, i.e., it expresses μ in terms of $f(x_1, x_2)$. In order to construct the inverse problem, i.e. in order to express μ in terms of an appropriate spectral function, we note that μ is an analytic function of k in the entire complex k-plane (including infinity) except for the unit circle. Thus in order to reconstruct μ, it is sufficient to compute the "jump" $\mu^+ - \mu^-$, where μ^+ and μ^- denote the limits of μ as k approaches the unit circle from inside and outside the unit disk. A simple computation yields [11]

$$\mu^\pm = \lim_{\varepsilon \to 0} \mu(x_1, x_2, (1 \mp \varepsilon)e^{i\theta}) = \mp P^\mp \hat{f}(\rho, \theta) - \int_\tau^\infty f ds, \tag{3.14}$$

where $P^\mp$ denote the usual projectors in the variable ρ defined in equation (3.9) and H denotes the Hilbert transform.

Equations (3.14) imply

$$\mu^+ - \mu^- = -\frac{H\hat{f}(\rho, \theta)}{i\pi}. \tag{3.15}$$

Substituting this expression in the equation

$$\mu = \frac{1}{2i\pi}\int_0^{2\pi} \frac{(\mu^+ - \mu^-)(e^{i\theta'})ie^{i\theta'}d\theta'}{e^{i\theta'} - k}, \quad |k| \neq 1, \tag{3.16}$$

we find μ in terms of $\hat{f}$,

$$\mu(x_1, x_2, k) = -\frac{1}{2i\pi^2}\int_0^{2\pi} \frac{e^{i\theta'}H\hat{f}d\theta'}{e^{i\theta'} - k}. \tag{3.17}$$

This equation provides the inverse map. Substituting this expression in equation (3.4) we find the inverse Radon transform formula (3.5) and (3.6).

We now present the derivation of the inverse attenuated Radon transform. Equation (3.7) can be rewritten in the form

$$\nu(k)\frac{\partial \mu}{\partial \bar{z}} + f(x_1, x_2)\mu = g(x_1, x_2), \tag{3.18}$$

where $f \in S(\mathbb{R}^2)$, $g \in S(\mathbb{R}^2)$, $k \in \mathbb{C}$, and $|k| \neq 1$. Equation (3.18) implies

$$\mu \exp\left[\partial_{\bar{z}}^{-1}\frac{f}{\nu}\right] = \partial_{\bar{z}}^{-1}\left(\frac{g}{\nu}\exp\left[\partial_{\bar{z}}^{-1}\frac{f}{\nu}\right]\right). \tag{3.19}$$

This equation provides the solution of the direct problem, i.e. it expresses μ in terms of f and g. Since μ is an analytic function of k in the entire complex k-plane except for the unit circle, it follows that μ satisfies the alternative representation (3.16). Thus, in order to express μ in terms of an appropriate spectral function, all that remains is to compute $\mu^+ - \mu^-$. But this can be easily derived using equation (3.14). Indeed, equation (3.11) can be rewritten in the form

$$\lim_{k\to k^\pm}\left\{\partial_{\bar{z}}^{-1}\left(\frac{f(x_1,x_2)}{\nu(k)}\right)\right\} = \mp P^{\mp}\hat{f}(\rho,\theta) - \int_{\tau}^{\infty} f ds. \tag{3.20}$$

Using this equation, equation (3.19) can be used to compute $\mu^\pm$, and then equation (3.16) provides an alternative representation of μ in terms of $\hat{g}_f(\rho,\theta)$ and of f. Substituting this representation in equation (3.7) we find the inverse attenuated Radon transform (3.5) and (3.8).

3.2. Integrals Arising in Moving Initial-Boundary Value Problems. It was shown in the introduction, that the characterization of the Dirichlet to Neumann map for equation (1.1) on the half-line involves the inversion of $\tilde{g}_1$, i.e. the inversion of the following integral

$$\mathcal{F}(k) = \int_0^T e^{ik^2 s} f(s) ds, \quad k \in \mathbb{C}, \tag{3.21}$$

where T is a positive constant. This integral can be inverted by a straightforward application of the inverse Fourier transform (after a suitable change of variables),

$$f(t) = \frac{1}{\pi}\int_{\partial I} e^{-ik^2 t} k\mathcal{F}(k) dk, \quad 0 < t < T,$$

where ∂I denotes the boundary of the first quadrant of the complex k-plane with the orientation shown in Fig. 3.2.

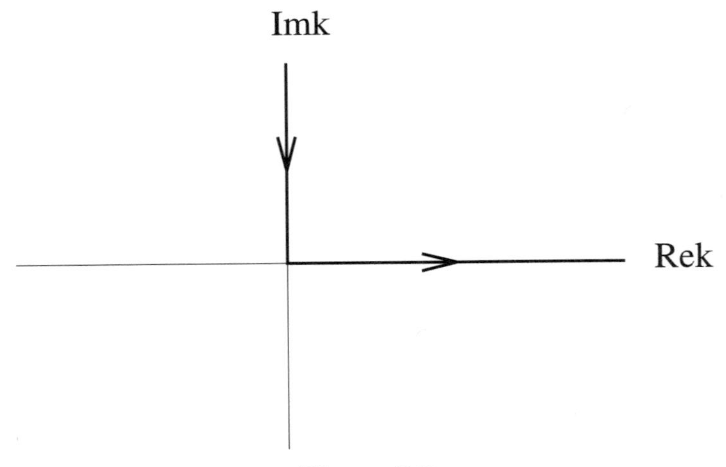

Figure 3.2

Alternatively, the integral in (4.1) can be inverted via the spectral analysis of the following eigenvalue equation

$$\mu_t + ik^2\mu = kf(t), \quad 0 < t < T, \quad k \in \mathbb{C}. \tag{3.22}$$

It is shown in [32] that the characterization of the Dirichlet to Neumann map for equation (1.1) formulated in the domain $\{l(t) < x < \infty, t > 0\}$, where $l(t)$ is a given smooth function, involves the inversion of the following integral

$$F(k) = \int_0^T e^{ik^2 s - ikl(s)} f(s)ds, \quad k \in \mathbb{C}. \tag{3.23}$$

This integral can be inverted via the spectral analysis of the following eigenvalue equation

$$\mu_t(t,k) + (ik^2 - ikl'(t))\mu(t,k) = kf(t), \quad 0 < t < T, \quad k \in \mathbb{C}. \tag{3.24}$$

There exists an important difference between equations (3.22) and (3.24): whereas there exists a solution $\mu(t,k)$ of equation (3.22) which is *sectionally analytic* in the entire complex k-plane, there does *not* exist such a solution for equation (3.24), i.e. the solution of equation (3.24) satisfies $\partial\mu/\partial\bar{k} \neq 0$. As a consequence of this difference, the‘ integral (3.23) cannot be inverted in closed form but is given through the solution of a linear Volterra integral equation. Indeed, the following result is derived in [12]:

Let $F(k)$ be defined in terms of $f(t)$ by equation (3.23), where $f(t)$ is a sufficiently smooth function and for simplicity the following restrictions are imposed on $l(t)$,

$$l(t) \in C^2[0,T], \quad l''(t) > 0, \quad l(0) = 0. \tag{3.25}$$

Then $f(t)$ satisfies the Volterra integral equation

$$f_1(t) = \frac{2}{3\pi}\int_0^t J(s,t) f_1(s)ds + \frac{2}{3\sqrt{\pi}}\mathrm{e}^{-i\pi/4}\left[\int_0^t \frac{\mathrm{e}^{i\frac{(l(t)-l(s))^2}{4(t-s)}}}{\sqrt{t-s}} f_0'(s)ds \right.$$

$$\left. + \frac{1}{\sqrt{t}}\int_0^\infty \mathrm{e}^{\frac{i(l(t)-x)^2}{4t}} q_0'(x)dx\right], \tag{3.26}$$

while $J(s,t)$ is defined by

$$J(s,t) = \frac{l(s)-l(t)}{2(s-t)}\left\{\int_{l'(s)/2}^\infty \mathrm{e}^{ik^2(s-t)-ik(l(s)-l(t))}dk - \right.$$

$$\left. i\,G(s,t)\int_0^\infty \mathrm{e}^{-ik^2(s-t)-k(s-t)\left[l'(s)-\frac{l(s)-l(t)}{s-t}\right]}dk\right\}, \tag{3.27}$$

$$G(s,t) = \mathrm{e}^{i\frac{l'(s)^2}{4}(s-t)-i\frac{l'(s)}{2}(l(s)-l(t))},$$

and $\partial\Omega_2^-(t)$ is given by

$$\partial\Omega_2^-(t) = \{k \in \mathbb{R} : \ k \leq l'(t)/2\} \cup \{k = k_R + ik_I : \ k_R = l'(t)/2, \ k_I \leq 0\} \tag{3.28}$$

with the orientation shown in figure 3.2.

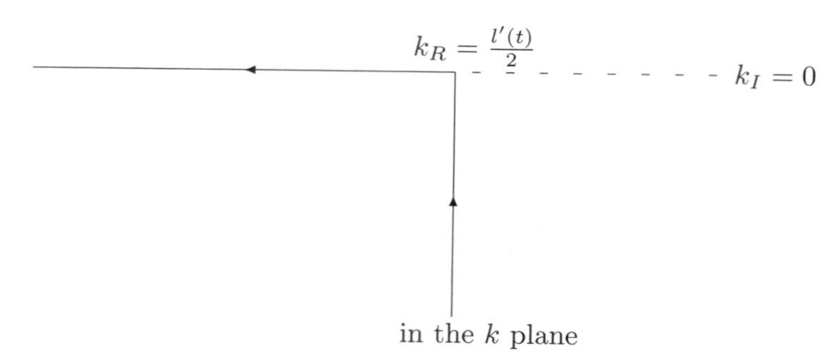

Figure 3.3: The curve $\partial\Omega_2^-(t)$ in the $k-$plane.

4. The Simultaneous Spectral Analysis of Lax pairs for Linear PDEs

We will derive the integral representation (1.17), as well as the corresponding formula for the modified Helmholtz equation in the interior of a bounded polygon via the *simultaneous* spectral analysis of both parts of the associated Lax pairs. A convenient way to implement this analysis is to rewrite the Lax pair as a single equation, and then to use the same steps discussed in section 3. The integral representation is derived under the assumption of existence. The solution of a concrete boundary value problem requires the analysis of the integral representation and of the associated global relation. Through this analysis it is possible to prove rigorously all relevant results *without* the assumption of existence.

4.1. Evolution PDEs on the Half-Line. The Lax pair associated with equation (1.16), i.e. equations (2.4), can be written in the form

$$d\left[e^{-ikx+w(k)t}\mu(x,t,k)\right] = e^{-ikx+w(k)t}\left[q(x,t)dx + \sum_{j=0}^{n-1} c_j(k)\partial_x^j q(x,t)dt\right]. \quad (4.1)$$

We consider equation (4.1) in the domain

$$0 < x < \infty, \quad 0 < t < T, \quad k \in \mathbb{C}.$$

Equation (4.1) implies

$$\mu_j(x,t,k) = \int_{(x_j,t_j)}^{(x,t)} e^{ik(x-\xi)-w(k)(t-\tau)}\left[q(\xi,\tau)d\xi + \sum_{j=0}^{n-1} c_j(k)\partial_\xi^j q(\xi,\tau)d\tau\right], \quad (4.2)$$

where μ_j depends only on (x_j,t_j) and not on the path of integration. It was shown in [17] that for a polygonal domain the proper choice of the set (x_j,t_j) is the set of the *corners* of this polygon. In the above particular example we take, see figure 4.1,

$$(x_1,t_1) = (0,T), \quad (x_2,t_2) = (0,0), \quad (x_3,t_3) = (\infty,t).$$

Thus, we define $\{\mu_j\}_1^3$ by the following equations,

$$\mu_3 = -\int_x^\infty e^{ik(x-\xi)}q(\xi,t)d\xi, \quad Imk \le 0, \quad (4.3a)$$

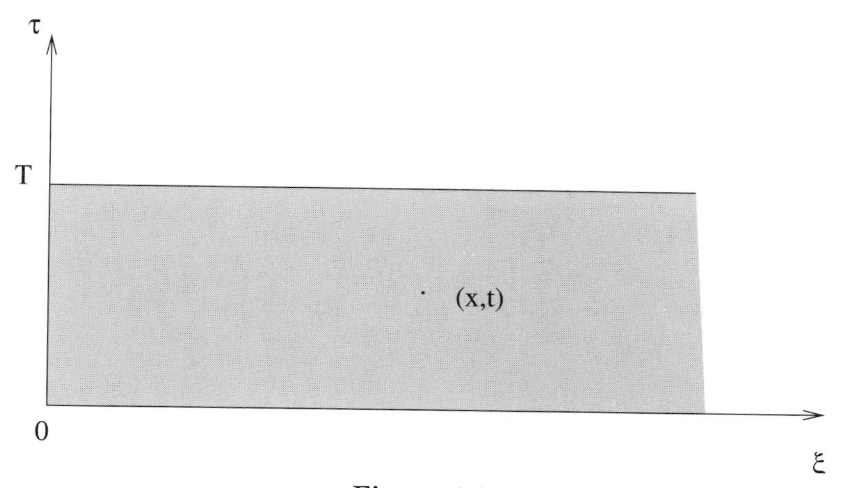

Figure 4.1

$$\mu_2 = \int_0^x e^{ik(x-\xi)} q(\xi,t) d\xi + e^{ikx} \int_0^t e^{-w(k)(t-\tau)} \sum_{j=0}^{n-1} c_j(k) \partial_\xi^j(0,\tau) d\tau, \; k \in \mathbb{C}, \quad (4,3b)$$

$$\mu_1 = \int_0^x e^{ik(x-\xi)} q(\xi,t) d\xi - e^{ikx} \int_t^T e^{-w(k)(t-\tau)} \sum_{j=0}^{n-1} c_j(k) \partial_\xi^j(0,\tau) d\tau, \; k \in \mathbb{C}. \quad (4.3c)$$

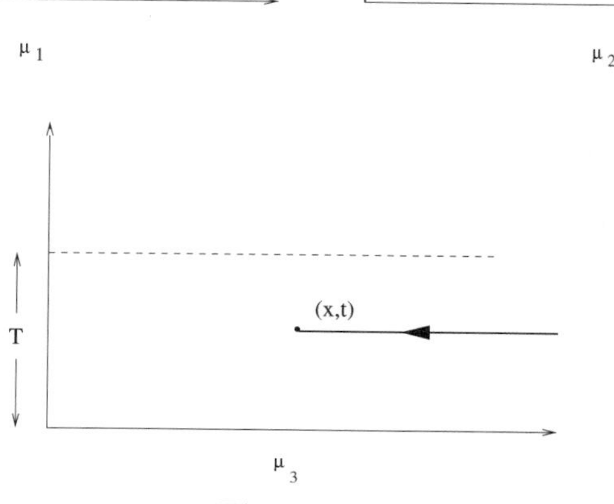

Figure 4.2

In these representations we have chosen the contours indicated in figure 4.2 . We note that the functions μ_1 and μ_2, which are entire functions in k are bounded in k in the following domains respectively,

$$\{Imk \geq 0\} \cap \{\mathrm{Re}w(k) \geq 0\}, \quad \{Imk \geq 0\} \cap \{\mathrm{Re}w(k) \leq 0\}. \tag{4.4}$$

Furthermore,

$$\mu = 0\left(\frac{1}{k}\right), \quad k \to \infty. \tag{4.5}$$

The advantage of performing the simultaneous spectral analysis is that, since μ_j is defined by (4.1), the difference of any two such μ_j's is proportional to $\exp[ikx - w(k)t]$. For example, in the domain where both μ_2 and μ_3 are valid,

$$\mu_2(x,t,k) - \mu_3(x,t,k) = e^{ikx-w(k)t}\rho_{23}(k);$$

evaluating this equation at $x = t = 0$, we find

$$\rho_{23}(k) = \hat{q}_0(k). \tag{4.6}$$

Also, subtracting equations (4.3b), (4.3c) we find

$$\mu_2 - \mu_1 = e^{ikx-w(k)t}\tilde{g}(k), \tag{4.7}$$

where $\tilde{g}(k)$ is defined by equation (1.17d).

Hence,

$$\mu_1 - \mu_3 = e^{ikx-w(k)t}[\hat{q}_0(k) - \tilde{g}(k)]. \tag{4.8}$$

Equations (4.5)-(4.8) imply

$$\mu = -\frac{1}{2i\pi}\left\{\int_{-\infty}^{\infty} e^{ilx-w(l)t}\hat{q}_0(l)\frac{dl}{l-k} - \int_{\partial D_+} e^{ilx-w(l)t}\frac{\tilde{g}(l)dl}{l-k}\right\}, \tag{4.9}$$

where ∂D_+ is defined by equation (1.17b).

Substituting equation (4.9) into equation (2.4a), we find equation (1.17a).

Remark.

Using Jordan's lemma, it is straightforward to show that equation (1.17a) is also valid if $\tilde{g}(k)$ is replaced by $\tilde{g}(k,t)$, where $\tilde{g}(k,t)$ is defined by equations similar to equations (1.17d) but with T replaced by t.

4.2. The Modified Helmholtz Equation in a Polygon. Using $z = x+iy$, the modified Helmholtz equation can be written in the form

$$q_{z\bar{z}} - \lambda q = 0, \quad \lambda > 0. \tag{4.10}$$

This equation can be rewritten in the following form

$$\left[e^{-(ikz+\frac{\lambda}{ik}\bar{z})}(q_z + ikq)\right]_{\bar{z}} + \left[e^{-(ikz+\frac{\lambda}{ik}\bar{z})}\left(q_{\bar{z}} + \frac{\lambda}{ik}q\right)\right]_z = 0. \tag{4.11}$$

Hence, the associated Lax pair is

$$\mu_z - ik\mu = q_z + ikq, \quad \mu_{\bar{z}} - \frac{\lambda}{ik}\mu = -q_{\bar{z}} - \frac{\lambda}{ik}q. \tag{4.12}$$

The simultaneous spectral analysis of this equation yields the following result.

Proposition 4.1. Let Ω be the interior of a convex bounded polygon in the complex $z-$plane with corners $z_1, ..., z_{n+1}$, see figure 4.3. Assume that there exists a solution $q(z,\bar{z})$ of the modified Helmholtz equation (4.10) valid in the interior of Ω and suppose that this solution has sufficient smoothness all the way to the boundary of the polygon. Then q can be expressed in the form

$$q(z,\bar{z}) = \frac{1}{4i\pi}\sum_{j=1}^{n}\int_{l_j} e^{ikz+\frac{\lambda}{ik}\bar{z}}\widehat{q}_j(k)\frac{dk}{k}, \quad k\in\Omega, \tag{4.13}$$

where the functions $\{\widehat{q}_j(k)\}_1^n$ are defined by

$$\widehat{q}_j(k) = \int_{z_{j+1}}^{z_j} e^{-ikz-\frac{\lambda}{ik}\bar{z}}\Big[iq_n - \Big(\frac{\lambda}{ik}\frac{d\bar{z}}{ds} - ik\frac{dz}{ds}\Big)q\Big]ds, \tag{4.14}$$

$$k\in C, \quad j=1,...,n, \quad z_{n+1}=z_1,$$

and $\{l_j\}_1^n$ are the rays on the complex $k-$ plane oriented towards infinity and defined by

$$l_j = \{k\in C : \arg(k) = -\arg(z_j - z_{j+1})\}, j=1,...,n, \quad z_{n+1}=z_1. \tag{4.15}$$

Furthermore, the following global relations are valid

$$\sum_{j=1}^{n}\widehat{q}_j(k) = 0, \quad \sum_{j=1}^{n}\widetilde{q}_j(k) = 0, \tag{4.16}$$

where $\{\widetilde{q}_j(k)\}_1^n$ are defined by

$$\widetilde{q}_j(k) = \int_{z_{j+1}}^{z_j} e^{ik\bar{z}+\frac{\lambda}{ik}z}\left[-iq_n + \left(\frac{\lambda}{ik}\frac{dz}{ds} - ik\frac{d\bar{z}}{ds}\right)q\right]ds, \tag{4.17}$$

$$k\in C, \quad j=1,...,n, \quad z_{n+1}=z_1.$$

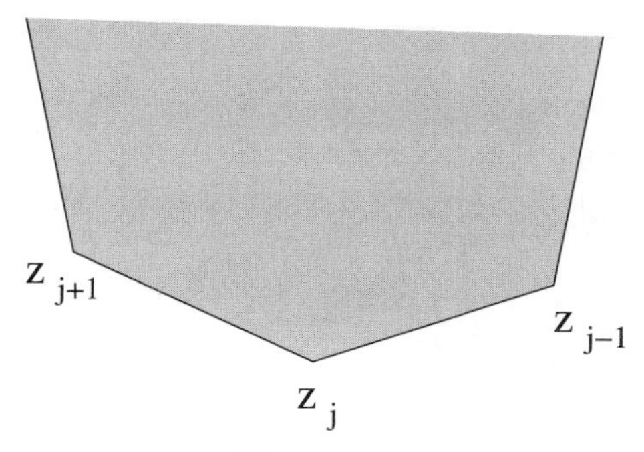

Figure 4.3

The global relations couple the function q and q_n on the boundary, i.e. they couple the Dirichlet and Neumann boundary values.

Proof.

Equations (4.12) imply

$$d\left[e^{-(ikz+\frac{\lambda}{ik}\bar{z})}\mu(z,\bar{z},k)\right] = e^{-\left(ikz+\frac{\lambda}{ik}\bar{z}\right)}\left[(q_z+ikq)dz - \left(q_{\bar{z}}+\frac{\lambda}{ik}q\right)d\bar{z}\right]. \tag{4.18}$$

Equation (4.18) with $z\in\Omega$ implies

$$\mu_j(z,\bar{z},k) = \int_{z_j}^{z} e^{ik(z-\zeta)+\frac{\lambda}{ik}(\bar{z}-\bar{\zeta})}\left[(q_\zeta+ikq)d\zeta - \left(q_{\bar{\zeta}}+\frac{\lambda}{ik}q\right)d\bar{\zeta}\right], \tag{4.19}$$

$$k\in\mathbb{C}.$$

This is an entire function of k, which is bounded in the sector S_j of the complex k-plane, which is defined by

$$S_j = \{k \in \mathbb{C} : \arg k \in [-\arg(z_{j-1} - z_j), \pi - \arg(z_{j+1} - z_j)]\}. \tag{4.20}$$

Indeed, the term $\exp[ik(z-\zeta)$ is bounded for

$$0 \leq \{\arg k + \arg(z-\zeta)\} \leq \pi. \tag{4.21}$$

Using the fact that

$$\arg(z-\zeta) \in [\arg(z_{j-1} - z_j), \arg(z_{j+1} - z_j)] \tag{4.22}$$

equation (4.21) implies that $\exp[ik(z-\zeta)]$ is bounded in S_j; similarly for the function $\exp[\lambda(\bar{z}-\bar{\zeta})/ik]$.

The sectors $\{S_j\}_1^n$ precisely cover the entire complex k-plane. Thus, the sectionally analytic function defined by

$$\mu = \mu_j, \quad k \in S_j, \quad j = 1, \cdots, n, \tag{4.23}$$

satisfies the following elementary RH problem:

$$\mu_j - \mu_{j+1} = e^{ikz + \frac{\lambda}{ik}\bar{z}} \hat{q}_j(k), \quad k \in l_j, \tag{4.24a}$$

$$\mu = -q + O\left(\frac{1}{k}\right), \quad k \to \infty, \tag{4.24b}$$

where $\hat{q}_j$ is defined by

$$\hat{q}_j(k) = \int_{z_{j+1}}^{z_j} e^{-(ikz + \frac{\lambda}{ik}\bar{z})} \left[(q_z + ikq)dz - \left(q_{\bar{z}} + \frac{\lambda}{ik}q\right)d\bar{z}\right], \tag{4.25}$$

$$k \in l_j, \quad j = 1, \cdots, n, \quad z_{n+1} = z_1;$$

equation (4.24b) is a direct consequence of the large k asymptotics of equation (4.12a).

The solution of the RH problem defined by equations (4.24) is given by

$$\mu = -q + \frac{1}{2i\pi} \sum_{j=1}^{n} \int_{l_j} e^{ilz + \frac{\lambda}{il}\bar{z}} \hat{q}_j(l) \frac{dl}{l-k}, \quad k \notin l_j. \tag{4.26}$$

Substituting this expression in equation (4.12b) and using

$$\frac{1}{l(l-k)} = \frac{1}{k}\left(\frac{1}{l-k} - \frac{1}{l}\right),$$

we find equation (4.13).

If a side of the polygon is parameterized by $z(s)$, then

$$q_z dz = \frac{1}{2}(q_s + iq_n)ds, \tag{4.27}$$

where q_s and q_n denote the derivatives of q along and normal to this side. Substituting this formula in the RHS of equation (4.25), we find that $\hat{q}_j(k)$ is given by equation (4.14).

The complex form of Green's theorem in the domain Ω and equation (4.11) imply the first of the global relations (4.16). The second global relation can be derived in a similar way using the transformation $k \to -\frac{\lambda}{k}$ in equation (4.18). **QED**

5. An Integral Representation for the Solution of the NLS on the Half-Line

The analysis of the initial-boundary value problems for integrable nonlinear PDEs requires the following steps.

Step 1. *Assuming existence: (a) Construct the integral representation of $q(x,t)$ in terms of appropriate spectral functions. This can be achieved via the simultaneous spectral analysis of the associated Lax pair. (b) Derive the global relation. This is a direct consequence of the Lax pair formulation.*

In order to implement this step for the defocusing NLS we first recall that if the operator $\hat{\sigma}_3$ is defined by equation (2.15), and A is a 2×2 matrix, then

$$e^{\hat{\sigma}_3} A = e^{\sigma_3} A e^{-\sigma_3}. \tag{5.1}$$

The two equations (1.12) with $r = \bar{q}$, which define the Lax pair of the defocusing NLS, can be rewritten in the form

$$d\left[e^{(ikx+2ik^2t)\hat{\sigma}_3} M(x,t,k)\right] = W(x,t,k), \quad k \in \mathbb{C}, \tag{5.2}$$

where M is a 2×2 matrix, and the differential form W is defined by

$$W = e^{(ikx+2ik^2t)\hat{\sigma}_3}\left[Q(x,t)M(x,t,k)dx + H(x,t,k)M(x,t,k)dt\right], \tag{5.3}$$

with

$$Q = \begin{pmatrix} 0 & q(x,t) \\ \bar{q}(x,t) & 0 \end{pmatrix}, \quad H = 2kQ - iQ_x\sigma_3 - i|q|^2\sigma_3. \tag{5.4}$$

The spectral analysis of equation (5.2) follows similar conceptual steps with those used in section 4.1. However, there exist the following technical differences: Since W involves M, the analogues of $\hat{q}_0(k)$ and $\tilde{g}(k)$ cannot be written in closed form but are characterized through linear Volterra integral equations uniquely defined in terms of $q_0(x)$ and $\{q(0,t) = g_0(t),\ q_x(0,t) = g_1(t)\}$, respectively. Similarly, although it is possible to define the matrix analogues of the scalar functions $\{\mu_j\}_1^3$ by the equations

$$M_j(x,t,k) = I_2 + e^{-(ikx+2ik^2t)\hat{\sigma}_3}\int_{(x_j,t_j)}^{(x,t)} W(\xi,\tau,k), \tag{5.5}$$

these functions yield a matrix, as opposed to a scalar, RH problem. Thus the integral representation of $q(x,t)$, in addition to spectral functions, it also involves the entries of the matrix M which is defined via a matrix RH problem.

Using equation (5.2), it is straightforward to derive the global relation.

In what follows we give the definitions of the spectral functions for the defocusing NLS:

Definition 5.1. The spectral functions $a(k)$ and $b(k)$ are defined in terms of $q_0(x)$ by

$$a(k) = \phi_2(0,k), \qquad b(k) = \phi_1(0,k), \qquad \operatorname{Im} k \geq 0, \tag{5.6}$$

where the vector $(\phi_1(x,k), \phi_2(x,k))$ is the solution of the linear Volterra integral equation

$$\phi_1(x,k) = \int_\infty^x e^{-2ik(x-x')} q_0(x')\phi_2(x',k)\,dx', \quad \operatorname{Im} k \geq 0 \tag{5.7a}$$

$$\phi_2(x,k) = 1 + \int_\infty^x \bar{q}_0(x')\phi_1(x',k)\,dx', \qquad \operatorname{Im} k \geq 0. \tag{5.7b}$$

The spectral functions $A(k)$ and $B(k)$ are defined in terms of $\{g_0(t), g_1(t)\}$ by

$$A(k) = \overline{\Phi_2(T, \bar{k})}, \qquad B(k) = -e^{4ik^2T}\Phi_1(T,k), \qquad k \in \mathbb{C}, \tag{5.8}$$

where the vector $(\Phi_1(t,k), \Phi_2(t,k))$ is the solution of the linear Volterra integral equation

$$\Phi_1(t,k) = \int_0^t e^{-4ik^2(t-t')}\left[(2kg_0(t')+ig_1(t'))\Phi_2(t',k) - i|g_0(t')|^2\Phi_1(t',k)\right] dt' \tag{5.9a}$$

$$\Phi_2(t,k) = 1 + \int_0^t \left[(2k\overline{g_0(t')} - i\,\overline{g_1(t')})\Phi_1(t',k) + i|g_0(t')|^2\Phi_2(t',k)\right] dt'. \tag{5.9b}$$

Step 2. *Existence under the assumption that the spectral functions satisfy the global relation.*

Given $q_0(x) \in S(\mathbb{R}^+)$, define $\{a(k), b(k)\}$. Define $\{A(k), B(k)\}$ in terms of the smooth functions g_0 and g_1. Define $q(x,t)$ through the solution of the RH problem formulated in step 1. Then prove that $q(x,t)$ solves the given nonlinear PDE and that $q(x,0) = q_0(x)$. *Assume* that there exist smooth functions g_0 and g_1 such that the spectral functions $\{a(k), b(k), A(k), B(k)\}$ satisfy the global relation. Then prove that $q(0,t) = g_0(t)$ and $q_x(0,t) = g_1(t)$.

In what follows we give the main theorem for the defocusing NLS. Details, as well as the analogous result for the focusing NLS are given in [20].

Theorem 5.1. Let $q_0(x) \in S(\mathbb{R}^+)$ and let $\{g_0(t), g_1(t)\}$ be smooth functions. Define $\{a(k), b(k)\}$ in terms of $q_0(x)$ by equations (5.6), (5.7), and $\{A(k), B(k)\}$ in terms of $\{g_0(t), g_1(t)\}$ by equations (5.8), (5.9).

Define $M(x,t,k)$ in terms of $\{a, b, A, B\}$ as the solution of the following RH problem:

(i) M is analytic in $k \in \mathbb{C} \setminus L$ where L denotes the union of the real and the imaginary axes.
(ii) $M = \operatorname{diag}(1,1) + O(1/k)$ as $k \to \infty$.
(iii) $M^- = M^+ J$, $k \in L$ (cf. Figure 5.1) where

$$J_1 = \begin{bmatrix} 1 & 0 \\ \Gamma(k)e^{2ikx+4ik^2t} & 1 \end{bmatrix}, \qquad J_2 = J_3 J_4^{-1} J_1, \qquad J_3 = \begin{bmatrix} 1 & \overline{\Gamma(\bar{k})}e^{-2ikx-4ik^2t} \\ 0 & 1 \end{bmatrix}$$

$$J_4 = \begin{bmatrix} 1 & -\gamma(k)e^{-2ikx-4ik^2t} \\ \overline{\gamma(\bar{k})}e^{i2kx+4ik^2t} & 1 - |\gamma|^2 \end{bmatrix}, \tag{5.10}$$

and

$$\gamma(k) = \frac{b(k)}{a(k)}, \quad \Gamma(k) = \frac{1}{a(k)}\left[\frac{\overline{A(\bar{k})}}{\overline{B(\bar{k})}}a(k) - b(k)\right]^{-1}. \tag{5.11}$$

Then M exists and is unique.

Define $q(x,t)$ in terms of $\{\gamma(k), \Gamma(k), M(x,t,k)\}$ by

$$q(x,t) = -\frac{1}{\pi}\Big\{\int_{\partial I_2} \overline{\Gamma(\bar{k})}e^{-2ikx-4ik^2t}M_{11}^+\, dk + \int_{-\infty}^{\infty} \gamma(k)e^{-2ikx-4ik^2t}M_{11}^+ dk$$
$$+ \int_0^\infty |\gamma(k)|^2 M_{12}^+ dk\Big\}, \tag{5.12}$$

where ∂I_2 denotes the boundary of the third quadrant of the complex k-plane. Then $q(x,t)$ solves the defocusing NLS and also satisfies $q(x,0) = q_0(x)$.

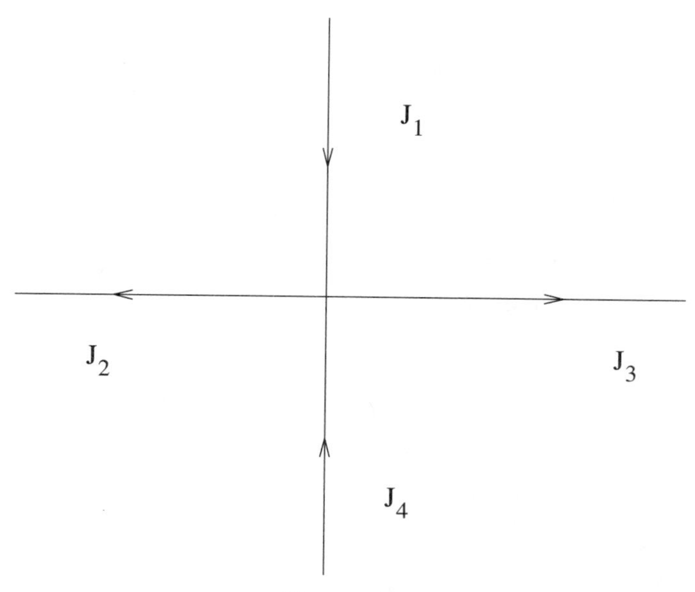

Figure 5.1

Assume that there exist smooth functions $\{g_0(t), g_1(t)\}$ such that

$$g_0(0) = q_0(0), \quad g_1(0) = \dot{q}_0(0), \tag{5.13}$$

and such that the following global relation is valid,

$$a(k)B(k) - b(k)A(k) = e^{4ik^2T}C_T(k), \quad \operatorname{Im} k \geq 0, \tag{5.14}$$

where $C_T(k)$ is a function analytic for Im $k > 0$ and of $O(1/k)$ as $k \to \infty$. Then

$$q(0,t) = g_0(t), \quad q_x(0,t) = g_1(t). \tag{5.15}$$

Remarks. 1. The above result reduces the solution of a given initial-boundary value problem for the defocusing NLS, to the problem of characterizing the Dirichlet to Neumann map. This problem will be discussed in sections 6 and 7.

2. The direct map defined by equations (5.6)-(5.9) is the nonlinearization of the corresponding direct map of equation (1.1). Indeed, if the initial condition $q_0(x)$ is small, equations (5.7) imply that $\phi_1 \sim 0$ and $\phi_2 \sim 1$, thus equations (5.6) imply that $a \sim 1$ and that $b(k)$ tends to the FT of $q_0(x)$. Similarly, if the boundary values $g_0(t) = q(0,t)$ and $g_1(t) = q_x(0,t)$ are small, equations (5.9) imply $\Phi_1 \sim 0$ and $\Phi_2 \sim 1$, thus equations (5.8) imply that $A \sim 1$ and that $B(k)$ tends to the time-transform of $2kg_0(t) + ig_1(t)$ (which is the spectral function associated with equation (1.1)). Similarly, the inverse map expressed by equation (5.12) is the nonlinearization of the integral representation (1.17a) for equation (1.1). Indeed, if the spectral functions are small, the RH problem defining M implies that $M \sim$ diag(1, 1), and then equation (5.12) implies that $q(x,t)$ tends to the expression defined by equation (1.17a) for the case of $\omega(k) = ik^2$.

6. The Dirichlet to Neumann Map and the Global Relation for the NLS on the Half-Line

For brevity of presentation we consider the particular case of $q_0(x) = 0$, which implies

$$a = 1, \quad b = 0. \tag{6.1}$$

In this case the global relation (5.14) becomes

$$B(k) = e^{4ik^2T}C_T(k), \quad \text{Im } k \geq 0; \quad B(k) = -e^{2ik^2T}\Phi_1(T,k). \tag{6.2}$$

Actually this equation follows from the evaluation at $t = T$ of the equation

$$-e^{2ik^2t}\Phi_1(t,k) = e^{4ik^2t}C(k,t), \quad \text{Im } k \geq 0, \quad 0 < t < T, \tag{6.3}$$

where $C(k,t)$ is an analytic functions of k for Im $k > 0$ and is of $O(1/k)$ as $k \to \infty$, for all $0 < t < T$.

Our aim is to solve equation (6.3) for $g_1(t)$ in terms of $g_0(t)$. The integral defining $\Phi_1(t,k)$, see equations (5.9), contains an exponential dependence on k, and also involves a complicated dependence on k through the functions $\Phi_1(t',k)$ and $\Phi_2(t',k)$. In order to have an integral which depends on k only through an exponential, we use the so-called Gel'fand-Levitan-Marchenko representation of the functions $(\Phi_1(t,k), \Phi_2(t,k))$. Namely, we employ the following result derived in [36]: Let the functions Φ_1 and Φ_2 satisfy the linear Volterra integral equations (5.9). Then, these functions can be represented in the form

$$\Phi_1(t,k) = 2e^{-2ik^2t}\int_0^t \left(L_1 - \frac{i}{2}g_0(t)M_2 + kM_1\right)e^{4ik^2\tau}d\tau, \tag{6.4a}$$

$$\Phi_2(t,k) = e^{2ik^2t} + 2e^{-2ik^2t}\int_0^t \left(L_2 + \frac{i\rho}{2}\bar{g}_0(t)M_1 + kM_2\right)e^{4ik^2\tau}d\tau, \tag{6.4b}$$

where L_1, L_2, M_1, M_2 are functions of $(t, 2\tau - t)$, and the functions $\{L_j(t,s), M_j(t,s)\}_1^2$ satisfy the following equations:

$$L_1(t,t) = \frac{i}{2}g_1(t), \quad M_1(t,t) = g_0(t), \quad L_2(t,-t) = M_2(t,-t) = 0, \tag{6.5a}$$

$$\begin{aligned}
L_{1_t} - L_{1_s} &= ig_1(t)L_2 + \alpha(t)M_1 + \beta(t)M_2,\\
L_{2_t} + L_{2_s} &= -i\rho\bar{g}_1(t)L_1 - \alpha(t)M_2 + \rho\bar{\beta}(t)M_1,\\
M_{1_t} - M_{1_s} &= 2g_0(t)L_2 + ig_1(t)M_2,\\
M_{2_t} + M_{2_s} &= 2\rho\bar{g}_0(t)L_1 - i\rho\bar{g}_1(t)M_1,
\end{aligned} \tag{6.5b}$$

and $\alpha(t)$, $\beta(t)$ are defined by the equations

$$\alpha(t) = \frac{\rho}{2}(g_0\bar{g}_1 - \bar{g}_0g_1), \quad \beta(t) = \frac{1}{2}(\dot{g}_0 - \rho|g_0|^2g_0). \tag{6.5c}$$

Using the first of equations (6.4), equation (6.3) becomes

$$\begin{aligned}
&\int_0^t e^{4ik^2\tau}\left[L_1(t,2\tau - t) - \frac{i}{2}g_0(t)M_2(t,2\tau - t) + kM_1(t,2\tau - t)\right]d\tau\\
&\qquad = -\frac{e^{4ik^2t}}{2}C(k,t), \quad \text{Im } k \geq 0, \quad 0 < t < T.
\end{aligned} \tag{6.6}$$

We multiply this equation by $k\exp[-4ik^2t']$, $t'<t$, and integrate along the boundary of the first quadrant of the complex k-plane, which we denote by ∂I (see figure 3.2).

The rhs of the resulting equation vanishes because $kC(k,t)$ is analytic and of $O(1)$ for Im $k>0$, and the oscillatory term $\exp[ik^2(t-t')]$ is bounded in the first quadrant.

The first two terms of the lhs of equation (6.6) give contributions which can be computed in closed form using the following identity

$$\int_{\partial I} k\left[\int_0^t e^{4ik^2(\tau-t')}K(\tau,t)d\tau\right]dk=\frac{\pi}{4}K(t',t),\quad t>0,\quad t>0,\quad t'<t \tag{6.7}$$

where $K(\tau,t)$ is a smooth function of its arguments. This identity follows from the usual Fourier transform identity after using the transformation $k^2\to l$ to map ∂I to the real axis. Using this identity, the first two terms of the lhs of equation (6.6) yield

$$\frac{\pi}{4}\left[L_1(t,2t'-t)-\frac{i}{2}g_0(t)M_2(t,2t'-t)\right]. \tag{6.8}$$

Before computing the contribution of the third term in the lhs of equation (6.6), we first use integration by parts:

$$\int_0^t ke^{4ik^2\tau}M_1(t,2\tau-t)d\tau=\frac{1}{4ik}\left[e^{4ik^2t}M_1(t,t)-M_1(t,-t)\right] \tag{6.9}$$

$$-\frac{1}{4ik}\int_0^t e^{4ik^2\tau}\frac{\partial M_1}{\partial\tau}(t,2\tau-t)d\tau.$$

Multiplying this term by $k\exp[-4ik^2t']$ and integrating along ∂I, we find three contributions. The first vanishes due to the fact that $\exp[4ik^2(t-t')]$ is bounded in the first quadrant of the complex k-plane. The k-integral of the second contribution can be computed in close form:

$$\int_{\partial I} e^{-4ik^2t'}dk=\int_{\partial I} e^{-4il^2}\frac{dl}{\sqrt{t'}}=\frac{c}{\sqrt{t'}}, \tag{6.10}$$

with

$$c=\int_{\partial I} e^{-4il^2}dl=\frac{1}{2}\int_{\partial I} e^{-il^2}dl=\frac{1}{2}\int_{-\infty}^{\infty} e^{-il^2}dl$$

$$=\int_0^\infty e^{-il^2}dl=\frac{1}{2}e^{-\frac{i\pi}{4}}\Gamma\left(\frac{1}{2}\right), \tag{6.11}$$

where in deforming ∂I to the real axis we have used the fact that $\exp(-il^2)$ is bounded in the second quadrant of the complex k-plane.

In order to compute the contribution of the third term in the rhs of equation (6.9) we split $\int_0^t$ into $\int_0^{t'}$ and $\int_{t'}^t$. The contribution of the second integral vanishes due to analyticity considerations and the contribution of the first integral yields a k-integral which equals $c/\sqrt{t'-t}$. Thus, equation (6.6) yields

$$\frac{\pi}{4}\left[L_1(t,2t'-t)-\frac{i}{2}g_0(t)M_2(t,2t'-t)\right]$$

$$-\frac{c}{4i}\left[\frac{M_1(t,-t)}{\sqrt{t'}}+\int_0^t\frac{\partial M_1}{\partial\tau}(t_1 2\tau-t)\frac{d\tau}{\sqrt{t-\tau}}\right]=0.$$

Letting $t' \to t$ and using $L_1(t,t) = \frac{i}{2}g_1(t)$, $M_1(\ t, -t) = g_0 = 0$, the above equation yields

$$g_1(t) = g_0(t)M_2(t,t) - \frac{e^{-\frac{i\pi}{4}}}{\sqrt{\pi}} \int_0^t \frac{\partial M_1}{\partial \tau}(t, 2\tau - t)\frac{d\tau}{\sqrt{t-\tau}}. \tag{6.12}$$

In summary: The solution of the global relation in the particular case of $q_0(x) = 0$ yields $g_1(t)$ in terms of $g_0(t)$ and the functions $\{M_1(t,s), M_2(t,s)\}$. Replacing in equations (6.5) g_1 by the rhs of equation (6.12), we obtain a nonlinear system for $\{L_j, M_j\}_1^2$ in terms of $g_0(t)$.

In the linear limit, equations (6.5) imply

$$L_1(t,s) \to L_1(t+s), \quad M_1(t,s) \to M_1(t+s), \quad L_2 \to 0, \quad M_2 \to 0.$$

Thus using the conditions at $s = t$ (equations (6.4a), we find

$$L_1(t,t) \to L_1(2t) = \frac{i}{2}g_1(t), \quad M_1(t,t) \to M_1(2t) = g_0(t),$$
$$M_1(t, 2\tau - t) \to M_1(2\tau) = g_0(\tau).$$

Hence, the linear limit of equation (6.12) yields

$$g_1(t) = -\frac{e^{-\frac{i\pi}{4}}}{\sqrt{\pi}} \int_0^t \frac{\partial g_0(\tau)}{\partial \tau}\frac{d\tau}{\sqrt{t-\tau}}, \tag{6.13}$$

which coincides with the Dirichlet to Neumann map of equation (1.1), with $q(x,0) = 0$, see equation (1.21).

7. Linearizable Boundary Value Problems

For several integrable nonlinear evolution PDEs, the t-part of the Lax pair evaluated at $x = 0$ takes the following form

$$\mu_t(t,k) + if_2(k)\hat{\sigma}_3\mu(t,k) = \tilde{Q}(t,k)\mu(t,k), \tag{7.1}$$

where μ is a 2×2 matrix, $\tilde{Q}(t,k)$ depends on the values of q and its derivatives evaluated at $x = 0$, and $f_2(k)$ is an analytic function of k. For example, for the NLS, $f_2 = 2k^2$, and

$$\tilde{Q}(t,k) = 2kQ(0,t) - i(Q_x(0,t) + Q^2(0,t))\sigma_3,$$
$$Q(0,t) = \begin{pmatrix} 0 & q(0,t) \\ \sigma\bar{q}(0,t) & 0 \end{pmatrix}, \quad \sigma^2 = 1. \tag{7.2}$$

The spectral functions $A(k)$ and $B(k)$ can be expressed in terms of the matrix $\mu(t,k)$ satisfying equation (7.1) and the initial condition $\mu(0,k) = I_2$.

Letting $M(t,k) = \mu(t,k)\exp[-if_2t\sigma_3]$, and using $\mu(0,k) = I$, we find

$$M_t + if_2(k)\sigma_3 M = \tilde{Q}(t,k)M, \quad M(0,k) = I_2. \tag{7.3}$$

If $\tilde{Q}(t,k)$ has certain symmetries, then M also has similar symmetries. We will assume that M possesses the following symmetry,

$$M(t,k) = \begin{pmatrix} \overline{M_2(t,\bar{k})} & M_1(t,k) \\ \rho\overline{M_1(t,\bar{k})} & M_2(t,k) \end{pmatrix}, \quad \rho^2 = 1. \tag{7.4}$$

Let $\nu(k)$ be defined by

$$f_2(\nu(k)) = f_2(k), \quad \nu(k) \neq k. \tag{7.5}$$

Suppose that given a subset of the boundary values $\{\partial_x^j q(0,t)\}_0^{n-1}$, it is possible to compute a nonsingular matrix $N(k)$ such that

$$U(t,\nu(k))N(k) = N(k)U(t,k); \quad U(t,k) = if_2(k)\sigma_3 - \tilde{Q}(t,k). \tag{7.6}$$

Then

$$M(t,\nu(k)) = N(k)M(t,k)N(k)^{-1}. \tag{7.7}$$

The spectral functions $A(k)$ and $B(k)$ can be expressed in terms of the entries of the matrix $M(t,k)$ by

$$A(k) = \overline{M_2(T,\bar{k})}e^{if_2(k)T}, \quad B(k) = -M_1(T,k)e^{if_2(k)T}. \tag{7.8}$$

Hence, equation (7.7) provides a relationship between $\{A(k),B(k)\}$ and $\{A(\nu(k)), B(\nu(k))\}$. This relationship, together with the global relation can be used to compute $\Gamma(k)$ in terms of $a(k)$, $b(k)$, and $N(k)$.

We note that equation (7.6) implies that a necessary condition for the existence of $N(k)$ is that the determinant of the matrix $U(t,k)$ depends on k only through $f_2(k)$.

Example.

In the case of the NLS, U depends only on k^2 iff

$$g_0(t)\overline{g_1}(t) - \overline{g_0}(t)g_1(t) = 0.$$

A particular solution of this equation is

$$g_1(t) = \chi g_0(t), \quad \chi \geq 0. \tag{7.9}$$

In this case, using $\nu(k) = -k$, it follows that if $(N)_{12} = (N)_{21} = 0$; then the diagonal part of equation (7.6) is satisfied identically and the off-diagonal parts yield

$$(2k - i\chi)N_{22} + (2k + i\chi)N_{11} = 0.$$

The second column vector of equation (7.7), yields

$$M_2(t,k) = M_2(t,-k), M_1(t,k) = -\frac{M_1(t,-k)}{f(k)}, \quad f(k) = \frac{2k - i\chi}{2k + i\chi}. \tag{7.10}$$

Hence,

$$A(k) = A(-k), \quad B(k) = -\frac{B(-k)}{f(k)}, \quad k \in \mathbb{C}. \tag{7.11}$$

We will now show that the global relation supplemented with these symmetry conditions yields $\Gamma(k)$ *explicitly* in terms of χ, $a(k)$, $b(k)$. For convenience we assume $T = \infty$. The function $\Gamma(k)$ for the NLS is defined by

$$\Gamma(k) = \frac{\sigma \overline{B(\bar{k})}}{a(k)d(k)}, \quad d(k) = a(k)\overline{A(\bar{k})} - \sigma b(k)\overline{B(\bar{k})}, \quad k \in D_2, \tag{7.12}$$

where $D_1, \cdots, D_4$ denote the first, $\cdots$, fourth quadrants of the complex k-plane. Letting $k \to -k$ in the definition of $d(k)$ and using (7.11) we find

$$A(k)\overline{a(-\bar{k})} + \sigma f(k)B(k)\overline{b(-\bar{k})} = \overline{d(-\bar{k})}, \quad k \in D_1. \tag{7.13}$$

This equation and the global relations which is also valid for $k \in D_1$, can be thought of as two algebraic equations for $A(k)$ and $B(k)$. Their solution yields,

$$A(k) = \frac{a(k)\overline{d(-\bar{k})}}{\Delta(k)}, \quad B(k) = \frac{b(k)\overline{d(-\bar{k})}}{\Delta(k)}, \quad k \in D_1. \tag{7.14a}$$

Letting $k \to -k$ and using (7.11) we find

$$A(k) = \frac{a(-k)\overline{d(-k)}}{\Delta(-k)}, \quad B(k) = -\frac{f(-k)b(-k)\overline{d(\bar{k})}}{\Delta(-k)}, \quad k \in D_3. \tag{7.14b}$$

Hence, equations (7.12) and (7.14) yield

$$\overline{\Gamma(\bar{k})} = \frac{\sigma}{\overline{a(\bar{k})}\left[\overline{a(\bar{k})}\frac{A(k)}{B(k)} - \sigma\overline{b(\bar{k})}\right]} = \frac{1}{\overline{a(\bar{k})}\left[\sigma\frac{\overline{a(\bar{k})}}{f(-k)}\frac{a(-k)}{b(-k)} + \overline{b(\bar{k})}\right]}, \quad k \in D_3.$$

In summary, the function $\Gamma(k)$ needed for the solution of the RH problem characterizing the solution of the NLS on th half-line with the boundary condition (7.9), is given by

$$\Gamma(\bar{k}) = \frac{1}{a(k)\left[\sigma\frac{a(k)}{f(-\bar{k})}\frac{a(-\bar{k})}{b(-\bar{k})} + b(k)\right]}. \tag{7.15}$$

Several linearizable problems are solved in [18] and [21].

8. Conclusions

The Cauchy problem on the infinite line for linear and integrable nonlinear evolution equations can be solved using the Fourier transform and a nonlinear Fourier transform respectively. These transforms can be constructed via the spectral analysis of *one* of the two equations forming the associated Lax pair. Boundary value problems can be solved by constructing appropriate integral representations and by analyzing the global relation. The integral representations can be constructed via the *simultaneous* spectral analysis of the associated Lax pair.

Following these developments, a new method has emerged for the analysis of boundary value problems. In what follows we discuss this method for both linear and integrable nonlinear PDEs.

The method for linear PDEs can be summarized as follows: Given a linear PDE for a scalar function q in a domain Ω construct an integral representation of the solution in terms of the values of q and its derivatives on the boundary of Ω denoted by $\partial\Omega$. Furthermore, formulate the associated global relation, which is an equation coupling the boundary values of q on $\partial\Omega$. By analyzing these two basic equations and in particular by utilizing the invariance properties of the global relation, eliminate the unknown boundary values appearing in the integral representation of q.

There exist two different types of integral representations, one formulated in the *physical space*, and one in the Fourier space, which for PDEs in two-dimensions is the complex k-plane.

It is shown in [33] that the classical Green's function formulation provides the integral representation in the physical spaces. Furthermore, it also immediately yields the associated global relation. The analysis of these two basic equations in the physical space provides *a simpler alternative as well as generalization of the method of images.* For example, the application of this method to the Laplace equation in the interior of the disc or of a sphere yields both the classical Poisson formula for the solution of the Dirichlet problem and the corresponding formulae for the solution of the Neumann problem.

For a large class of boundary value problems the physical space formulation is *not* effective. For such more complicated problems it is necessary to analyze the *representation of the basic equations in the Fourier space.* For example, for equation (1.16) these basic equations are equations (1.17a) and (1.18). The analysis of these equations in the Fourier space provides a *simpler alterative as well as a generalization of the method of the classical transforms.* For example, the classical sine- and cosine-transforms associated with the solution of the Dirichlet and Neumann problems of equation (1.1) can be derived from the global relation (1.19): Replacing k by $-k$ in (1.19) and adding or subtracting the resulting equation to equation (1.19) (with T replaced by t), we find

$$\int_0^\infty \sin(kx)q(x,t)dx = e^{-ik^2t}\left[\int_0^\infty \sin(kx)q(x,0)dx + ik\int_0^t e^{ik^2s}q(0,s)ds\right],$$

$$\int_0^\infty \cos(kx)q(x,t)dx = e^{-ik^2t}\left[\int_0^\infty \cos(kx)q(x,0)dx - i\int_0^\infty e^{ik^2s}q_x(0,s)ds\right],$$

$$k \quad \text{real}.$$

It is shown in [33] that the formulation of the method in the complex k-plane provides a unified, simple, and powerful approach to the analysis of a great variety of boundary value problems.

For linear PDEs the basic equations in the complex k-plane can be derived through a variety of approaches one of which is based on the Lax pair formalism. Apparently, it is only this approach that can be extended to integrable nonlinear PDEs. In other words, although for linear PDEs there exist two different formulations, one in the physical space and one in the Fourier space, it is only the latter that can be nonlinearized. Furthermore, among the various formalisms that exist for the Fourier space formulation for linear PDEs, it is only the Lax pair formalism that nonlinearizes.

In summary, the simultaneous spectral analysis of a Lax pair and the analysis of the associated global relation provides a unified method for the analysis of boundary value problems for both linear and integrable nonlinear PDEs. For a variety of boundary value problems, the analysis of the global relation requires the inversion of certain integrals. This can be achieved by using a new method based on the spectral analysis of an eigenvalue equation and on the formulation of either a RH or a $\bar{\partial}$ problem.

We conclude with some remarks:

1. The representation (1.17a) is consistent with the Ehrenpreis-Palamodov integral representation [34], [35]. Indeed, applying this abstract result to equation (1.16) it follows that there exists a measure $\rho(k)$ such that $q(x,t)$ is given by

$$q(x,t) = \int_L e^{ikx-iw(k)t}d\rho(k).$$

The limitation of this result is that it does *not* provide an algorithm for constructing $\rho(k)$ in terms of the given initial and boundary conditions. The new method shows that the Ehrenpreis-Palamodov representations can be made effective and can be nonlinearized. For example, the expression for $q(x,t)$ in (5.12) involves an explicit exponential (x,t) dependence, as well as the entries of the 2×2-matrix $M(x,t,k)$. Although M has complicated (x,t) dependence, it nevertheless is characterized through the solution of a RH problem whose jump matrix has explicit exponential (x,t) dependence.

2. The explicit (x,t) dependence of the RH problem mentioned above has important implications: There exist an elegant nonlinearization of the steepest descent

method introduced by Deift and Zhou [37], [38] that yields rigorous long time asymptotic results precisely for these types of RH problems. Also, by employing an essential extention of the Deift-Zhou method intoduced by Deift, Venakides and Zhou (the so-called g-function mechanism) [39],[40], it is possible to calculate asymptotically fully nonlinear waveforms such as those occuring in the zero-dispersion limit of integrable PDEs. Using these methods it is possible to investigate both the long time asymptotics and the zero-dispersion limit of integrable nonlinear PDEs solved via the simultaneous spectral analysis of the Lax pair.

3. For linear PDEs, the physical space formulation can be implemented in any number of dimensions. Furthermore, the global relation in the Fourier space formulation can be derived in any number of dimensions. The derivation of integral representations using the simultaneous spectral analysis for PDEs in n-dimensions, $n \geq 3$, is work in progress.

For nonlinear PDEs, the implementation of the new method for elliptic PDEs in 2 and 3 dimensions, and of evolution PDEs in $2+1$ dimensions is also work in progress.

References

[1] C. S. Gardner, J. M. Greene, M. D. Kruskal and R. M. Miura, Method for Solving the Korteweg-de-Vries Equation, Phys. Rev. Lett. **19**, 1095 (1967).

[2] D.J. Korteweg and G. de Vries, On the change of form of long waves advancing in a rectangular canal, and on a new type of long stationary waves, Philos. Mag. **539**, 422 (1895).

[3] J.Boussinesq, Theorie de l' intumescence liquid appelee onde solitaire ou de translation, se propagente dans un canal rectangulaire, Compte Rendus Acad. Sci. Paris, **72**, pp. 755-759 (1871)

[4] P.D. Lax, Integrals of nonlinear equations of evolution and solitary waves, CPAM **21**, 467 (1968).

[5] M. J. Ablowitz, D. J. Kaup, A. C. Newell and H. Segur, Method for solving the sine-Gordon equation, Phys. Rev. Lett. **30**, 1262 (1973)

[6] A. S. Fokas and I. M. Gel'fand, Integrability of Linear and Nonlinear Evolution Equations, and the Associated Nonlinear Fourier Transforms, Lett. Math. Physics, **32**, 189-210 (1994).

[7] V. E. Zakharov and A. B. Shabat, A Scheme for Integrating the Nonlinear Equations of Mathematical Physics Part I by the method of Inverse Scattering Problem, Funct. Annal. Appl. **8**, 43 (1974).

[8] V. E. Zakharov and A. B. Shabat, A Scheme for Integrating the Nonlinear Equations of Mathematical Physics by the Method of Inverse Scattering Problem,Part II, Funct. Annal. Appl. **13**, 13 (1979).

[9] V. E. Zakharov and S. V. Manakov, Construction of multidimensional nonlinear integrable systems and their solutions, **19**, 11 (1985).

[10] R. G. Novikov, An Inversion Formula for the Attenuated X-ray Trasnformation, Ark. Mat. **40**, 145 (2002).

[11] A.S. Fokas, A. Iserles and V. Marinakis, Reconstruction Algorithm for Single Photon Emission Computed Tomography and its numerical implementation, J. R. Soc. Interface **3** , 6, 45 (2006).

[12] A.S. Fokas and B. Pelloni, The Generalised Dirichlet to Neumann Map for Moving Initial-Boundary Value Problems (preprint)

[13] S. Delillo and A.S. Fokas, The Dirichlet to Neumann Map for the Heat Equation on a Moving Boundary (in preparation)

[14] A.S. Fokas, A Unified Transform Method for Solving Linear and Certain Nonlinear PDE's. Proc. R. Soc. Lond. A, **453**, 1411-1443 (1997).

[15] A.S. Fokas and L.Y. Sung, Initial-Boundary Value Problems for Linear Dispersive Evolution Equations on the Half-Line (preprint)

[16] A.S. Fokas, A New Transform Method for Evolution PDEs, IMA J. Appl. Math. **67**, 1-32 (2002).

[17] A.S. Fokas, Two Dimensional Linear PDEs in a Convex Polygon, Proc. R. Soc. Lond. A, **457**, 371-393 (2001).

[18] A.S. Fokas, Integrable Nonlinear Evolution Equations on the Half-Line, Comm. Math. Phys. **230**, 1-39 (2002).

[19] A.S. Fokas and L.Y. Sung, Generalised Fourier Transforms, their Nonlinearisation and the Imaging of the Brain, Notices of the AMS Feature Article, 52, 1176-1190 (2005).

[20] A.S. Fokas, A.R. Its and L.Y. Sung, The Nonlinear Schroedinger Equation on the Half-Line, Nonlinearity, **18**, 1771-1822 (2005).

[21] A.S. Fokas, Linearizable Initial-Boundary Value Problems for the sine-Gordon Equation on the Half-Line, Nonlinearity **17**, 1521-1534 (2004).

[22] A. Boutet de Monvel, A.S. Fokas and D. Shepelsky, The Modified KdV Equation on the Half-Line, J. of the Inst. of Math. Jussieu **3**, 139-164 (2004).

[23] A.S. Fokas and A.R. Its, The Nonlinear Schrödinger Equation on the Interval, J. Phys. A: Math. Gen. **37**, 6091-6114 (2004).

[24] A. Boutet de Monvel, A.S. Fokas and D. Shepelsky, Integrable Nonlinear Evolution Equations on the Interval, Comm. Math. Phys. **263** , 133-172 (2006).

[25] A. Boutet de Monvel and D. Shepelsky, The mKdV equation on the finite interval, C.R. Acad. Sci. Paris Ser. I Math., 337, 8, 517-522 (2003).

[26] A. Boutet de Monvel and V. Kotlyarov,Generation of asymptotic solitons of the nonlinear Schrdinger equation by boundary data, J. Math. Phys., **44**, 3185 (2003)

[27] A. Boutet de Monvel, A.S. Fokas and D. Shepelsky, The Analysis of the Global Relation for the Nonlinear Schrödinger Equation on the Half-Line, Lett. Math. Phys. **65**, 199-212 (2003).

[28] A.S. Fokas, A Generalised Dirichlet to Neumann Map for Certain Nonlinear Evolution PDEs, Comm. Pure Appl. Math. LVIII, 639-670 (2005).

[29] P.A. Treharne and A.S. Fokas, The Generalised Dirichlet to Neumann Map for the KdV Equation on the Half Line (preprint)

[30] V. E. Zakahrov and A. B. Shabat, Exact Theory of Two-Dimensional Self-Focusing and One-Dimensional Self Modulation of Waves in Nonlinear Media, Sov. Phys. JHEP **34**, 62 (1972).

[31] A. S. Fokas and R. G. Novikov, Discrete Analogues of the Dbar Equation and of Radon Transform, C.R. Acad. Sci., Paris, 313, 75-80 (1991).

[32] A. S. Fokas and B. Pelloni, Method for Solving Moving Boundary Value Problems for Linear Evolution Equations, Phys. Rev. Lett. bf 84, 4785 - 4789 (2000);A. S. Fokas and B. Pelloni, Inverse Problems **17** 919-935 (2001).

[33] A. S. Fokas and J. B. Keller, A New Method for Boundary Value Problems (preprint).

[34] L. Ehrenpreis, *Fourier Analysis in Several Complex Variables*, Wiley-Interscience, New York (1970).

[35] V. P. Palamodov, *Die Grundlehren der Mathematischen Wissenschaft*, Vol 168, Springer, Berlin (1970).

[36] Boutet de Monvel, A.; Kotlyarov, V. Scattering problem for the Zakharov-Shabat equationson the semi-axis. *Inverse Problems* **16**, no. 6, 1813-1837 (2000).

[37] P. Deift and X. Zhou, A steepest descent method for oscillatory Riemann-Hilbert problems, Bull. Am. Math. Soc., New Ser. **26**,1, 119 (1992).

[38] P. Deift and X. Zhou, A steepest descent method for oscillatory Riemann Hilbert problems. Asymptotics for the mkdv equation. Annals of Math., **137**, 295, (1993).

[39] P. Deift, S. Venakides, X. Zhou, The Collisionless Shock Region for the Long Time Behavior of the Solutions of the KdV Equation, CPAM. vol. 47, pp. 199-206, (1994).

[40] P. Deift, S. Venakides, X. Zhou, New Results in the Small-Dispersion KdV by an Extension of the Method of Steepest Descent for Riemann-Hilbert Problems, IMRN, N0. 6, 285-299 (1997).

[41] P. C. Sabatier, On elbow potential scattering and Korteweg-de Vries. *Inverse Problems* **18**, no. 3,611-630, (2002).

[42] P. C. Sabatier, Lax equations scattering and KdV. Integrability, topological solitons and beyond. *J. Math. Phys.* **44** , no. 8, 3216-3225, (2003).

Department of Applied Mathematics and Theoretical Physics, University of Cambridge, Cambridge, CB30WA, UK

E-mail address: `t.fokas@damtp.cam.ac.uk`

Proceedings of Symposia in Applied Mathematics
Volume **65**, 2007

UNTANGLING WALL TURBULENCE THROUGH DIRECT SIMULATIONS

Javier Jiménez

Abstract. The study of turbulence near walls has experienced a renaissance in the last decade, largely because of the availability of high-quality numerical simulations. The viscous and buffer layers over smooth walls are essentially independent of the outer flow, and there is a family of numerically-exact non-linear structures that account for most of the energy production and dissipation. Many of the best-known characteristics of the wall layer, such as the dimensions of the dominant structures, are well predicted by those models, which were essentially completed at the end of the last century. It is argued that this was made possible by the increase in computer power that made the kinematic simulations of the late 1980s cheap enough to undertake dynamic experiments. We are today at the early stages of simulating the logarithmic layer. A kinematic picture of the various cascades present in that part of the flow is beginning to emerge. Dynamical understanding can be expected in the next decade.

1. Introduction

Some of the first systems in which turbulence was studied scientifically were wall-bounded flows [**Hag39, Dar54**], in spite of which, wall turbulence remains to this day worse understood than homogeneous or free-shear flows. Part of the reason is that we are interested in different things in both cases. Turbulence is a multiscale phenomenon. Energy resides in the largest eddies, but it cannot be dissipated until it is transferred to smaller scales where viscosity can act. The classical conceptual framework for how this process takes place is the self-similar cascade proposed by [**Ric20**], which basically assumes that the transfer is local in scale space, with no significant interaction between eddies of very different sizes.

In homogeneous turbulence there are no spatial fluxes, and this multiscale transfer is everything. Starting from the cascade idea, and using energy conservation arguments, Kolmogorov [**Kol41**] was able to derive how energy is distributed among the inviscid eddies in the 'inertial' range of scales. He also computed the size of the eddies where the energy is finally dissipated, and where the cascade ends. The resulting energy spectrum, although now recognized as only an approximation, describes well the experimental observations, not only for isotropic turbulence, but also for small-scale turbulence in general.

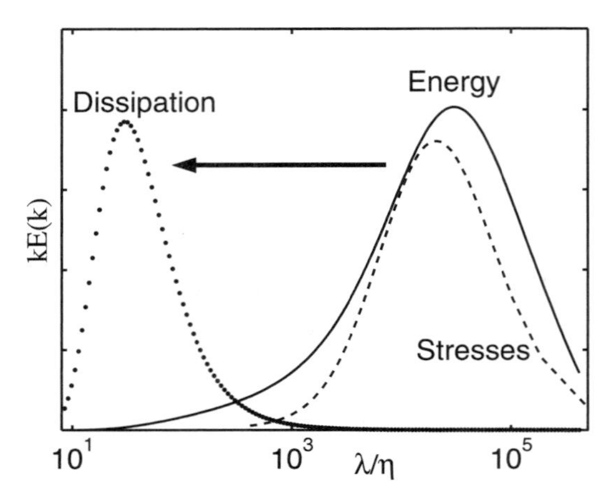

FIGURE 1. Premultiplied spectra, $kE(k)$, of the kinetic energy $|\boldsymbol{u}'|^2$, the dissipation $\nu|\boldsymbol{\omega}'|^2$, and the Reynolds stresses $-\langle u'v'\rangle$, where u and v are the streamwise and wall-normal velocity components. They are plotted as functions of the eddy size $\lambda = 2\pi/|\boldsymbol{k}|$, in a generic shear flow away from walls. The maxima for the energy and for the Reynolds stresses are near the integral length scale, while the dissipation peaks near the viscous Kolmogorov scale η. The energy transport, indicated by the arrow, is predominantly from large to small scales at the same spatial location.

Kolmogorov's theory gives no indication of how energy is fed into the cascade, which is a much less universal process than the cascade itself. In shear flows, the energy source is the gradient of the mean velocity. The mechanism is the interaction between that gradient and the Reynolds stresses, $-\langle u_i'u_j'\rangle$, which are the average momentum fluxes due to the velocity fluctuations [**TL72**]. In free shear flows, such as jets or mixing layers, this leads to a large-scale instability of the mean velocity profile [**BR74**], and to large-scale eddies with sizes of the order of the flow thickness. Those 'integral' scales contain most of the energy and of the Reynolds stresses. The subsequent transfer to the smaller eddies is thought to be essentially similar to the isotropic case. A sketch can be found in figure 1.

The mean velocity profiles of wall-bounded flows, such as pipes or boundary layers, are not subject to the same global instabilities as the free shear cases. Wall-bounded turbulence is consequently a weaker phenomenon. While the velocity fluctuations in a jet can easily reach 15-20% of the mean velocity differences, they rarely exceed 5% in a boundary layer. Wall-bounded flows are however of huge technological importance. Probably half of the energy being spent worldwide to move fluids through pipes and canals, or to move vehicles through air or through water, is spent in overcoming turbulent friction.

Wall-bounded flows are also scientifically tougher than isotropic of free-shear turbulence, because they are essentially inhomogeneous and anisotropic. The eddy sizes containing most of the energy at one wall distance are in the midst of the inertial cascade when they are observed farther away from the wall. The Reynolds number, defined as the scale disparity between energy and dissipation at some given location, also changes continuously with wall distance. The main emphasis in

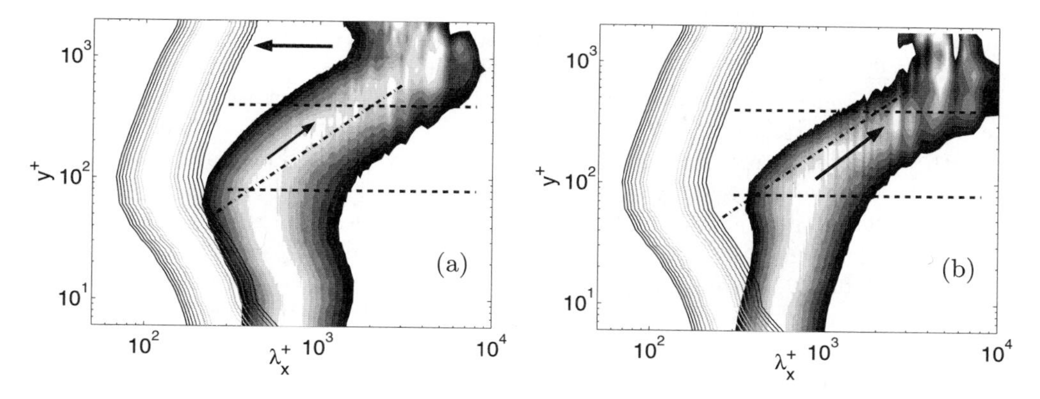

FIGURE 2. Premultiplied spectra, $kE(k)$. Shades are: (a) the kinetic energy $|\boldsymbol{u}'|^2$, and (b) the Reynolds stress $-\langle u'v' \rangle$. The lines are the premultiplied enstrophy spectrum, $|\boldsymbol{\omega}'|^2$. All are plotted as functions of the streamwise wavelength $\lambda_x = 2\pi/k_x$, and of the wall distance y. At each y the lowest contour is 0.86 times the local maximum. The horizontal lines are $y^+ = 80$ and $y/h = 0.2$, and represent limits of the logarithmic layer. The diagonal is $\lambda_x = 5y$. The arrows indicate the implied energy and momentum cascades. Data from a numerical channel with half-width $h^+ = 2000$ [**HJ06**].

studying wall turbulence is not on understanding the inertial energy cascade, but on the interplay between different scales at different distances from the wall.

Models for wall-bounded turbulence also have to deal with an extra cascade that is not present in the homogeneous case. Axial momentum is fed into a pipe by a mean pressure gradient that acts over its whole cross section. It is removed from the pipe only at the wall, by viscous friction. Momentum has to flow from the wall to the centre, carried that way by the Reynolds stresses mentioned above. We will see in the next section that this implies a cascade, because the size of the eddies carrying the stresses changes as a function of the wall distance by as much as the scale of the energy changes across the inertial cascade. This momentum transfer introduces an extra constraint to wall-bounded turbulence that complicates the problem with respect to free shear flows.

This paper is organized as follows. In the next section we outline the classical theory of wall-bounded flows, and define the different flow regions. In section 3 we review the recent work on equilibrium solutions for the near-wall layer, and how they are related to turbulence. In section 4 we discuss briefly the present status of our understanding of the logarithmic layer, and the likely impact of computers in future developments. Finally some conclusions are offered.

2. The classical theory of wall-bounded turbulence

The spectra of the energy and of the Reynolds stresses are plotted separately in figure 2, where both are compared with the dissipation. Each horizontal section of these figures is equivalent to figure 1. The energy and the Reynolds stresses are again at similar large scales, while the dissipative eddies are smaller. In this case, however, the size of the energy- and stress-containing eddies changes with the distance to the wall, and so does the range of scales over which the energy has to cascade. This divides the flow into several distinct regions.

Wall-bounded turbulence over smooth walls can be described by two sets of scaling parameters [**TL72**] . Viscosity is important near the wall, and the units for length and velocity in that region are constructed with the kinematic viscosity ν and with the friction velocity $u_\tau = (\tau_w/\rho)^{1/2}$, which is based on the shear stress at the wall τ_w, and on the fluid density ρ. Magnitudes expressed in these 'wall units' are denoted by $^+$ superscripts. There is no scale disparity in this region, as seen in figure 2, because most large eddies are excluded by the presence of the impermeable wall. The energy, the stresses, and the dissipation, are all at similar sizes. If y is the distance to the wall, y^+ is a Reynolds number for the size of the structures, and it is never large within this region. The near-wall layer extends at most to $y^+ = 150$ [**ÖJNH00**]. It is conventionally divided into a viscous sublayer, $y^+ \lesssim 5$, where viscosity is dominant, and a 'buffer' layer in which neither viscosity nor inertial effects can be neglected.

Away from the wall the velocity also scales with u_τ, because the momentum equation requires that the Reynolds stress, $-\langle u'v' \rangle$, can only change slowly with y to compensate the pressure gradient. This uniform velocity scale is the extra constraint introduced to wall-bounded flows by the momentum transfer. The length scale in the outer region far from the wall is the flow thickness h. Between the inner and the outer regions there is an intermediate layer where the only available length scale is the wall distance y.

Both the constant velocity scale across the intermediate region, and the absence of a length scale other than y, are only approximations. It will be seen below that large-scale eddies of size $O(h)$ penetrate to the wall, and that the velocity does not scale strictly with u_τ even in the viscous sublayer. However, if those approximation as accepted, it follows from simple symmetry arguments that the mean velocity in this 'logarithmic' layer is given by

$$U^+ = \kappa^{-1} \log y^+ + A. \tag{2.1}$$

The Kármán constant $\kappa \approx 0.4$ is approximately universal. Equation (2.1) does not extend to the wall, and the intercept constant A depends on the details of the viscous near-wall region. For smooth walls, $A \approx 5$.

The buffer, viscous and logarithmic layers are the most characteristic features of wall-bounded flows, and they constitute the main difference between those flows and other types of turbulence.

The viscous inner layer is extremely important for the flow as a whole. The ratio between the inner and the outer length scales is the friction Reynolds number, h^+, which ranges from 200 for barely turbulent flows to 5×10^5 for large water pipes. In the latter, the near-wall layer is only about 3×10^{-4} times the pipe radius, but it follows from (2.1) that, even in that case, 40% of the velocity drop takes place below $y^+ = 50$. Turbulence is characterized by the expulsion towards the small scales of the energy dissipation, away from the large energy-containing eddies. In wall-bounded flows that separation occurs not only in the scale space for the velocity fluctuations, but also in the shape of the mean velocity profile. The singularities are expelled both from the large scales, and from the centre of the flow towards the walls.

Because of this singular nature, the near-wall layer is not only important for the rest of the flow, but it is also essentially independent from it. That was for example shown by numerical experiments in [**JP99**], where the outer flow was artificially removed above a certain wall distance δ. The near-wall dynamics was essentially unaffected as long as $\delta^+ \gtrsim 60$.

The near-wall layer, among turbulent flows, is relatively easy to simulate numerically, because the local Reynolds numbers are low, but it is difficult to study experimentally because it is usually very thin in laboratory flows. Its modern study began experimentally in the 1970's [**KKR71, MBK71**], but it got its strongest impulse with the advent of high-quality direct numerical simulations [**KMM87**] in the late 1980's and in the 1990's. We will see below that it is one of the turbulent systems about which most is known. This knowledge has practical implications. Not only is this layer responsible for most of the friction drag of turbulent boundary layers, but most schemes to control wall friction can only act from the wall, and they interact directly with the viscous region.

The logarithmic law is located just above the near-wall layer, and it is also unique to wall turbulence. Most of the velocity difference that does not reside in the near-wall region is concentrated in the logarithmic layer, which extends experimentally up to $y \approx 0.2h$ (figure 2). It follows from (2.1) that the velocity difference above the logarithmic layer is only 20% of the total when $h^+ = 200$, and that it decreases logarithmically as the Reynolds number increases. In the limit of infinite Reynolds number, all the velocity drop is in the logarithmic layer.

The logarithmic layer is an intrinsically high-Reynolds number phenomenon. Its existence requires at least that its upper limit should be above the lower one, so that $0.2h+ \gtrsim 150$, and $h^+ \gtrsim 750$. The local Reynolds numbers y^+ of the eddies are also never too low. The logarithmic layer has been studied experimentally for a long time, but numerical simulations with an appreciable logarithmic region have only recently become available [**dÁJZM04, HJ06, FJ06**]. It is also much worse understood than the viscous layer.

2.1. The classical model for the buffer layer. The region below $y^+ \approx 100$ is dominated by coherent streaks of the streamwise velocity and by quasi-streamwise vortices. The former are an irregular array of long ($x^+ \approx 1000$) sinuous alternating streamwise jets superimposed on the mean shear, with an average spanwise separation of the order of $z^+ \approx 100$ [**SM83**]. The quasi-streamwise vortices are slightly tilted away from the wall [**JHSK97**], and stay in the near-wall region only for $x^+ \approx 200$. Several vortices are associated with each streak [**JM91**], with a longitudinal spacing of the order of $x^+ \approx 400$. Most of them merge into disorganized vorticity outside the immediate neighbourhood of the wall [**Rob91**].

It was proposed soon after they were discovered that streaks and vortices were involved in a regeneration cycle in which the vortices were the results of an instability of the streaks [**SB87**], while the streaks were caused by the advection of the mean velocity gradient by the vortices [**BL67, KKR71**]. Both processes have been documented and sharpened by numerical experiments. For example, disturbing the streaks inhibits the formation of the vortices, but only if it is done between $y^+ \approx 10$ and $y^+ \approx 60$ [**JP99**], suggesting that it is predominantly between those two levels that the regeneration cycle works. There is a substantial body of numerical [**HKW95, Wal97, SH02**] and analytic [**RSBH98, KJUP03**] work on the linear instability of model streaks. It shows that streaks are unstable to sinuous perturbations associated with inflection points of the distorted velocity profile, whose eigenfunctions correspond well with the shape and location of the observed

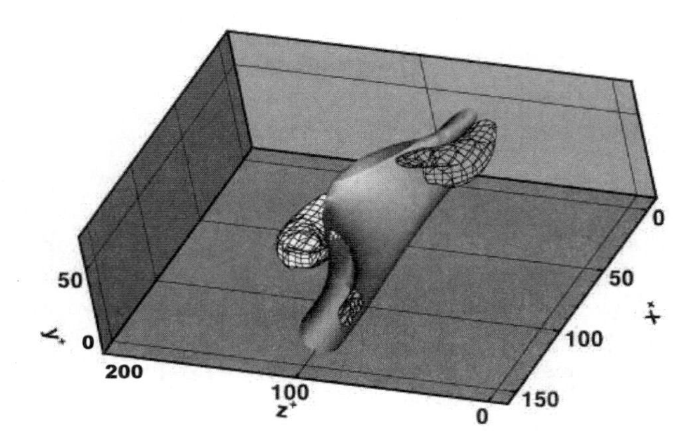

FIGURE 3. Exact permanent-wave solution for the Navier-Stokes equations in an 'autonomous' domain below $\delta^+ = 40$. The flow is from top-right to bottom-left. The central object is an isosurface of the streamwise perturbation velocity, $\widetilde{u}^+ = -3.5$, and defines the streak. It is flanked by two staggered streamwise vortices of opposite signs, $\omega_x^+ = \pm 0.18$, whose effect is to create an upwash that maintains the streak [**JS01**].

vortices. The model implied by these instabilities is a time-dependent cycle in which streaks and vortices are created, grow, generate each other, and eventually decay. Reference [**JP99**] discusses other unsteady models of this type, and gives additional references.

3. Exact solutions for the buffer layer

A slightly different point of view is that the regeneration cycle is organized around a nonlinear travelling wave, a fixed point in some phase space, which represents a nonuniform streak. This is actually not too different from the previous model, which essentially assumes that the undisturbed streak is a fixed point in phase space, and that the cycle is an approximation to an orbit along its unstable manifold. The new model however considers fixed points which are non-trivially perturbed streaks, and therefore separates the dynamics of turbulence from that of transition.

The organization of the buffer layer does not require the chaos observed in fully-turbulent flows. Simulations in which the flow is substituted by an ordered 'crystal' of identical single structures [**JM91**] reproduce the correct statistics (figure 4). In a further simplification, that occurred at roughly the same time as the previous one, nonlinear equilibrium solutions of the three-dimensional Navier–Stokes equations were obtained numerically, with characteristics that suggested that they could be useful in a dynamical description of the near-wall region [**Nag90**]. Other such solutions were soon found for plane Couette flow [**Nag90, Wal03**], plane Poiseuille flow [**TI01, Wal01, Wal03**], and for an autonomous wall flow [**JS01**]. All those solutions look qualitatively similar [**Wal98, KJUP03**], and take the form

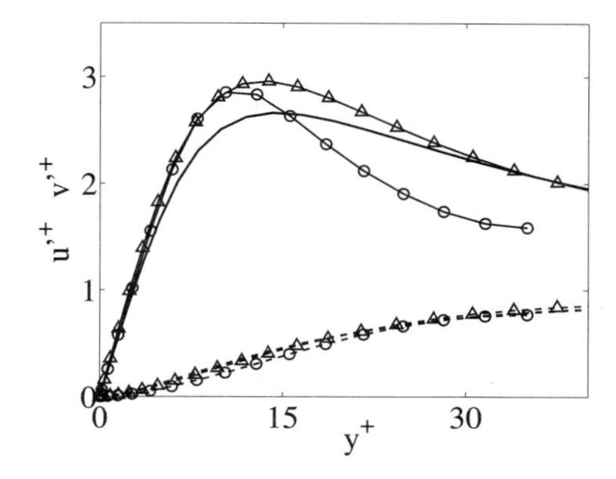

FIGURE 4. Profiles of the root-mean-square velocity fluctuations. Simple lines are a full channel with $h^+ = 180$ [**KMM87**]; —△ —, a minimal channel with $h^+ = 180$ [**JM91**]; —○—, the permanent-wave autonomous solution in figure 3. ——, streamwise velocity; - - - -, wall-normal velocity.

of a wavy low-velocity streak flanked by a pair of staggered quasi-streamwise vortices of alternating signs, closely resembling the spatially-coherent objects educed from the near-wall region of true turbulence. An example is shown in figure 3. Their mean and fluctuation intensity profiles are reminiscent of experimental values [**JS01, Wal03**], as shown in figure 4. Other properties are also suggestive of real turbulence. For example, the range of spanwise wavelengths in which the non-linear solutions exist is always in the neighbourhood of the observed spacing of the streaks of the sublayer [**JKS⁺05**].

In those cases in which the stability of the equilibrium solutions has been investigated, they have been found to be saddles in phase space, with few unstable directions. They are not therefore expected to exist as such in real turbulence, but the system could spend a substantial fraction of its lifetime in their neighbourhood, because it would move slowly in phase space in the neighbourhood of a fixed point. Exact limit cycles and heteroclinic orbits based on these fixed points have been found numerically [**KK01, TI03**], and several reduced dynamical models of the near-wall region have been formulated in terms of low-dimensional projections of such solutions [**AHLS88, SZ94, Wal97**].

The fixed-point and limit-cycle solutions found by different authors were recently reviewed and extended in [**JKS⁺05**]. It turns out that they can be classified into 'upper' and 'lower' branches in terms of their mean wall shear, and that both branches have very different profiles of their fluctuation intensities. The 'upper' solutions have relatively weak sinuous streaks flanked by strong vortices. They consequently have relatively weak root-mean-square streamwise-velocity fluctuations, and strong wall-normal ones, at least when compared to those in the lower branch. The solution in figure 3 belongs to the upper branch, and we already saw in figure 4 that its r.m.s. velocity fluctuations agree well with those of a full channel. 'Lower' solutions have stronger and essentially straight streaks, and much weaker vortices. Their statistics are very different from turbulence.

The statistics of full turbulent flows, when compiled over scales corresponding to a single streak and to a single vortex pair, agree reasonably well with those of

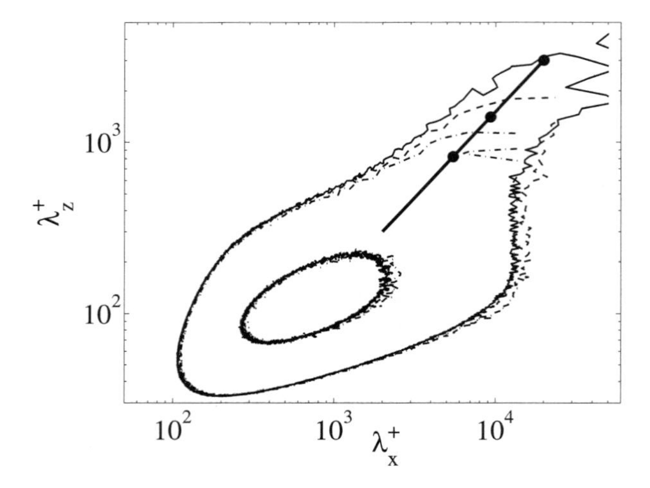

FIGURE 5. Two-dimensional spectral energy density of the streamwise velocity in the near-wall region ($y^+ = 15$), in terms of the streamwise and spanwise wavelengths. Numerical channels [**dÁJ03, dÁJZM04, HJ06**]. —·— , $h^+ = 547$; ---- , 934; —— , 2003. Spectra are normalized in wall units, and the two contours for each spectrum are 0.125 and 0.625 times the maximum of the spectrum for the highest Reynolds number. The heavy straight line is $\lambda_z = 0.15\lambda_x$, and the heavy dots are $\lambda_x = 10h$ for the three cases.

the fixed points, although there is a noticeable contribution from unsteady bursting [**JKS$^+$05**]. When the statistics are compiled over much larger boxes, however, the intensity of the fluctuations does not scale well in wall units, even very near the wall [**DE00**]. That effect is due to large outer-flow velocity fluctuations reaching the wall [**dÁJ03, JdÁF04, HJ06**], and is unrelated to the structures being considered here.

This is shown in figure 5, which contains two-dimensional premultiplied energy spectra of the streamwise velocity, $k_x k_z E_{uu}(k_x, k_z)$ in the buffer layer, displayed as functions of the streamwise and spanwise wavelengths. The three spectra in the figure correspond to large numerical turbulent channels at different Reynolds numbers. They differ from each other almost exclusively in the long and wide structures represented in the upper-right corner of the spectrum, whose sizes are of the order of $\lambda_x \times \lambda_z = 10h \times h$. Those spectra are fairly well understood [**dÁJ03, JdÁF04, HJ06**]. The lower-left corner contains the structures discussed in this section, which are very approximately universal and local to the near-wall layer. The larger structures in the upper edge of the spectra, and specially those in the top-right corner, extend into the logarithmic layer, scale in outer units, and correspond approximately to the 'attached eddies' proposed by Townsend [**Tow76**].

4. The logarithmic layer

We noted in section 2 that the logarithmic layer is expensive to compute. The first simulations with an appreciable logarithmic range have only recently appeared, and even in them the range of wall distances is short. In [**HJ06**], for example,

$h^+ = 2000$, and the upper and lower logarithmic limits are approximately $y^+ = 400$ and $y^+ = 150$. Some of the results of those simulations have been used in this article. The linear range in the lengths of the energy- and stress-containing eddies in figure 2, and the linear behaviour in the large-scale tails of the two-dimensional spectra in figure 5, are logarithmic-layer results.

These simulations, as well as corresponding advances in experimental methods, have greatly improved our kinematic understanding of the structures in this region. It is for example known that there is a self-similar hierarchy of compact ejections extending from the buffer layer into the outer flow, within which the vorticity is more intense than elsewhere. They are associated with extremely long, conical, low velocity regions in the logarithmic layer [**dÁJZM06**] – "wakes" – which agree well with the energy-containing structures of the streamwise velocity spectrum. The overall arrangement is reminiscent of the association of vortices and streaks in the buffer layer, but at a much larger scale. It is also known that the low-velocity regions are almost identical to the transient-growth structures forced on the mean velocity profile by a concentrated ejection [**dÁJ06**], but that whatever is causing them does not grow directly from the buffer layer. The lifetimes of the observed ejections are too short for that [**dÁJZM06**], and the spectra of flows in which the buffer region has been purposefully destroyed are essentially identical to those over smooth walls [**FJ06**].

We know much less about how the ejections are created, although there are indications that their association with the low-velocity regions goes both ways. Not only are the wakes seen when the statistics are conditioned to the ejections, but viceversa. The knowledge that we are gaining from the simulations is however essentially kinematic.

The problem is one of cost, and was shared by the original low-Reynolds number simulations that eventually led to the understanding of the buffer layer. The simulation in [**HJ06**] took six months on 2000 supercomputer processors. It took a similar time to run the simulation in [**KMM87**] at $h^+ = 180$. As long as each numerical experiment takes such long times, it is only possible to observe the results, and the simulations are little more than better-instrumented laboratory experiments.

As computers improve, however, other things become possible. When the low-Reynolds number simulations of the 1980s became roughly 100 times cheaper in the 1990s, it became possible to experiment with them in ways that were not possible in the laboratory. The series of 'conceptual' simulations that led to the results in section 3 were of this kind. The perturbed-wall simulations cited above [**FJ06**] are one of the first examples of this type of simulations for the logarithmic layer, but their Reynolds numbers are still only marginal, and they are in any case conceptually similar to flows over rough walls.

There is however no reason to believe that computer improvements have stopped. The next decade will bring the cost of the simulations of the logarithmic layer to the level at which dynamical experiments should become commonplace. It is only then that we can expect a dynamical theory for this part of the flow to emerge from simulations.

5. Conclusions

We have briefly reviewed the present state of the understanding of the dynamics of turbulent flows near smooth walls. This is a subject that, like most others in turbulence, is not completely closed, but which has evolved in the last two decades from empirical observations to relatively coherent theoretical models. It is also one of the first cases in turbulence, perhaps together with the structure of small-scale vorticity in isotropic turbulence, in which the key technique for cracking the problem has been the numerical simulation of the flow. The reason is that the Reynolds numbers of the important structures are low, and therefore accessible to computation, while experiments are difficult. For example the spanwise Reynolds number of the streaks is only of the order of $z^+ = 100$, which is less than a millimetre in most experiments, but we have seen that it is well predicted by the range of parameters in which the associated equilibrium solutions exist. We have seen that the larger structures coming from the outside flow interfere only weakly with the near-wall region, because the local dynamics are intense enough to be always dominant. The spacing of the streaks just mentioned has been observed up to the highest Reynolds numbers of the atmospheric boundary layer [**KMKT95**].

On the other hand the thinness of the layer in which the dynamics takes place makes the flow very sensitive to small perturbations at the wall. Roughness elements with heights of the order of a few wall units, microns in a large pipe, completely destroy the delicate cycle that has been described here, and can increase the friction coefficient by a factor of two or more [**Jim04**]. Conversely it only takes a concentration of polymers of a few parts per million in the near-wall region [**McC90**] to decrease the drag coefficient by 40%.

The next few years will probably be dominated by modelling efforts for the logarithmic layer similar to those described here for the viscous ones. The cost of simulations is higher in that case, but it is beginning to be within the reach of modern computers. The motivation is both theoretical and technological. The cascade of momentum across the range of scales in the logarithmic layer will probably be the first three-dimensional self-similar cascade to become accessible to computational experiments. Its simplifying feature is the alignment of most of the net transfer along the direction normal to the wall. The main practical drive is probably large-eddy simulation, in which the momentum transfer across scales in the inertial range has to be modelled for the method to be practical [**JM00**]. Only by understanding the structures involved would we be sure of how to accomplish that.

Acknowledgments

The preparation of this paper was supported in part by the CICYT grant TRA2006–08226. I am deeply indebted to J.C. del Álamo, O. Flores, S. Hoyas, G. Kawahara and M.P. Simens for providing most of the data used in the figures.

References

[AHLS88] N. Aubry, P. Holmes, J. L. Lumley, and E. Stone, *The dynamics of coherent structures in the wall region of a turbulent boundary layer*, J. Fluid Mech. **192** (1988), 115–173.

[BL67] H. P. Bakewell and J. L. Lumley, *Viscous sublayer and adjacent wall region in turbulent pipe flow*, Phys. Fluids **10** (1967), 1880–1889.

[BR74] G.L. Brown and A. Roshko, *On the density effects and large structure in turbulent mixing layers*, J. Fluid Mech. **64** (1974), 775–816.

[dÁJ03] J. C. del Álamo and J. Jiménez, *Spectra of very large anisotropic scales in turbulent channels*, Phys. Fluids **15** (2003), L41–L44.

[dÁJ06] ———, *Linear energy amplification in turbulent channels*, J. Fluid Mech. **559** (2006), 205–213.

[dÁJZM04] J. C. del Álamo, J. Jiménez, P. Zandonade, and R. D. Moser, *Scaling of the energy spectra of turbulent channels*, J. Fluid Mech. **500** (2004), 135–144.

[dÁJZM06] ———, *Self-similar vortex clusters in the logarithmic region*, J. Fluid Mech. **561** (2006), 329–358.

[Dar54] H. Darcy, *Recherches expérimentales rélatives au mouvement de l'eau dans les tuyeaux*, Mém. Savants Etrang. Acad. Sci. Paris **17** (1854), 1–268.

[DE00] D. B. DeGraaf and J. K. Eaton, *Reynolds number scaling of the flat-plate turbulent boundary layer*, J. Fluid Mech. **422** (2000), 319–346.

[FJ06] O. Flores and J. Jiménez, *Effect of wall-boundary disturbances on turbulent channel flows*, J. Fluid Mech. **566** (2006), 357–376.

[Hag39] G. H. L. Hagen, *Über den Bewegung des Wassers in engen cylindrischen Röhren*, Poggendorfs Ann. Physik Chemie **46** (1839), 423–442.

[HJ06] S. Hoyas and J. Jiménez, *Scaling of the velocity fluctuations in turbulent channels up to* $Re_\tau = 2003$, Phys. Fluids **18** (2006), 011702.

[HKW95] J. M. Hamilton, J. Kim, and F. Waleffe, *Regeneration mechanisms of near-wall turbulence structures*, J. Fluid Mech. **287** (1995), 317–248.

[JdÁF04] J. Jiménez, J. C. del Álamo, and O. Flores, *The large-scale dynamics of near-wall turbulence*, J. Fluid Mech. **505** (2004), 179–199.

[JHSK97] J. Jeong, F. Hussain, W. Schoppa, and J. Kim, *Coherent structures near the wall in a turbulent channel flow*, J. Fluid Mech. **332** (1997), 185–214.

[Jim04] J. Jiménez, *Turbulent flows over rough walls*, Ann. Rev. Fluid Mech. **36** (2004), 173–196.

[JKS+05] J. Jiménez, G. Kawahara, M. P. Simens, M. Nagata, and M. Shiba, *Characterization of near-wall turbulence in terms of equilibrium and 'bursting' solutions*, Phys. Fluids **17** (2005), 015105.

[JM91] J. Jiménez and P. Moin, *The minimal flow unit in near-wall turbulence*, J. Fluid Mech. **225** (1991), 221–240.

[JM00] J. Jiménez and R. D. Moser, *LES: where are we and what can we expect?*, AIAA J. **38** (2000), 605–612.

[JP99] J. Jiménez and A. Pinelli, *The autonomous cycle of near wall turbulence*, J. Fluid Mech. **389** (1999), 335–359.

[JS01] J. Jiménez and M. P. Simens, *Low-dimensional dynamics in a turbulent wall flow*, J. Fluid Mech. **435** (2001), 81–91.

[KJUP03] G. Kawahara, J. Jiménez, M. Uhlmann, and A. Pinelli, *Linear instability of a corrugated vortex sheet – a model for streak instability*, J. Fluid Mech. **483** (2003), 315–342.

[KK01] G. Kawahara and S. Kida, *Periodic motion embedded in plane Couette turbulence: regeneration cycle and burst*, J. Fluid Mech. **449** (2001), 291–300.

[KKR71] H. T. Kim, S. J. Kline, and W. C. Reynolds, *The production of turbulence near a smooth wall in a turbulent boundary layer*, J. Fluid Mech. **50** (1971), 133–160.

[KMKT95] J. C. Klewicki, M. M. Metzger, E. Kelner, and E.M. Thurlow, *Viscous sublayer flow visualizations at* $R_\theta \approx 1,500,000$, Phys. Fluids **7** (1995), 857–863.

[KMM87] J. Kim, P. Moin, and R. D. Moser, *Turbulence statistics in fully developed channel flow at low Reynolds number*, J. Fluid Mech. **177** (1987), 133–166.

[Kol41] A. N. Kolmogorov, *The local structure of turbulence in incompressible viscous fluids a very large Reynolds numbers*, Dokl. Akad. Nauk. SSSR **30** (1941), 301–305, Reprinted in *Proc. R. Soc. London.* A **434**, 9–13 (1991).

[MBK71] W. R. B. Morrison, K. J. Bullock, and R. E. Kronauer, *Experimental evidence of waves in the sublayer*, J. Fluid Mech. **47** (1971), 639–656.

[McC90] W.D. McComb, *The physics of fluid turbulence*, Oxford U. Press, 1990.

[Nag90] M. Nagata, *Three-dimensional finite-amplitude solutions in plane Couette flow: bifurcation from infinity*, J. Fluid Mech. **217** (1990), 519–527.

[ÖJNH00] J. M. Österlund, A. V. Johansson, H. M. Nagib, and M. Hites, *A note on the overlap region in turbulent boundary layers*, Phys. Fluids **12** (2000), 1–4.

[Ric20] L. F. Richardson, *The supply of energy from and to atmospheric eddies*, Proc. Roy. Soc. A **97** (1920), 354–373.

[Rob91] S. K. Robinson, *Coherent motions in the turbulent boundary layer*, Ann. Rev. Fluid Mech. **23** (1991), 601–639.

[RSBH98] S. C. Reddy, P. J. Schmid, J. S. Baggett, and D. S. Henningson, *On stability of streamwise streaks and transition thresholds in plane channel flows*, J. Fluid Mech. **365** (1998), 269–303.

[SB87] J. D. Swearingen and R. F. Blackwelder, *The growth and breakdown of streamwise vortices in the presence of a wall*, J. Fluid Mech. **182** (1987), 255–290.

[SH02] W. Schoppa and F. Hussain, *Coherent structure generation in near-wall turbulence*, J. Fluid Mech. **453** (2002), 57–108.

[SM83] C. R. Smith and S. P. Metzler, *The characteristics of low speed streaks in the near wall region of a turbulent boundary layer*, J. Fluid Mech. **129** (1983), 27–54.

[SZ94] L. Sirovich and X. Zhou, *Dynamical model of wall-bounded turbulence*, Phys. Rev. Lett. **72** (1994), 340–343.

[TI01] S. Toh and T. Itano, *On the regeneration mechanism of turbulence in the channel flow*, Proc. Iutam Symp. on Geometry and Statistics of Turbulence (T. Kambe, Nakano T., and T. Muiyauchi, eds.), Kluwer, 2001, pp. 305–310.

[TI03] ______, *A periodic-like solution in channel flow*, J. Fluid Mech. **481** (2003), 67–76.

[TL72] H. Tennekes and J. L. Lumley, *A first course in turbulence*, MIT Press, 1972.

[Tow76] A.A. Townsend, *The structure of turbulent shear flow*, second ed., Cambridge U. Press, 1976.

[Wal97] F. Waleffe, *On a self-sustaining process in shear flows*, Phys. Fluids **9** (1997), 883–900.

[Wal98] ______, *Three-dimensional coherent states in plane shear flows*, Phys. Rev. Lett. **81** (1998), 4140–4143.

[Wal01] ______, *Exact coherent structures in channel flow*, J. Fluid Mech. **435** (2001), 93–102.

[Wal03] ______, *Homotopy of exact coherent structures in plane shear flows*, Phys. Fluids **15** (2003), 1517–1534.

School of Aeronautics, Universidad Politécnica, 28040 Madrid SPAIN, and Centre for Turbulence Research, Stanford University, Stanford CA, 94305 USA

E-mail address: `jimenez@torroja.dmt.upm.es`

Proceedings of Symposia in Applied Mathematics
Volume **65**, 2007

Defects, Singularities and Waves

Luis L. Bonilla and Ana Carpio

This paper is dedicated to Peter Lax and Louis Nirenberg on the occasion of their 80th birthdays.

ABSTRACT. Crystal defects such as dislocations are the basis of macroscopic properties such as the strength of materials and control their mechanical, optical and electronic properties. In recent times, advances in electronic microscopy have allowed imaging of atoms and therefore to visualize the core of dislocations, cracks, and so on. In continuum mechanics, dislocations are treated as source terms proportional to delta functions supported on the dislocation line. Cores and crystal structure are not properly considered and it is hard to describe the motion of crystal defects. Unlike defects in fluids (such as vortices), dislocations move only within glide planes, not in arbitrary directions, and they move only when the applied stress surpasses the Peierls stress, which is not infinitesimal.

We have proposed a discrete model describing defects in crystal lattices with cubic symmetry and having the standard linear anisotropic elasticity (Navier equations) as its continuum limit. Moving dislocations are traveling waves which become stationary solutions if the applied stress falls below the Peierls value. The corresponding transition is a global bifurcation of the model equations similar to that observed in simpler one-dimensional Frenkel-Kontorova models. Discrete models can also be used to study the interaction of dislocations and the creation of dislocations under sufficient applied stress.

1. Introduction

Defects such as dislocations, vacancies and cracks control mechanical properties of materials, including crystal plasticity, creep, fatigue, ductility and brittleness, hardness and friction. Crystal growth, radiation damage of materials, and their optical and electronic properties are also strongly affected by defects, particularly dislocations. These phenomena occur over many different scales of length and time and the properties at each scale are influenced by the others. In a certain sense, these multiscale phenomena are reminiscent of fluid turbulence as indicated in Table 1. Aspects of complex physical phenomena such as fluid turbulence may be captured by the dynamics of entangled vortices, whereas dislocations play a paramount role in plasticity and strength of materials. Line vortices are solutions

The first author was supported in part by MEC grant MAT2005-05730-C02-01.
The second author was supported in part by MEC grant MAT2005-05730-C02-02.

of the Euler equations with singular sources supported on lines and dislocations are solutions of the Navier equations of elasticity with singular sources supported on lines. However, vortices may move in any direction under infinitesimal stress whereas dislocations may move only in glide planes provided they are subject to a stress larger than the *Peierls stress* (defined as the minimal stress necessary for a straight dislocation to move at zero temperature). Both the Euler and Navier equations are approximations of theories that describe the outer structure of vortices and dislocations, respectively. In the case of fluids, the more detailed theory consists of the Navier-Stokes equations, whereas the inner core of dislocations may be described by atomic models or by discrete model equations as explained in this paper.

TABLE 1. Comparison between fluid vortices and crystal dislocations

Vortices	**Dislocations**
move in any direction	move only in glide planes
move under infinitesimal stress	move for stress above threshold
Euler equations	Navier eqs + line singularities (Volterra)
Navier-Stokes equations	discrete/atomic model equations
Turbulence: many entangled vortices	Plasticity: many entangled dislocations

The mathematical study of atomistic models (such as those based on molecular dynamics; see below) is not developed at the present time. For example, it has not been proved that a perfect crystal is a minimizer of the energy or, equivalently, a stable equilibrium of the equations of motion. We do not know which is the functional framework for studying solutions of the equations of motion corresponding to crystal defects such as dislocations.

It seems desirable that dislocation models should be simple enough to allow developing a rigorous existence theory of solutions with crystal defects. Numerical simulations of these models should have a moderate cost and pave the way to a detailed theoretical understanding of dislocation behavior. For example, we would like to analyze dislocation cores and pinning of dislocations due to the crystal lattice. Dislocation models should allow to understand and calculate the Peierls stress and the motion of dislocations under stress, the generation of dislocations and the dynamics of a finite number of dislocations. Moreover, these models could in principle be used to understand the coupled evolution of the dislocation density and the elastic field in the continuum limit. In this paper, we present discrete elasticity models of dislocations in cubic crystals, and carry out parts of the sketched program in simplified equations.

2. Basic notions on dislocations

A simple example of crystal dislocation is obtained by inserting an extra half-plane of atoms in a simple cubic crystal. As shown in Fig. 1, the added half-plane is the upper part of the yz plane. The edge of this half-plane (i.e., the z axis) is the *dislocation line* of this *edge dislocation*. Near the dislocation line, the lattice is greatly distorted but, as we move away from it, the planes of atoms fit almost regularly. The distortion is recognized if we form a closed circuit (called Burgers circuit) of lattice points around the dislocation line, as in Fig. 1. Let the dislocation line point outside the page (the positive z axis) and let the circuit be oriented counterclockwise, following the right-hand rule [**HL**]. Let r_0 be an arbitrary lattice point which we take as the initial point of the circuit, and let $v_i, i = 1, \ldots, N$ be vectors of length equal to one lattice period comprising the Burgers circuit in the undistorted lattice, such that $\sum_{i=1}^{N} v_i = 0$. Each $v_i \in \{(1,0), (-1,0), (0,1), (0,-1)\}$. The points of the Burgers circuit in the distorted lattice are such that r_i is the lattice point closest to $r_{i-1} + v_i$. A circuit that does not enclose a dislocation line ends at the initial point, $r_N = r_0$. A circuit containing a dislocation line is not closed, and we define *the Burgers vector* as [**BC**]

$$b = r_N - r_0 = \sum_{i=1}^{N} \Delta u_i, \quad \Delta u_i = r_i - (r_{i-1} + v_i). \tag{2.1}$$

In our example, $b = (1, 0, 0)$, one period in the positive x direction.

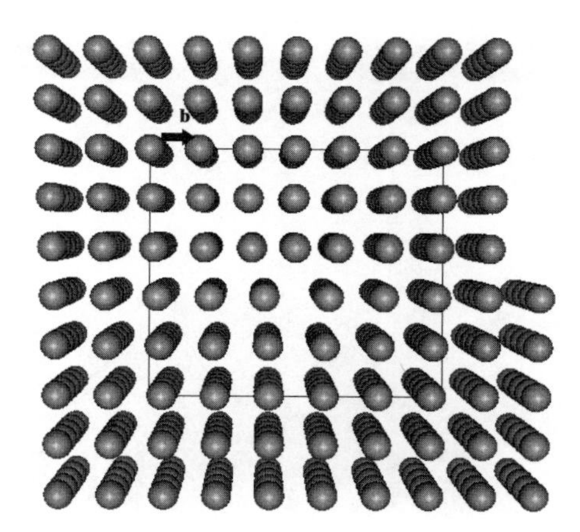

FIGURE 1. Edge dislocation and Burgers circuit about it.

The Burgers vector is the same for any deformation of the Burgers circuit as long as a dislocation line is not crossed during the deformation process. If n dislocation lines merge at a lattice point and we take their directions to flow out of the common point, then conservation of the Burgers vector implies $b_1 + \ldots + b_n = 0$.

An edge dislocation (at low temperature) may move along the *glide plane* formed by its line and its Burgers vector (the plane xz in Fig. 1). For edge dislocations, the Burgers vector and the dislocation line are orthogonal. In a finite crystal, the motion of a dislocation to the crystal boundary leaves a permanent deformation as indicated in Fig. 2. Notice that the final configuration corresponds to cutting the perfect crystal along the glide plane and sliding the upper half-crystal by the Burgers vector.

Another simple dislocation is shown in Fig. 3. This dislocation is obtained by cutting the cubic crystal along one plane and sliding half the crystal along the dislocation line. The Burgers vector of the resulting *screw dislocation* is parallel to the dislocation line and any crystal plane containing it is a glide plane.

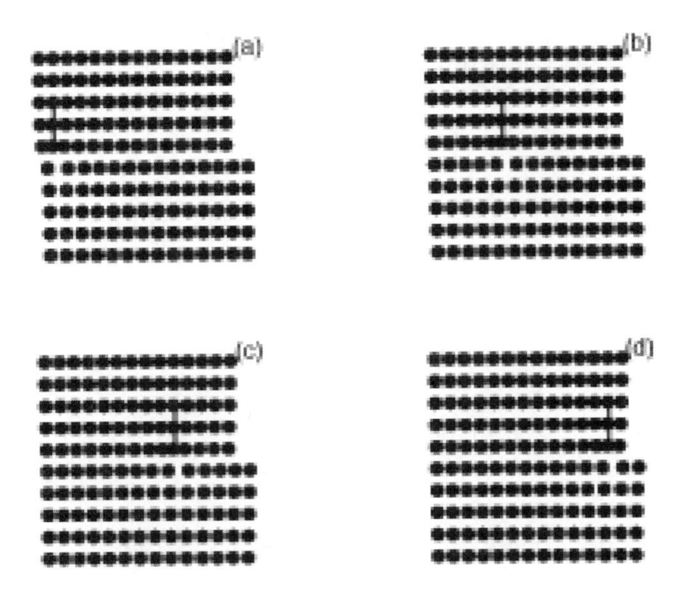

FIGURE 2. Motion of an edge dislocation from left to right.

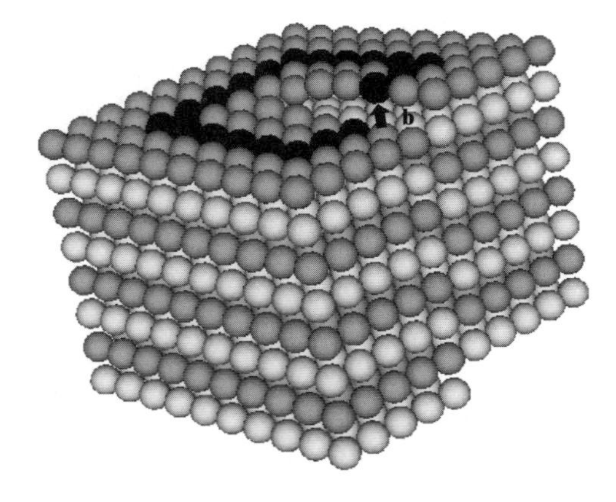

FIGURE 3. Burgers circuit about an screw dislocation.

3. Discrete elasticity and dislocations

3.1. Atomistic calculations. A direct calculation of dislocations and their dynamics can be carried out by simulating the motion of atoms in the crystal. The resulting atomistic models vary widely in accuracy and computational cost, from first-principles models such as density functional theory to molecular dynamics with empirical potentials. For example, molecular dynamics consists of solving the

Newton equations of motion for all the atoms in a sample, i.e,

$$m\,\frac{d^2 r_i}{dt^2} = -\frac{\partial V}{\partial r_i}, \tag{3.1}$$

$$V(\{r_j\}) = \sum_{i,j,i<j} V(|r_i - r_j|) + \sum_i F\left(\sum_{j\neq i} \rho_e(|r_i - r_j|)\right). \tag{3.2}$$

The interatomic potential of Eq. (3.2) is called the embedded-atom model (EAM) which is widely used for metals [**BC**]. It is a sum of two-body potentials and a "glue-potential" term containing an embedding function F that represents the energy needed to embed the atom i into an environment with a given electron density $\rho_e(r)$.

Molecular dynamics calculations are used to simulate creation and motion of dislocations, crack propagation, etc., but they have several drawbacks:

- Uncertainty about the potentials, which are derived by fitting empirical data or first-principles descriptions such as density functional theory for some key properties of the material.
- Huge computational cost which forces simulations to be restricted to small samples during time intervals that may not be enough to display the desired material behavior.
- Numerical artifacts such as numerical chaos and spurious oscillations have to be eliminated and technical issues as, for example, time discretization for long time computations and elimination of reflected waves from boundary conditions have to be solved.
- Qualitative information may be hard to extract from the simulations.

3.2. Continuum mechanics. The classical way to describe dislocations in continuum mechanics was initiated by Volterra. It consists of using the Navier equations of linear elasticity with singular source terms supported on dislocation lines. For a static dislocation in an isotropic medium, these equations are [**HL, LL**]:

$$\Delta u + \frac{1}{1-2\nu}\nabla \mathrm{div} u = b \times \xi\,\delta_\Gamma, \tag{3.3}$$

in which u, b, ξ and ν are the displacement vector from a lattice site, the Burgers vector, the unit tangent vector to the dislocation line Γ and the Poisson ratio, respectively. The displacement vector has a jump discontinuity of magnitude $|b|$ at a branch cut which, for the edge dislocation of Fig. 1, is the positive x axis:

$$\oint_c \nabla u \cdot dx = b, \tag{3.4}$$

where the integral is over any closed Burgers circuit c encircling the dislocation line. Note that (3.4) is the continuum version of (2.1). Then the gradient of u and the stress have a singularity along the dislocation line: they are inversely proportional to the distance to the line, $1/r$. This unphysical behavior indicates that modeling a dislocation by (3.3) is a reasonable approximation of the elastic stress and strain at the far field of a dislocation, but this description breaks down at the dislocation core, near the origin. Moreover, elasticity gives no information about the motion of dislocations.

3.3. Classical theory of dislocation density. A widely used theory of plasticity is to consider eigendistortions [**M**] (chapter 6). To include dislocation lines, instead of adding delta sources to the Navier equations, we consider that the gradient of the displacement vector (the distortion tensor) is a sum of elastic and plastic distortions:

$$\frac{\partial u_i}{\partial x_j} = \beta_{ij} + \beta^*_{ij}, \tag{3.5}$$

and no source. The equations of motion are found from the stress-strain relation $\sigma_{ij} = c_{ijkl}\beta_{kl}$:

$$\rho\, \ddot{u}_i - \frac{\partial}{\partial x_j}\left(c_{ijkl}\frac{\partial u_k}{\partial x_l}\right) = -\frac{\partial}{\partial x_j}(c_{ijkl}\beta^*_{kl}). \tag{3.6}$$

For constant ρ, the solution of this equation is

$$u_i(x,t) = -\int c_{jlmn}\beta^*_{mn}(x',t')\frac{\partial G_{ij}}{\partial x_l}(x-x',t-t')\, dx'dt', \tag{3.7}$$

$$\rho\, \ddot{G}_{im} - c_{ijkl}\frac{\partial^2 G_{km}}{\partial x_l \partial x_j}(x,t) = \delta_{im}\delta(x)\,\delta(t), \tag{3.8}$$

where the elastic Green function satisfies the previous equation for $t > 0$ in $\mathbb{R}^3$ and $G_{km} = 0$ if $t < 0$. The plastic distortion is connected to the *dislocation density tensor* α_{ij} by the Kröner formula [**M**]:

$$\alpha_{ik} = -\epsilon_{ilm}\frac{\partial \beta^*_{mk}}{\partial x_l}, \quad \text{so that} \quad \frac{\partial \alpha_{ik}}{\partial x_i} = 0, \tag{3.9}$$

in which ϵ_{ijk} is the completely antisymmetric unit tensor such that $\epsilon_{123} = 1$ and summation over repeated indices is implied. The Nye dislocation density tensor α_{ik} is the x_k component of the total Burgers vector of dislocations threading the unit surface perpendicular to the x_i direction. According to (3.4), the total Burgers vector of dislocations threading the surface S encircled by the x_k component of the Burgers circuit c obeys

$$\int_S \alpha_{ik}\xi_i dS = -\oint_c \beta^*_{jk}dx_j = -\int_S \epsilon_{ilj}\frac{\partial \beta^*_{jk}}{\partial x_l}\xi_i dS.$$

This gives (3.9). For the planar edge dislocation directed along the positive z axis ($\xi = (0,0,1)$ is its tangent vector) with Burgers vector directed along the positive x axis we have been considering,

$$\alpha_{ik} = \xi_i b_k \delta_\Gamma \Longrightarrow \alpha_{31} = b\delta(x)\delta(y), \quad \beta^*_{21} = -b\, H(x)\delta(y). \tag{3.10}$$

All other components of α_{ij} and β^*_{ij} are zero. For a single dislocation, the condition in (3.9) that the divergence of the dislocation density is zero means that its Burgers vector is constant along the dislocation line. In the case of continuously distributed dislocations,

$$\alpha_{ik} = \sum_\Gamma \xi_i b_k \delta_\Gamma \Longrightarrow \alpha_{ik} d\mathbf{x} = \sum_\Gamma \xi_i b_k dl, \tag{3.11}$$

where the summation is taken over all dislocation segments of length dl contained in the infinitesimal cube $d\mathbf{x}$. The growth rate of the total Burgers vector of the dislocations threading the surface S encircled by the Burgers circuit c is

$$\int_S \dot{\alpha}_{ji}\xi_j dS = \oint_c \epsilon_{klj}V_{kli}dx_j, \quad V_{ijk} = \sum V_i \xi_j b_k, \tag{3.12}$$

where V_i is the dislocation velocity, V_{ijk} is Mura's dislocation flux tensor and the sum is over all dislocations threading a surface dS. The Stokes theorem and (3.9) imply

$$\dot{\alpha}_{ji} = \epsilon_{jlk}\epsilon_{mnk}\frac{\partial V_{mni}}{\partial x_l}, \quad \dot{\beta}^*_{ij} = -\epsilon_{ikl}V_{klj}. \tag{3.13}$$

Other authors [**LL**] use the Kosevich dislocation flux tensor, $j_{ij} = \epsilon_{ikl}V_{klj}$ instead of V_{ijk}. Knowing the dislocation flux tensor, the dislocation density and the plastic distortion can be found from (3.13).

In the classical continuum theory of dislocations, the plastic distortion β^*_{ij} is supposed to be known as a function of space and time (equivalently, the dislocation density and the dislocation flux are known), the displacement vector is calculated according to (3.7) and the elastic distortion is related to the distortion tensor by (3.5). In terms of the dislocation density and the dislocation flux, the elastic distortion is given by [**M**]

$$\beta_{ji} = \epsilon_{jnh}\int\left(c_{klmn}\frac{\partial G_{ik}}{\partial x_l}\alpha_{hm} + \rho\,\dot{G}_{mi}V_{nhm}\right)dx'dt', \tag{3.14}$$

and stresses can be calculated from the linear stress-strain relation. If we know the dislocation density and the dislocation flux, the following local equations for the Mura potentials allow us to find the elastic distortion:

$$c_{ijkl}\frac{\partial^2 B_{kpnm}}{\partial x_l \partial x_j} - \rho\ddot{B}_{ipnm} = -\delta_{ip}\dot{\beta}^*_{nm}, \tag{3.15}$$

$$c_{ijkl}\frac{\partial^2 \phi_{kpnm}}{\partial x_l \partial x_j} - \rho\ddot{\phi}_{ipnm} = -\delta_{ip}\alpha_{nm}, \tag{3.16}$$

$$c_{ijkl}\frac{\partial^2 A_{kpsnm}}{\partial x_l \partial x_j} - \rho\ddot{A}_{ipsnm} = -\delta_{ip}V_{snm}. \tag{3.17}$$

The Mura potentials are related to the plastic distortion, the dislocation density and flux by

$$B_{kmji}(x,t) = \int G_{km}(x-x',t-t')\beta^*_{ji}(x',t')\,dx'\,dt', \tag{3.18}$$

$$\phi_{kmji} = \int G_{km}(x-x',t-t')\alpha_{ji}(x',t')\,dx'\,dt', \tag{3.19}$$

$$A_{kmsji} = \int G_{km}(x-x',t-t')V_{sji}(x',t')\,dx'\,dt'. \tag{3.20}$$

The solutions of (3.16) and (3.17) (that are reminiscent of Maxwell equations) should satisfy the Lorentz condition

$$\dot{\phi}_{kpnm} = \frac{\partial A_{kpnlm}}{\partial x_l} - \frac{\partial A_{kplnm}}{\partial x_l}. \tag{3.21}$$

For constant ρ, the elastic distortion is calculated in terms of the Mura potentials by means of

$$\beta_{ji} = \epsilon_{jnh}\left(c_{klmn}\frac{\partial \phi_{ikhm}}{\partial x_l} + \rho\,\dot{A}_{minhm}\right). \tag{3.22}$$

The trouble with the classical theory is that the information on dislocation density and flux (or dislocation velocity) should be known in advance [**M**]. Alternatively, equations for these magnitudes should be postulated. See for example the

Boltzmann-like kinetic equations of Groma and Bakó [**GB**]. Moreover, the evolution of the dislocation density should also include nucleation and destruction of dislocations. A different approach to track the simultaneous evolution of a large number of dislocations in a continuum is to divide a given dislocation network into small straight segments (each having its own Burgers vector) and to consider their energies, forces acting on them, empirical dislocation velocity depending on the acting force, and mechanisms of dislocation creation. The resulting theory is called *line dislocation dynamics*, and it can be used to numerically simulate and discuss more complex situations involving billions of dislocation segments. See for example Bulatov and Cai's book [**BC**].

3.4. Peierls-Nabarro model. There are models that eliminate the singularity of the elastic strain at the dislocation line without having to consider the crystal lattice as a regularization of elasticity. Among them, the Peierls-Nabarro (PN) model is widely used. This model solves the equations of linear elasticity for a configuration with a straight dislocation such as that in Fig. 1. Here we shall consider only this one-dimensional situation with Burgers vector b directed along the positive x axis. There is a cut at the positive x axis such that $u(x) \equiv u^+(x) - u^-(x)$ is the *disregistry* across the cut plane. In a Volterra dislocation, $u(x)$ is the step function $b\,\theta(x)$, equal to b if $x > 0$ and zero otherwise, and the disregistry density $\rho_{\text{dis}}(x) \equiv du/dx = b\,\delta(x)$. In the PN model, ρ_{dis} is a smooth function and the force on the dislocation due to the elastic stress can be calculated to be the left hand side of the following equilibrium equation defining the model [**BC**]

$$-\frac{\mu}{2\pi\,(1-\nu)} \int_{-\infty}^{\infty} \frac{\rho_{\text{dis}}(x')}{x - x'}\, dx' = \frac{U\pi}{b} \sin\left(\frac{2\pi u(x)}{b}\right). \tag{3.23}$$

Here μ and U are the shear modulus and the energy per unit area of the cut plane, respectively. The right hand side of (3.23) is the force acting on the atoms as a result of having uniformly slided the upper half of the crystal a distance $u(x)$ over the lower half. This force is a b-periodic function of u that can be calculated from an atomistic model, but a particularly simple choice is the sine form in (3.23), which comes from the *misfit potential* $U\,[1 - \cos(2\pi u/b)]/2$. Peierls found the following simple analytical solution

$$u(x) = \frac{b}{\pi} \tan^{-1}\left(\frac{x}{d}\right) - \frac{b}{2}, \quad d = \frac{\mu b^2}{4\pi^2 U\,(1-\nu)}, \tag{3.24}$$

in which d can be interpreted as the width of the dislocation core. The stress field of this dislocation along the x axis is proportional to $x/(x^2 + d^2)$, and the singularity at $x = 0$ has been removed.

The PN model does not take into account the discrete nature of the crystal lattice because it is invariant under arbitrary translations. Then it does not describe properly the motion of dislocations and the Peierls stress. Different modifications have improved the model, without changing its hybrid nature, that an atomistic force in the right hand side of (3.23) is balanced with the continuum response of the elastic medium on the left hand side of this equation.

3.5. Discrete models. Discrete models of dislocations have been analyzed since 1938, when Frenkel and Kontorova (FK) studied a model of interconnected harmonic springs in a periodic potential governed by the equations [**FK, N**]

$$m\,\frac{d^2u_n}{dt^2} = \kappa\,(u_{n+1}+u_{n-1}-2u_n) - \frac{U\pi}{b}\,\sin\left(\frac{2\pi u_n}{b}\right). \tag{3.25}$$

For a static dislocation, the equations describing the model are (3.23) except that $u = u_n$ (n is an integer) in the right hand side, and the left hand side is now proportional to $(u_{n+1}+u_{n-1}-2u_n)$. Dislocations are kink solutions of these equations [**N**], but the distortion $(u_{n+1}-u_n)$ decays exponentially far from the dislocation core, not as $1/n$ like the elastic far field of a true dislocation. Related models such as the discrete models of crystal growth by Frank and van der Merwe [**FM**] share the same unrealistic exponential decay of the distortion far from dislocation cores. Other discrete models such as the Suzuki model of moving screw dislocations in terms of sliding chains [**S**] and the Landau-Kovalev-Kondratiuk model of interacting atomic chains (IAC) [**LKK**] change the FK model so as to achieve algebraic decay of the distortion far from dislocation cores. They are all related to the FK model of idealized springs on a periodic substrate and are limited to simplified geometries. Here we describe our own proposal [**CB3**] which can be used to describe the core and far field of dislocations and it coincides with anisotropic elasticity far from dislocation cores.

3.5.1. *Discrete elasticity.* Let us consider a simple cubic crystal with one atom per lattice point and having a unit cell of side length a. Extensions to face centered cubic and body centered cubic crystals can be found in [**CB3**] and to cubic crystals with a two-atom basis in [**BCP**]. We want to ensure that the equations of our model become those of anisotropic linear elasticity far from dislocations, i.e., they should become the Cauchy equations for the displacement vector $\tilde{u}_i$ ($i = 1,2,3$) [**HL, LL**],

$$\rho\,\frac{\partial^2\tilde{u}_i}{\partial t^2} = \sum_{j,k,l}\frac{\partial}{\partial x_j}\left(c_{ijkl}\,\frac{\partial\tilde{u}_k}{\partial x_l}\right) + \tilde{f}_i, \tag{3.26}$$

$$\begin{aligned} c_{ijkl} = {} & \lambda\,\delta_{ij}\delta_{kl} + \mu\,(\delta_{ik}\delta_{jl}+\delta_{il}\delta_{jk}) + 2(C_{44}-\mu) \\ & \times\left(\frac{\delta_{ik}\delta_{jl}+\delta_{il}\delta_{jk}}{2} - \delta_{1i}\delta_{1j}\delta_{1k}\delta_{1l} - \delta_{2i}\delta_{2j}\delta_{2k}\delta_{2l} - \delta_{3i}\delta_{3j}\delta_{3k}\delta_{3l}\right), \end{aligned} \tag{3.27}$$

in which summation over repeated indices is understood. Here $\tilde{f}_i$ is a body force, ρ is the mass density of the crystal, and $\lambda = C_{12}$, $\mu = (C_{11}-C_{12})/2$, where C_{ij} are the stiffness constants of a cubic crystal. If $C_{44} = \mu$, the strain energy is isotropic and λ and μ are the usual Lamé coefficients. Equations (3.26) are the Euler-Lagrange equations associated to a potential energy $-W - \int \tilde{f}_i\tilde{u}_i$, where W (strain energy) is

$$W = \frac{1}{2}\int c_{ijkl}e_{ij}e_{kl}dx_1dx_2dx_3. \tag{3.28}$$

Here the strain tensor e_{ij} is the symmetrized gradient of the displacement vector. To obtain a discrete model from (3.26), we proceed as follows:

(1) Discretize space along the primitive vectors defining the unit cell of the crystal: $x = x_1 = (l+\epsilon_1)\,a$, $y = x_2 = (m+\epsilon_2)\,a$, $z = x_3 = (n+\epsilon_3)\,a$, where l, m and n are integer numbers. The numbers $\epsilon_i \in (0,1)$ are chosen different from zero so as to avoid that any lattice site coincide with the origin $x_i = 0$. For a simple cubic crystal it is convenient to set $\epsilon_i = 1/2$,

so that the origin is at the center of a unit cell. The discrete displacement vector from lattice sites $u_i(l,m,n;t)$ is a nondimensional vector such that $\tilde{u}_i(la,ma,na,t) = a\, u_i(l,m,n;t)$.

(2) Define the *discrete* distortion tensor as

$$w_i^{(j)} = g(D_j^+ u_i), \tag{3.29}$$

where $g(x)$ is an odd periodic function of period one satisfying $g(x) \sim x$ as $x \to 0$, [1] and D_j^+ and D_j^- represent the standard forward and backward difference operators, so that $D_1^\pm u_i(l,m,n;t) = \pm\,[u_i(l \pm 1,m,n;t) - u_i(l,m,n;t)]$, and so on.

(3) Replace the strain energy in (3.28) by

$$W(\{u_i\})(l,m,n;t) = \frac{a^3}{2} \sum_{l,m,n} c_{ijkl} e_{ij} e_{kl}, \tag{3.30}$$

$$e_{ij} = \frac{1}{2}\,(w_i^{(j)} + w_j^{(i)}) = \frac{g(D_j^+ u_i) + g(D_i^+ u_j)}{2}, \tag{3.31}$$

(4) Then the Euler-Lagrange equations for the potential energy $V = W + a^3 \sum_{l,m,n} f_i u_i$ (with $f_i = a\,\tilde{f}_i$) yield the equations of motion:

$$M\,\ddot{u}_i = \sum_{j,k,l} D_j^- \left[c_{ijkl}\, g'(D_j^+ u_i)\, g(D_l^+ u_k)\right] + f_i. \tag{3.32}$$

$$M = \rho a^2. \tag{3.33}$$

Here $\ddot{u}_i \equiv \partial^2 u_i/\partial t^2$, M has units of mass per unit length (because ρ is the mass density) and the displacement vector is dimensionless, so that both sides of Eq. (3.32) have units of force per unit area. To recover the continuum limit of our discrete model, we restore dimensional units to Eq. (3.32) and use

$$u_i(l,m,n;t) = \frac{\tilde{u}_i\left((l+\epsilon_1)\,a, (m+\epsilon_2)\,a, (n+\epsilon_3)\,a; t\right)}{a}, \tag{3.34}$$

where $\tilde{f}_i = f_i/a$, then let $a \to 0$, use Eq. (3.33) and that $g(x) \sim x$ as $x \to 0$. The resulting equations are (3.26). Viscosity and fluctuation effects can be added as explained in Ref. [**BCP**].

3.5.2. *Dislocation solutions.* To illustrate how our model produces dislocations, we shall construct an edge dislocation in an isotropic simple cubic crystal ($C_{44} = (C_{11} - C_{12})/2$) with planar discrete symmetry, so that $(u_1(l,m;t), u_2(l,m;t), 0)$ is independent of $z = na$. The construction proceeds as follows:

- Calculate the corresponding singular solution of the static Navier equations of linear elasticity. Use (3.34) to obtain a time-independent nondimensional displacement vector $(U_1(l,m), U_2(l,m))$ and notice that the singularity at $x_i = 0$ does not coincide with any lattice site.
- Use $U_i(l,m)$, $i = 1,2$, as the boundary condition at the crystal borders ($l = \pm X$, $m = \pm Y$) and also as an initial condition to solve an overdamped discrete model in which $M\gamma\dot{u}_i$ replaces the left hand side of the equations

[1] This function can be chosen so that the Peierls stress calculated using the model agrees with experimentally measured values or with results of molecular dynamics simulations [**CB3**].

of motion (3.32):

$$M\gamma\, \dot{u}_i - \sum_{j,k,l} D_j^- [c_{ijkl}\, g'(D_j^+ u_i)\, g(D_l^+ u_k)] = f_i. \qquad (3.35)$$

- The solution of the overdamped equations of motion tends to the stationary dislocation solution of the discrete model as time goes to infinity.

The solution of the static Navier equations corresponding to an edge dislocation as in Fig. 1 under a dimensionless shear stress F equals the corresponding solution under zero stress plus $(Fy, Fx, 0)$. Then we can use as initial and boundary displacement vector in the previous procedure $(U_1 + Fm, U_2 + Fl, 0)$, provided U_i is the displacement vector under zero shear stress. The resulting edge dislocation is depicted in Fig. 4.

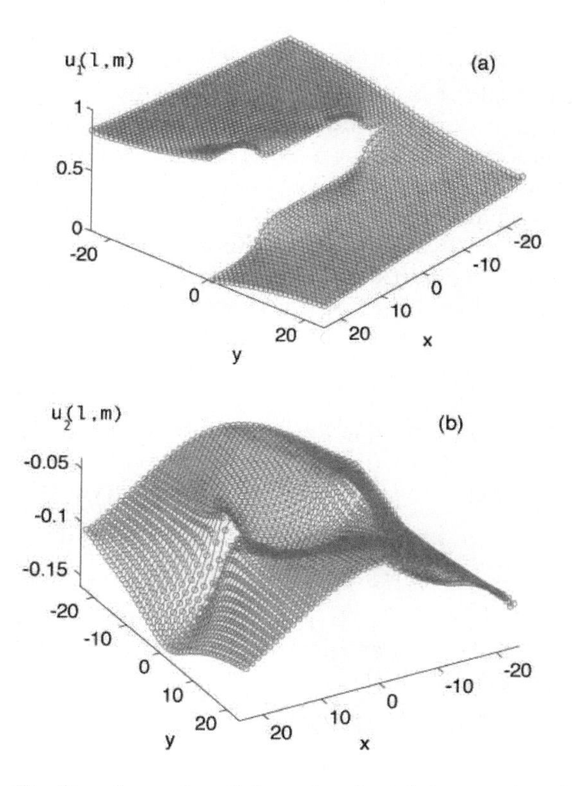

FIGURE 4. Profile of an edge dislocation for $g(x) = \sin(2\pi x)/(2\pi)$: (a) $u_1(l,m)$, (b) $u_2(l,m)$.

3.5.3. *Mathematical setting.* For our discrete models, it is immediate to prove that the perfect crystal is a minimizer of the energy and a stable solution of the equations of motion. We have also identified the class of functions in which existence of dislocation solutions can be established. The idea is suggested by our previous numerical tests.

Consider the class of functions S that are sequences

$$(u_1(l,m,n), u_2(l,m,n), u_3(l,m,n)) \quad \text{in} \quad \mathbb{Z}^3$$

behaving at infinity as the corresponding singular solution of the Navier equations with $x = (l + 1/2)a$, $y = (m + 1/2)a$, $z = (n + 1/2)a$. Static dislocations solve the equation

$$(\mathbb{L}u)_i \equiv \sum_{j,k,l} D_j^- [c_{ijkl}\, g'(D_j^+ u_i)\, g(D_l^+ u_k)] = -f_i \tag{3.36}$$

in S. We have two options to prove their existence:

(1) Minimize the following potential energy in S:

$$\frac{1}{2} \sum_{l,m,n,i,j,a,b} c_{ijab} \frac{g(D_i^+ u_j) + g(D_j^+ u_i)}{2} \frac{g(D_a^+ u_b) + g(D_b^+ u_a)}{2}.$$

This energy is not globally convex because of the periodicity of the function $g(x)$, but it has local convexity properties in S.

(2) Compute the long time limit of the overdamped equations (3.35) in S using the singular solution of the Navier equations as an initial datum.

The operator $\mathbb{L}$ in (3.36) is a discrete elliptic operator in the class S provided an applied stress is below a certain value called the Peierls stress. If the applied stress surpasses the Peierls stress, $\mathbb{L}$ changes type, and the dislocation moves (see below).

3.5.4. *Simplified model for edge dislocations.* In order to analyze how dislocations move when a sufficient stress is applied, it is convenient to use a simplified version of the model equations (3.32). If we consider planar edge dislocations as in Fig. 4, they move when an applied shear stress parallel to the Burgers vector surpasses the Peierls stress. If the Burgers vector is directed along the positive x axis, the extra half-plane of atoms finishing at the dislocation line glides along the x direction but the atoms do not climb in the y direction. Then $D_2^+ u_1$ can be larger than 1 (the period of $g(x)$) as the dislocation line moves, but $D_1^+ u_1$ remains always smaller than 1. Therefore, no qualitative features of the model are lost if we set $u_1 = u$, $u_2 = u_3 = 0$, $g'(x) = 1$ (its far field value), and replace $g(D_1^+ u)$ by $D_1^+ u$ in the equations. We obtain the following *nondimensional* equations instead of (3.32):

$$\begin{aligned} m\, \ddot{u}_{i,j} + \dot{u}_{i,j} = {} & u_{i+1,j} - 2u_{i,j} + u_{i-1,j} \\ & + A\,[g(u_{i,j+1} - u_{i,j}) + g(u_{i,j-1} - u_{i,j})]. \end{aligned} \tag{3.37}$$

Here $m = C_{11}/(M\gamma^2)$, $A = C_{44}/C_{11}$, provided we added a damping term $\gamma M \dot{u}_i$ to the left hand side of (3.32) and ignore the body force. Note we have used the notation $u_{i,j}(t)$ instead of $u(i,j;t)$, and the nondimensional time is $C_{11}t/(M\gamma) \to t$. Selecting $g(x) = \sin(2\pi x)/(2\pi)$, (3.37) becomes

$$\begin{aligned} m\, \ddot{u}_{i,j} + \dot{u}_{i,j} = {} & u_{i+1,j} - 2u_{i,j} + u_{i-1,j} \\ & + A\, \frac{\sin[2\pi(u_{i,j+1} - u_{i,j})] + \sin[2\pi(u_{i,j-1} - u_{i,j})]}{2\pi}. \end{aligned} \tag{3.38}$$

Equations (3.38) are the same as in the IAC model [**LKK**].

4. Analysis of pinning and nucleation of dislocations in simplified models

4.1. Edge dislocations. We shall now study the structure of a static edge dislocation of the simple model given by Eq. (3.38) under shear stress, the critical stress needed to set it in motion and its subsequent speed. The Burgers vector of the dislocation will be directed along the positive x axis and its length will be one lattice period. According to the procedure explained in Section 3, first we need to obtain the corresponding static solutions of the model in the continuum limit. They obey the following equation:

$$\frac{\partial \tilde{u}^2}{\partial x^2} + A\,\frac{\partial \tilde{u}^2}{\partial y^2} = 0, \tag{4.1}$$

and satisfy (3.4) for the Burgers circuit around the dislocation line. Notice we are using dimensional quantities $\tilde{u} = a\,u$, with $x = ia$, $y = ja$. Edge dislocations are generated from the zero-stress solution

$$\tilde{u} = \frac{a}{2\pi}\,\theta\left(x, \frac{y}{\sqrt{A}}\right), \tag{4.2}$$

where $\theta = \tan^{-1}(y/x) \in [0, 2\pi)$ is the angle function and a the length of the Burgers vector. The corresponding strain made out of the gradient of u decays as $1/r$ as $r = \sqrt{x^2 + y^2/A} \to \infty$, same as the elastic strain about an edge dislocation. From (4.2), we obtain the discrete static solution:

$$U_{i,j} = \frac{1}{2\pi}\,\theta\left(a\left(i + \frac{1}{2}\right), \frac{a}{\sqrt{A}}\left(j + \frac{1}{2}\right)\right) \equiv \frac{\theta^A_{i,j}}{2\pi}, \tag{4.3}$$

which is the only nonzero component of the nondimensional displacement vector. Under a shear stress F, the static solution is $(jF + \theta^A_{i,j})/(2\pi)$.

4.1.1. *Stationary edge dislocation under shear stress.* Next, we solve numerically Eq. (3.38) with $m = 0$ in a large lattice, using

$$u_{i,j} = \frac{\theta^A_{i,j} + F\,j}{2\pi}, \quad \text{for } i = \pm N \text{ and for } j = \pm N, \tag{4.4}$$

$$u_{i,j}(0) = \frac{\theta^A_{i,j}}{2\pi}, \tag{4.5}$$

Provided $|F| < F_{\text{cs}}(A)$ (F_{cs} is the *static Peierls stress*, to be defined and calculated below), the system relaxes to the stationary configuration which we denote by $\hat{U}_{i,j}(F, A)$. In Fig. 5, we depict the zero-stress dislocation $\hat{U}_{i,j}(0, A)$, whose core is clearly visible. The core width is a decreasing function of A. Nonlinear stability

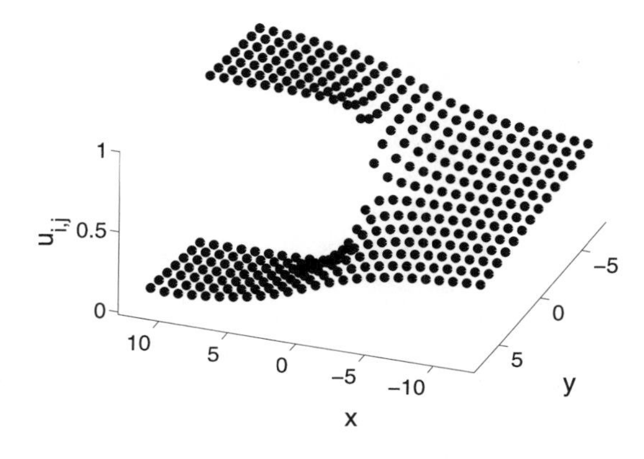

FIGURE 5. Displacement field profile for the stationary edge dislocation for $F = 0$, with $A = 1$ and $N = 50$.

of the stationary edge dislocation for $|F| < F_{cs}$ was proven in Ref. [**C**]. If $|F|$ is large enough, the dislocation is observed to glide in the x direction: to the right if $F > 0$, and to the left if $F < 0$.

4.1.2. *Linear stability analysis of the edge dislocation for* $m = 0$. To calculate $F_{\rm cs}(A)$, we perform a linear stability analysis of $\hat{U}(F, A)$. We insert $u_{i,j}(t) = \hat{U}_{i,j}(F, A) + v_{i,j}(t)$ in Eq. (3.38) with $m = 0$ and expand the result in powers of $v_{i,j}$, *about the stationary state* $\hat{U}_{i,j}(A, F)$ keeping up to quadratic terms. Subscripts in the resulting quadratic equation can be numbered with a single one starting from the point $i = j = -N$: $\hat{U}_{i,j} = \hat{U}_k$ and $v_{i,j} = v_k$, $k = i + N + 1 + (j + N)(2N + 1)$ for $i, j = -N, \ldots, N$. The quadratic equation can be written formally as

$$\frac{d\mathbf{v}}{dt} = \mathcal{M}(F)\mathbf{v} + \mathcal{B}(\mathbf{v}, \mathbf{v}; F), \tag{4.6}$$

where the vector $\mathbf{v}$ has components v_k. The linear stability of the stationary state $\hat{U}_k(A, F)$ depends on the eigenvalues of the matrix $\mathcal{M}(F)$. These eigenvalues are all real negative for $|F| < F_{\rm cs}$ whereas one of them vanishes at $|F| = F_{\rm cs}$.[2] This criterion allows us to numerically determine $F_{\rm cs}$ as a function of A; see Fig. 6(a). Notice that the critical stress increases with A. Thus narrow core dislocations (A large) are harder to move.

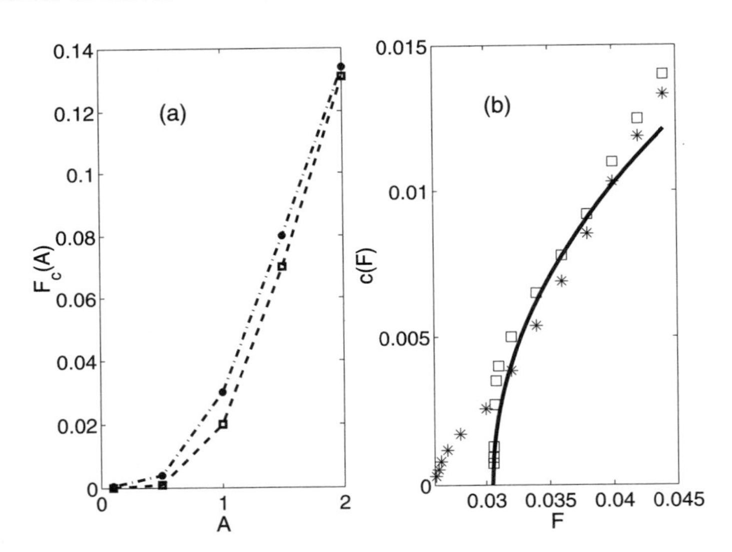

FIGURE 6. (a) Static (squares, $m = 0$) and dynamic (asterisks, $m = 0.5$) critical stresses F_{cs} and F_{cd} versus A. (b) Theoretical (solid line, $m = 0$) and numerical (squares, $m = 0$; asterisks, $m = 0.5$) dislocation velocity vs. F ($A = 1$, $N = 25$).

4.1.3. *The transition from stationary to moving dislocations as a global bifurcation* ($m = 0$). Let us assume that $F > F_{cs}$ (the case $F < -F_{cs}$ is similar). The motion of dislocations depends on whether m is zero or not. Consider first the overdamped case, $m = 0$. Let $\mathbf{l}$ and $\mathbf{r}$ (with components l_{ij} and r_{ij}) be the left and right eigenvectors of the matrix $\mathcal{M}(F_{cs})$ corresponding to its zero eigenvalue

[2]At this applied stress, the operator $\mathbb{L}$ in (3.36) ceases to be elliptic.

(its largest one). Then $v_{i,j} = [(F - F_c)j + \phi(t)\, r_{ij}]/(2\pi)$ (plus terms that decay exponentially fast in time) satisfies the BC $v_{i,j} = (F - F_c)j/(2\pi)$ for $j = \pm N$, and $v_{i,j} = 0$ for $i = \pm N$. Inserting $u_{i,j} = \hat{U}_{i,j}(F_{\text{cs}}) + v_{i,j}$ in (3.35) with $m = 0$, we obtain

$$\text{(4.7)} \qquad \begin{aligned} \dot{\phi}\, r_{ij} &= A\,(F - F_{\text{cs}})\, D_2^-\{\cos[2\pi(D_2^+\hat{U}_{i,j})]\} \\ &- \frac{A\phi^2}{2}\, D_2^-\{\sin[2\pi(D_2^+\hat{U}_{i,j})\,(D_2^+ r_{ij})^2]\}, \end{aligned}$$

plus terms of orders $(F - F_{\text{cs}})\phi$ and $(F - F_{\text{cs}})^2$. The cubic terms that were ignored when (4.6) was written are of order ϕ^3. The scalar product of this equation by $\mathbf{l}$ yields the following amplitude equation

$$\text{(4.8)} \qquad \frac{d\phi}{dt} = \alpha\,(F - F_{\text{cs}}) + \beta\phi^2,$$

$$\text{(4.9)} \qquad \alpha = \frac{A\sum_{ij} l_{ij} D_2^-\{\cos[2\pi(D_2^+\hat{U}_{i,j})]\}}{\sum_{ij} l_{ij} r_{ij}},$$

$$\text{(4.10)} \qquad \beta = \frac{\mathbf{l}\cdot\mathcal{B}(\mathbf{r},\mathbf{r};F_{cs})}{(\mathbf{l}\cdot\mathbf{r})} = -\frac{A\sum_{ij} l_{ij} D_2^-\{\sin[2\pi(D_2^+\hat{U}_{i,j})\,(D_2^+ r_{ij})^2]\}}{2\sum_{ij} l_{ij} r_{ij}}.$$

The two terms in the right hand side of (4.5) are of the same order, $(F - F_{\text{cs}})$, whereby ϕ is of order $(F - F_{\text{cs}})^{1/2}$. Then the neglected error terms are of order $(F - F_{\text{cs}})^{3/2}$, and they tend faster to zero as $F \to F_{\text{cs}}+$ than the terms in (4.7). Numerical evidence shows that $\alpha > 0$ and $\beta > 0$.

Equation (4.7) is the normal form of a saddle-node bifurcation. Its solution is

$$\text{(4.11)} \qquad \phi(t) = \sqrt{\frac{\alpha\,(F - F_{\text{cs}})}{\beta}}\, \tan\left(\sqrt{\alpha\beta\,(F - F_{\text{cs}})}\,(t - t_0)\right),$$

which remains close to $\phi = 0$ for long time periods, but it blows up at the large times $(t - t_0) = \pm 1/(2c)$, where

$$\text{(4.12)} \qquad c(A, F) = \frac{1}{\pi}\sqrt{\alpha\beta\,(F - F_{\text{cs}})}.$$

What happens after the blow up time $(t - t_0) = 1/(2c)$? The approximation $u_{i,j}(t) \sim \hat{U}_{i,j}(F_{\text{cs}}, A) + [(F - F_{\text{cs}})\, j + \phi(t)\, r_{i,j}]/(2\pi)$ breaks down because $\phi(t)$ is no longer of order $(F - F_{\text{cs}})^{1/2}$. We have to solve an inner problem consisting of Equations (3.38) at $F = F_{\text{cs}}$ with the matching conditions $u_{i,j} \sim \hat{U}_{i,j}(F_{\text{cs}}, A) + r_{ij}/[\pi^2\sqrt{\beta/[\alpha\,(F - F_c)]} - 2\pi\beta\,(t - t_0)]$, as $(t - t_0) \to -\infty$. This inner solution evolves towards $\hat{U}_{i-1,j}$ in a time of order 1 after $(t - t_0) = 1/(2c)$. Then another jump of the whole dislocation profile to the right occurs when $(t - t_0) = 1/c$, and so on. The dislocation is moving to the right with velocity $c(A, F)$ given by (4.12) and its profile is $u_{i,j}(t) = u(\zeta, j)$, $\zeta = i - ct$, which has been depicted in Fig. 7. Notice that the wave front profiles exhibit many smoothed steps for F slightly larger than F_{cs}. In the flat part of these steps, $u(\zeta, j)$ takes on the stationary values $\hat{U}_{i,j}(F_{\text{cs}})$ because $\phi(t)$ in (4.11) is almost zero for $|c(t - t_0)| < 1/2$. The steep parts of the profiles between steps correspond to the solutions of the inner problem.

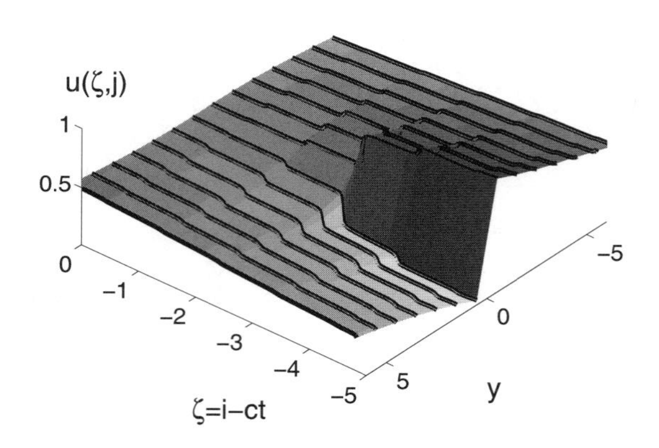

FIGURE 7. Wave front profiles, $u_{i,j}(t) = u(\zeta, j)$, $\zeta = i - ct$, $c > 0$, near $F = F_{cs}$ for $A = 3$, $m = 0$ and $N = 25$.

4.1.4. *Dislocation velocity near the Peierls stress.* $c(A, F)$ is the approximate dislocation velocity. Numerically measured and theoretically predicted dislocation velocities are compared in Fig. 6(b). Calculations in lattices of different sizes yield similar results.

How do we calculate numerically the dislocation velocity? This is an important point for using the calculated dislocation velocity as a function of stress in mesoscopic theories and a few comments are in order. If we solve numerically the initial-boundary value problem (IBVP) (3.38) and (4.4)-(4.5) for $|F| > F_{\text{cs}}$, the velocity of the dislocation decreases as it moves towards the boundary. The dislocation decelerates because we are using the far field of a steady dislocation as boundary condition, instead of the (more sensible) far field of a moving dislocation. However, the latter is in principle unknown because we do not know the dislocation speed. We will assume nevertheless that the dislocation moves at constant speed c once it starts moving, as it would in a stressed infinite system. Then the correct dislocation far field is $(\theta^A_{i-ct,j} + Fj)/(2\pi)$. With this far field as boundary condition, Eq. (3.38) has traveling wave solutions $u_{i,j}(t)$ whose velocity can be calculated self-consistently. How? By an iterative procedure that adopts as initial trial velocity that of a dislocation subject to static BC (4.4) as it starts moving. Figure 6(b) compares the numerically calculated velocity with the theoretical prediction (4.12). As explained above, step-like profiles are observed near threshold (see Fig. 7), that become smoother as F increases. Notice that the wave front profiles are kinks for $j < 0$ and antikinks for $j \geq 0$.

4.1.5. *Existence of traveling waves.* To prove existence of the traveling waves representing moving dislocations is an open problem. In principle, two possible ways to tackle this problem for the general discrete equations (3.32) are as follows:

(1) *Consider the class of functions* $\mathbb{S}_1$ *defined as*

$$(u_1(\zeta, m, n), u_2(\zeta, m, n), u_3(\zeta, m, n)) \quad \text{in} \quad \mathbb{R} \times \mathbb{Z}^2$$

$$\int_{-\infty}^{\infty} \sum_{m,n} \sum_{i} \left| \frac{\partial u_i}{\partial \zeta}(\zeta, m, n) \right|^2 d\zeta = 1,$$

behaving at infinity as the singular solution of the Navier equations corresponding to the dislocation we want to study with $x = \zeta$, $y = (m+1/2)a$, $z = (n+1/2)a$. Find

$$\min_{\mathbb{S}_1} \frac{1}{2} \int_{-\infty}^{\infty} \sum_{m,n} \sum_{ijkl} c_{ijkl}\, e_{ij}(\zeta, m, n)\, e_{kl}(\zeta, m, n)\, d\zeta,$$

$$e_{ij} = \frac{1}{2} [g(D_j^+ u_i) + g(D_i^+ u_j)].$$

(2) *Consider the class of functions $\mathbb{S}_2$ defined as*

$$(u_1(\zeta, m, n), u_2(\zeta, m, n), u_3(\zeta, m, n)) \quad \text{in} \quad \mathbb{R} \times \mathbb{Z}^2$$

$$\int_{-\infty}^{\infty} \sum_{m,n} \sum_{ijkl} c_{ijkl}\, e_{ij}(\zeta, m, n)\, e_{kl}(\zeta, m, n)\, d\zeta = 1,$$

behaving at infinity as the singular solution of the Navier equations corresponding to the dislocation we want to study with $x = \zeta$, $y = (m+1/2)a$, $z = (n+1/2)a$. Find

$$\min_{\mathbb{S}_2} \frac{1}{2} \int_{-\infty}^{\infty} \sum_{m,n} \sum_{i} \left| \frac{\partial u_i}{\partial \zeta}(\zeta, m, n) \right|^2 d\zeta.$$

For the simplified model (3.37), $(u_1(\zeta, m, n), u_2(\zeta, m, n), u_3(\zeta, m, n))$ should be replaced by $(u(\zeta, j), 0, 0)$. Related results in one-dimensional chains,

$$m\ddot{u}_n = -V'(u_{n+1} - u_n) + V'(u_n - u_{n-1}),$$

were obtained by Friesecke and Wattis [**FW**] assuming that the potential $V(x)$ is convex and grows faster than quadratic at infinity. In our case, the strain energy is periodic, therefore not convex.

4.1.6. *Effects of inertia ($m \neq 0$).* Numerical simulations of (3.38) show that inertia changes the previous picture of dislocation motion in one important aspect: the dislocations keep moving for an interval of stresses below the static Peierls stress, $F_{cd} < |F| < F_{cs}$. On this stress interval, stable solutions representing static and moving dislocations coexist: to depin a static dislocation, we need $|F| > F_{cs}$. However if $|F|$ decreases below F_{cs}, a moving dislocation keeps moving until $|F| < F_{cd}$; see Fig. 6(b). Thus F_{cd} represents the *dynamic Peierls stress* of the dislocation[**N**]. Our theory therefore yields the static and the dynamic Peierls stresses and the velocity of a dislocation.

Depinning of dislocations is a supercritical global bifurcation for $m = 0$ which becomes subcritical for $m \neq 0$. The effects of inertia on the Peierls stress have been investigated more thoroughly in one-dimensional chains related to the Frenkel-Kontorova model (3.25) [**AC, CB2, KT**]. For example, the wave fronts of the following model with a piecewise linear source term,

$$\frac{d^2 u_n}{dt^2} + \upsilon \frac{du_n}{dt} = u_{n+1} + u_{n-1} - 2u_n + F - A\, g(u_n), \tag{4.13}$$

$$g(u) = \begin{cases} u+1 & \text{for} \quad u < 0, \\ u-1 & \text{for} \quad u > 0, \end{cases} \tag{4.14}$$

have an explicit integral representation [**AC, CB2**]. Using this representation for $\upsilon \geq 0$, it is possible to find the values of F for which a given wave front moving with velocity c exists, $F = F(c)$. Exchanging the coordinate axes, the corresponding graph yields a bifurcation diagram which, for large enough v, is like that in Fig. 6(b). For $v = 0$ in (4.13), $F(c)$ has infinitely many vertical asymptotes, a minimum between each two consecutive asymptotes and a global minimum at $F = F_{\rm cd}$. For sufficiently small $v > 0$, the asymptotes become local maxima and $F(c)$ has infinitely many extrema that accumulate at $F_{\rm cs}$ as $c \to 0$. Exchanging the coordinate axes, the bifurcation diagram is conjectured to have infinitely many limit points (which are saddle-node bifurcations) that accumulate at $F_{\rm cs}$ [**CB2**]. Except for the solution branch with larger c, the wave front solutions corresponding to branches between two limit points are unstable. Kresse and Truskinovsky [**KT**] used the Wiener-Hopf technique to construct wave fronts for (4.13) with $v = 0$ and a non-symmetric piecewise linear function $g(u)$ having $B\,(u-1)$ instead of $(u-1)$ in (4.14). They found that the vertical asymptotes in the graph of $F(c)$ for $B = 1$ became maxima for $B \neq 1$. As in the symmetric case with damping, the bifurcation diagram in the (c, F) plane exhibits infinitely many limit points and again an accumulation point at $F = F_{\rm cs}$, $c = 0$ [**KT**]. Whether this picture persists for the other models considered here merits further investigation.

4.1.7. *Homogeneous nucleation of edge dislocations.* The discrete models exhibit homogeneous nucleation of dislocations as illustrated in Figure 8. Let us solve the overdamped model (3.37) with $m = 0$ under shear boundary conditions

$$u_{i,j} = \frac{Fj}{2\pi}, \quad \text{for } i = 0, N_x \text{ and for } j = 0, N_y, \tag{4.15}$$

$$u_{i,j}(0) = 0. \tag{4.16}$$

For $F = 0$, the initial and boundary conditions correspond to a perfect lattice. As $F > 0$ takes on larger and larger values, the lattice is deformed until a critical stress F_c is surpassed. At $F = F_c$, the largest eigenvalue of the linearized problem about the stationary configuration becomes zero. For $F > F_c$ and depending on the size of the lattice, one or more edge dipoles (each dipole consists of two edge dislocations with Burgers vectors $\pm(1, 0)$) may be created. Then these dipoles split and dislocations with Burgers vectors directed along the positive and negative x axis move to opposite boundaries of the lattice. Clearly the sum of the Burgers vectors of all newly created edge dislocations add up to zero, the same Burgers vector of the originally undistorted lattice.

Figure 8 shows the final configuration with the dislocations at the boundaries for a small lattice such that only one dipole is created. As the size of the lattice increases, two and more dipoles may be nucleated. The evidence for nucleation of edge dislocations is purely numerical. Using the AUTO program to numerically continue stationary solutions, we have observed that at $F = F_c$ there is a subcritical pitchfork bifurcation corresponding to a solution branch whose configuration contains edge dislocations that result from dipole creation and splitting. This solution branch becomes stable at a smaller stress, $F_{c0} < F_c$. There is also evidence that stationary solutions whose configurations contain two and four edge dislocations are simultaneously stable for certain ranges of F. These scenarios are analyzed in a forthcoming publication [**PCB**]. Further work is under way.

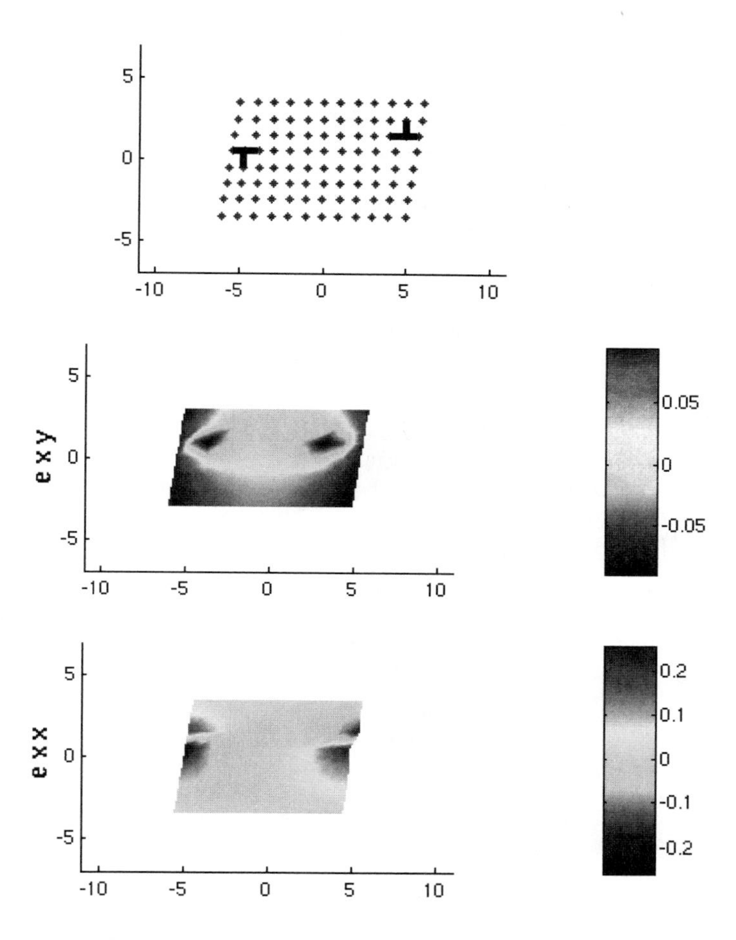

FIGURE 8. Final configuration with two edge dislocations for $F = 0.16$, $N_x = 12$, $N_y = 8$ and an odd period-1 piecewise linear function $g(x) = x$, for $|x| < 0.5-\alpha$, and $g(x) = (1-2\alpha)(1-2x)/(4\alpha)$ for $0.5 - \alpha < x < 0.5$, $\alpha = 0.15$. Here $e_{xx} = g(D_1^+ u)$, $e_{xy} = g(D_2^+ u)/2$ are the non-zero components of the strain tensor illustrating where the edge dislocations are.

5. Concluding remarks

We have presented several discrete models of crystal defects for which mathematical analysis can be carried out and economic numerical solutions can be obtained. Phenomena such as pinning of defects by the lattice and their motion for stresses surpassing a critical value can be understood as global bifurcations of traveling waves from stationary solutions of the model. Homogeneous nucleation of dislocations has been observed by numerically solving a simplified discrete model.

Acknowledgments

We thank Ignacio Plans for calculating Figure 8. This work was supported by the Spanish MEC grants MAT2005-05730-C02-01 and MAT2005-05730-C02-02.

References

[AC] W. Atkinson and N. Cabrera, *Motion of a Frenkel-Kontorowa dislocation in a one-dimensional crystal.* Phys. Rev. **138** (1965), A763-A766.

[BCP] L.L. Bonilla, A. Carpio and I. Plans, *Dislocations in cubic crystals described by discrete models.* Physica A **376** (2007), 361-377.

[BC] V.V. Bulatov and W. Cai, *Computer simulations of dislocations*, Oxford U.P., Oxford, UK 2006.

[C] A. Carpio, *Wavefronts for discrete two-dimensional nonlinear diffusion equations.* Appl. Math. Lett. **15**(4) (2002), 415-421.

[CB1] A. Carpio and L.L. Bonilla, *Edge dislocations in crystal structures considered as traveling waves of discrete models.* Phys. Rev. Lett. **90** (2003), 135502 (4 pages). Erratum: Ibid. **91** (2003), 029901.

[CB2] A. Carpio and L.L. Bonilla, *Oscillatory wave fronts in chains of coupled nonlinear oscillators.* Phys. Rev. E **67** (2003), 056621 (11 pages).

[CB3] A. Carpio and L. L. Bonilla, *Discrete models of dislocations and their motion in cubic crystals.* Phys. Rev. B **71** (2005), 134105 (10 pages).

[FK] J. Frenkel and T. Kontorova, *On the theory of plastic deformation and twinning.* J. Phys. Moscow **1** (1939), 137-149.

[FM] F.C. Frank and J.H. van der Merwe, *One dimensional dislocations. II. Misfitting monolayers and oriented overgrowth.* Proc. Roy. Soc. (London) A **198** (1949), 216-225.

[FW] G. Friesecke and J.A.D. Wattis, *Existence theorem for solitary waves on lattices.* Commun. Math. Phys. **161** (1994), 391-418.

[GB] I. Groma and B. Bakó, *Dislocation patterning: From micro- to mesoscale description.* Phys. Rev. Lett. **84** (2000), 1487-1490.

[HL] J.P. Hirth and J. Lothe, *Theory of Dislocations*, 2nd ed. Wiley, New York, 1982.

[KT] O. Kresse and L. Truskinovsky, *Lattice friction for crystalline defects: from dislocations to cracks.* J. Mech. Phys. Solids **52** (2004), 2521-2543.

[LKK] A. I. Landau, A.S. Kovalev and A. D. Kondratyuk, *Model of interacting atomic chains and its application to the description of the crowdion in an anisotropic crystal.* Phys. stat. sol. (b) **179** (1993), 373-381.

[LL] L.D. Landau and E.M. Lifshitz, *Theory of elasticity*, 3rd ed., Pergamon Press, London, UK, 1986.

[M] T. Mura, *Micromechanics of defects in solids*, 2nd. rev. ed., Martinus Nijhoff Publs., Dordrecht, 1987.

[N] F. R. N. Nabarro, *Theory of Crystal Dislocations*, Dover, N. Y. 1987.

[Neu] J.C. Neu, *Vortices in complex scalar fields.* Physica D **43** (1990), 385-406.

[PCB] I. Plans, A. Carpio and L.L. Bonilla, *Homogeneous nucleation of dislocations as bifurcations in a discrete elasticity model.* Preprint.

[S] H. Suzuki, *Motion of dislocations in body-centered cubic crystals*, in *Dislocation Dynamics*, ed. by A. H. Rosenfield et al. MacGraw Hill, New York, 1967, pp. 679–700.

Modelización, Simulación y Matemática Industrial, Universidad Carlos III de Madrid, E-28911 Leganés, Spain

E-mail address: bonilla@ing.uc3m.es

Departamento de Matemática Aplicada, Universidad Complutense de Madrid, E-28040 Madrid, Spain

E-mail address: ana_carpio@mat.ucm.es

Proceedings of Symposia in Applied Mathematics
Volume **65**, 2007

Fluid Dynamics from Boltzmann Equations

C. David Levermore

Dedicated to Peter Lax and Louis Nirenberg on the fete of their eightieth birthdays.

ABSTRACT. Maxwell (1867) and Boltzmann (1872) developed a recipe to go from certain Newtonian laws of molecular dynamics to the Euler and Navier-Stokes systems of gas dynamics. The problem of mathematically justifying this recipe remains largely open. We give a brief history of this problem. Recent significant advances start with the DiPerna-Lions (1990) theory of global solutions to the Boltzmann equation. Thses solutions are not known to be unique, and are not known to satisfy all the relations one would formally expect of classical solutions. To work with them one must use derivations of fluid approximations that are moment-based rather than traditional expansion-based ones. We survey results for the acoustic limit, for incompressible Stokes, Navier-Stokes, and Euler limits, and for weakly compressible Stokes and Navier-Stokes approximations.

1. Introduction

Euler published derivations of his continuity and motion equations in 1757, thereby laying the cornerstone for the theories of both incompressible and compressible fluid dynamics [**Eul**]. In the absence of external forces his motion equation reduces to

$$\rho(\partial_t u + u\cdot\nabla_x u) + \nabla_x p = 0\,, \tag{1.1}$$

where $\rho(x,t)$ is the mass density, $u(x,t)$ is the bulk velocity, and $p(x,t)$ is the pressure of the fluid at position x and time t. This was the first equation to express Newton's law for any three-dimensional continuum. His continuity equation is

$$\partial_t \rho + u\cdot\nabla_x \rho + \rho\nabla_x\cdot u = 0\,. \tag{1.2}$$

Taken together, these equations express the then known conservation laws of mass and momentum. Euler's derivations are essentially the same ones used by all books on fluid mechanics today.

For incompressible fluids Euler supplimented these equations with the incompressibility condition

$$\nabla_x\cdot u = 0\,, \tag{1.3}$$

in which case (1.2) is satisfied when ρ is a constant — an excellent approximation for liquids. For compressible fluids Euler supplimented them with with an equation of state $p = p(\rho)$ or $p = p(\rho, T)$ where $T(x,t)$ was a known temperature profile.

Euler realized that these theories of fluids were incomplete. He knew about viscosity (then commonly called internal friction), but did not see how to include it in his motion equation. He also knew about the ideal gas law (and even discussed corrections to it), but did not see how to derive an equation for the temperature. Indeed, the very nature of heat was not understood at the time. Heat and temperature were associated with a quantity called caloric. Because caloric was understood to be massless or nearly massless, it was not clear how to develope a theory for its dynamics. The theory that heat was a form of energy and that total energy should be conserved would not be formulated until almost a century later.

Fourier developed his theory of heat in 1807, however controversy over his methods kept it from being published until 1822. Fourier viewed his heat equation as govening the motion of caloric in solids. Inspired by Fourier, in 1823 Navier introduced additional terms into Euler's motion equation to model viscosity. His derivation was not widely accepted because (contrary to conventional wisdom) he adopted a molecular perspective. However, experiments showed that his theory had merit, so Poisson (1829), Saint-Venant (1843), and Stokes (1845) developed continuum-based derivations.

The notions that heat is a form of energy and that energy is conserved became widely accepted between 1840 and 1850 due to work by Mayer, Joule, Helmholtz, and others [**Br1**]. However, the nature of heat was still not understood. The theory that heat is the kinetic energy associated with the microscopic dynamics of molecules began to take firm root around 1850, and was consequenctly dubbed kinetic theory. There were earlier kinetic theories of gases, most notably one by Daniel Bernoulli in 1738. However, these theories did not explain more than continuum theories, as so were largely forgotten or ignored. In 1858 Clausius introduced the concept of "mean-free-path" into kinetic theory [**Br1**]. This caught the attention of Maxwell, who would elevate kinetic theory to another level.

Maxwell produced two kinetic theories; the first appeared in 1860 and the second in 1867. In the first [**Max1**] he introduces the so-called Maxwellian distribution for the molecular velocities in a gas. He then uses this equilibrium distribution and the concept of mean-free-path to derive viscosity coefficients for a gas of hard spheres. In the second kinetic theory [**Max2**] he derives a kinetic equation from microscopic Newtonian physics. He then uses this equation to derive a fluid dynamical system for a gas of so-called Maxwell molecules. This was the first such fluid system to include an energy equation and terms that model both viscosity and thermal conductivity. The resulting formulas for the viscosity and thermal conductivity coefficients in terms of Newtonian microscopic physics were one of the earliest examples of what we now call up-scaling.

Maxwell realized that kinetic theory had to address the fundamental question of how reversible microscopic Newtonian mechanics could lead to irreversible macroscopic dynamics. He introduced the famous "Maxwell demon" as a device to illustrate the statistical nature of macroscopic irreversibility in an 1867 private letter to Kelvin. The demon made its first public appearance in Maxwell's "Theory of Heat" published in 1871.

In 1868 Kirchhoff proposed that the dynamics of every gas is governed by a system of the type Maxwell had found. Such a system is now called a *compressible Navier-Stokes system* of gas dynamics. This is the earliest such proposal that I know. In fact, I do not know of any reference to a full *compressible Euler system* (one with an energy equation) that predates Maxwell's work. If true, the full fluid theory of gas dynamics followed from the kinetic one!

In 1872 Boltzmann reformulated Maxwell's kinetic theory and discovered what came to be called the H-Theorem [**Bolt**]. Their theory was controversial at the time because the notion of molecule was not generally accepted, and because their arguments had many gaps — many of which still have to be bridged. Boltzmann became the sole champion of the theory after Maxwell died in 1879 and the kinetic equation subsequently became known as the Boltzmann equation. Mathematicians came down on both sides of this controversy, which became very intense in the 1890's. Klein and Hilbert sided with Boltzmann. Poincare and Zermelo took the other side [**Br2**]. They claimed that the Boltzmann equation is inconsistent with the Poincare Recurrence Theorem, which had appeared in 1890.

Experimental evidence for the existence of atoms and molecules (based on Einstein's 1905 theory of Brownian motion) became generally accepted around 1906. That changed, but did not end the debate over the validity of the Boltzmann equation. (Poincare did switch sides.) It was only in the 1970's that the combination of careful experiments and careful simulations led to a general belief in its validity. The Boltzmann equation has been mathematically justified (for finite time) only for the case of elastic hard spheres by Landford (1975) in the so-called Boltzmann-Grad limit [**Lan**]. Here we will assume the validity of the Boltzmann equation, and use it as a starting point to establish fluid dynamical systems. Hilbert (1900 ICM) called for this problem to be addressed as part of his sixth problem.

Maxwell's 1867 derivation of the Navier-Stokes system from the Boltzmann equation rests on arguments about how various terms in the "Boltzmann equation" balance each other. These balance arguments seemed arbitrary to some, so Hilbert (1912) proposed that such derivations should be based on a systematic expansion in a small nondimensional parameter, which we now call the Knudsen number [**Hil**]. Later Enskog (1916) proposed a somewhat different systematic expansion, now often called the Chapman-Enskog expansion, in the same small parameter [**Ensk**].

Full justification of traditional compressible fluid dynamical approximations based on a formal Hilbert or Chapman-Enskog expansion has proven difficult because the basic well-posedness and regularity questions remain open for both the compressible fluid systems and the Boltzmann equation. The problem is exacerbated by the fact that to bound the error of these asymptotic expansions requires control of successively higher order spatial derivatives of the fluid variables. By the later half of the 20th century theories of local in time well-posedness for classical solutions had been developed by Grad, Ukai, Nishida, Caflisch, and others [**Caf2, Grad2, KMN, Nsh, Uk**]. The resulting justifications thereby where restricted to a meager subset of all physically natural initial data and usually to finite times.

In 1988 DiPerna and Lions gave an theory of global solutions to the Boltzmann equation for all physically natural initial data [**DiPL**]. This theory rests on a compactness argument, and therefore, does not yield uniqueness. Their theory is similar in spirit to that of Leray (1934) regarding global solutions to the incompressible

Navier-Stokes system [**Lry**]. It is natural to ask

"Does the Leray theory follow from that of DiPerna-Lions?"

It was shown by Golse-Saint Raymond in 2004 that the answer to this question is "yes" [**GStR3**]. This result is part of a program that was begun in 1989 when Bardos-Golse-L gave a new formal derivation of the incompressible Navier-Stokes system from the Boltzmann equation [**BGL1, BGL2**]. Rather than use traditional expansions, they gave a moment-based derivation, which puts fewer demands on the well-posedness and regularity theory. More generally, this program seeks to:

- study derivations of linear or weakly nonlinear fluid dynamical systems, such as the acoustic system and the incompressible systems, for which global theories exist;
- use moment-based formal derivations, which put fewer demands on the well-posedness and regularity theory;
- work within the framework of DiPerna-Lions solutions, thereby within the class of all physically natural initial data.

This article presents a survey of this program.

2. Boltzmann Equation Preliminaries

Here we will introduce the Boltzmann equation only so far as to set our notation, which is essentially that of [**BGL3**]. More complete introductions to the Boltzmann equation can be found in [**Cer, CIP, Gla, Gol**].

The state of a fluid composed of identical point particles confined to a spatial domain $\Omega \subset \mathbb{R}^D$ is described at the kinetic level by a mass density F over the single-particle phase space $\mathbb{R}^D \times \Omega$. More specifically, $F(v,x,t)\,\mathrm{d}v\,\mathrm{d}x$ gives the mass of the particles that occupy any infinitesimal volume $\mathrm{d}v\,\mathrm{d}x$ centered at the point $(v,x) \in \mathbb{R}^D \times \Omega$ at the instant of time $t \geq 0$. To remove complications due to boundaries, we take Ω to be the periodic domain $\mathbb{T}^D = \mathbb{R}^D/\mathbb{L}^D$, where $\mathbb{L}^D \subset \mathbb{R}^D$ is any D-dimensional lattice. Here $D \geq 2$.

The evolution of $F = F(v,x,t)$ is governed by the Boltzmann equation:

$$\partial_t F + v\cdot\nabla_x F = \mathcal{B}(F,F)\,, \qquad F(v,x,0) = F^{in}(v,x) \geq 0\,. \tag{2.1}$$

The collision operator $\mathcal{B}$ models binary collisions and acts only on the v argument of F. It is formally given by

$$\mathcal{B}(F,F) = \iint_{\mathbb{S}^{D-1}\times\mathbb{R}^D} (F_1'F' - F_1F)\, b(\omega, v_1 - v)\,\mathrm{d}\omega\,\mathrm{d}v_1\,, \tag{2.2}$$

where v_1 ranges over $\mathbb{R}^D$ endowed with its Lebesgue measure $\mathrm{d}v_1$ while ω ranges over the unit sphere $\mathbb{S}^{D-1} = \{\omega \in \mathbb{R}^D : |\omega| = 1\}$ endowed with its rotationally invariant measure $d\omega$. The F_1', F', F_1, and F appearing in the integrand designate $F(\cdot,x,t)$ evaluated at the velocities v_1', v', v_1, and v respectively, where the primed velocities are defined by

$$v_1' = v_1 - \omega\,\omega\cdot(v_1 - v)\,, \qquad v' = v + \omega\,\omega\cdot(v_1 - v)\,, \tag{2.3}$$

for any given $(\omega, v_1, v) \in \mathbb{S}^{D-1}\times\mathbb{R}^D\times\mathbb{R}^D$.

The unprimed and primed velocities are possible velocities for a pair of particles either before and after, or after and before, they interact through an elastic binary collision. Conservation o momentum and energy for particle pairs during collisions is expressed as

$$v + v_1 = v' + v_1' \,, \qquad |v|^2 + |v_1|^2 = |v'|^2 + |v_1'|^2 \,.$$

Equation (2.3) represents the general nontrivial solution of these $D+1$ equations for the $4D$ unknowns v_1', v', v_1, and v in terms of the $3D-1$ parameters (ω, v_1, v).

Quadratic operators like $\mathcal{B}$ are extended to be bilinear and symmetric by polarization formula

$$\mathcal{B}(F,G) = \tfrac{1}{4}\big[\mathcal{B}(F+G,F+G) - \mathcal{B}(F-G,F-G)\big] \,.$$

2.1. Collision Kernels. The collision kernel b is positive almost everywhere. The Galilean invariance of the collisional physics implies that b has the classical form

$$b(\omega, v_1 - v) = |v_1 - v| \, \Sigma(|\omega \cdot n|, |v_1 - v|) \,,$$

where $n = (v_1 - v)/|v_1 - v|$ and Σ is the specific differential cross-section.

Maxwell gave a recipe for the collision kernel in terms of the microscopic intermolecular potential [**Max2**]. For hard spheres of mass m and radius r_o it has the form

$$b(\omega, v_1 - v) = |\omega \cdot (v_1 - v)| \frac{(2r_o)^{D-1}}{2m} \,.$$

For a repulsive intermolecular potential of the form c/r^k with $k > 2\frac{D-1}{D+1}$ it has the form

$$b(\omega, v_1 - v) = \hat{b}(\omega \cdot n) \, |v_1 - v|^\beta \quad \text{with } \beta = 1 - 2\tfrac{D-1}{k} \,,$$

where $\hat{b}(\omega \cdot n)$ is positive almost everywhere and has even symmetry in ω. The condition $k > 2\frac{D-1}{D+1}$ is equivalent to $\beta > -D$, which insures that $b(\omega, v_1 - v)$ is locally integrable with respect to $v_1 - v$. The cases $\beta > 0$, $\beta = 0$, and $-D < \beta < 0$ are called the "hard", "Maxwell", and "soft" potential cases.

The function $\hat{b}(\omega \cdot n)$ devived by Maxwell's 1867 recipe for potentials of the form c/r^k has a singularity that is not locally integrable at $\omega \cdot n = 0$. This was not a problem for Maxwell because he used a weak form of the Boltzmann equation that regularizes this singularity [**Max2**]. Hilbert (1912) avoided this problem by studying the Boltzmann equation for only the hard sphere case [**Hil**]. Grad (1954) was able to extended some of Hilbert's analysis by introducing a small deflection cutoff that requires $\hat{b}(\omega \cdot n)$ to vanish like $|\omega \cdot n|$ as $\omega \cdot n \to 0$ — the so-called *Grad cutoff* condition [**Grad2**]. More generally one can assume [**LS**] that $\hat{b}(\omega \cdot n)$ satisfies the less restrictive small deflection cutoff condition

$$\int_{\mathbb{S}^{D-1}} \hat{b}(\omega \cdot n) \, \mathrm{d}\omega < \infty \,.$$

This so-called *weak cutoff* condition is required to have the gain and loss terms of the Boltzmann collision operator make sense separately.

2.2. Conservation, Dissipation, and Equilibria. Maxwell showed that the collision operator has the following property related to the conservation laws of mass, momentum, and energy [**Max2**]. For every measurable ζ the following are equivalent:

- $\zeta \in \mathrm{span}\{1, v_1, \cdots, v_D, \frac{1}{2}|v|^2\}$;
- $\langle \zeta \, \mathcal{B}(F,F) \rangle = 0$ for "every" F;
- $\zeta_1' + \zeta' - \zeta_1 - \zeta = 0$ for every (ω, v_1, v).

Here we have introduced the notation $\langle \cdot \rangle = \int \cdot \, \mathrm{d}v$.

If F is a classical solution of the Boltzmann equation then F satisfies local conservation laws of mass, momentum, and energy:

$$\partial_t \langle F \rangle + \nabla_x \cdot \langle v\, F \rangle = 0 \,,$$
$$\partial_t \langle v\, F \rangle + \nabla_x \cdot \langle v \otimes v\, F \rangle = 0 \,,$$
$$\partial_t \langle \tfrac{1}{2}|v|^2 F \rangle + \nabla_x \cdot \langle v \tfrac{1}{2}|v|^2 F \rangle = 0 \,.$$

Boltzmann's H-Theorem (1872) states that the collision operator has the following property related to the dissipation of entropy and equilibrium [**Bolt**].

$$\langle \log(F)\, \mathcal{B}(F,F) \rangle \leq 0 \quad \text{for "every" } F \,.$$

Moreover, for "every" F the following are equivalent:

- $\langle \log(F)\, \mathcal{B}(F,F) \rangle = 0$;
- $\mathcal{B}(F,F) = 0$;
- F is a Maxwellian,

where Maxwellians have the form

$$F = \mathcal{M}(v;\rho,u,\theta) = \frac{\rho}{(2\pi\theta)^{\frac{D}{2}}} \exp\left(-\frac{|v-u|^2}{2\theta} \right),$$

with ρ, u, and θ given by

$$\rho = \langle F \rangle \,, \qquad \rho u = \langle v\, F \rangle \,, \qquad \rho\theta = \tfrac{1}{D}\langle |v-u|^2 F \rangle \,.$$

Here ρ is the mass density, u is the bulk velocity, and $\theta = \mathrm{k_B} T/m$ where $\mathrm{k_B}$ is Boltzmann's constant, T is the temperature of the gas, and m is the molecular mass. In 1860 Maxwell argued these were the local equilibrium kinetic densities of gas dynamics [**Max1**]. In 1866 he showed they were equilibria of what we now call the collision operator [**Max2**]. Both times he gave an argument that they were the only such equilibria, but there were gaps in each argument. In 1872 Boltzmann filled the gap in Maxwell's second argument with his H-Theorem [**Bolt**].

If F is a classical solution of the scaled Boltzmann equation then F satisfies local entropy dissipation law:

$$\begin{aligned}
\partial_t \langle (F\log(F) - F) \rangle &+ \nabla_x \cdot \langle v\,(F\log(F) - F) \rangle \\
&= \langle \log(F)\, \mathcal{B}(F,F) \rangle \\
&= -\iiint \tfrac{1}{4} \log\left(\frac{F_1' F'}{F_1 F} \right) (F_1' F' - F_1 F)\, b \, \mathrm{d}\omega \, \mathrm{d}v_1 \, \mathrm{d}v \\
&\leq 0 \,.
\end{aligned}$$

3. Formal Relations to Fluid Systems

The Boltzmann equation can be brought into the nondimensional form

$$\partial_t F + v \cdot \nabla_x F = \frac{1}{\epsilon} \mathcal{B}(F,F) \,,$$

where ϵ is the Knudsen number. The Knudsen number is the ratio of the scale of mean-free-paths to the macroscopic length scale. Fluid dynamical regimes are ones in which collisions dominate — i.e. ones characterized by the Knudsen number being small.

3.1. Compressible Euler System. The H-Theorem suggests that collisions will then drive F towards a so-called local Maxwellian — namely that

$$F(v,x,t) \approx \mathcal{M}\big(v;\rho(x,t),u(x,t),\theta(x,t)\big)\,,$$

where $\rho(x,t)$, $u(x,t)$, and $\theta(x,t)$ are given by

$$\rho = \langle F\rangle\,, \qquad \rho u = \langle v\,F\rangle\,, \qquad \rho\theta = \tfrac{1}{D}\langle |v-u|^2 F\rangle\,.$$

A local Maxwellian is generally not a solution of the Boltzmann equation. However in fluid dynamical regimes the solution of the Boltzmann equation is almost a local Maxwellian.

When the local Maxwellian approximation is placed into the local conservation laws one obtains the compressible Euler system of gas dynamics

$$\begin{aligned} \partial_t\rho + \nabla_x\!\cdot(\rho u) &= 0\,,\\ \partial_t(\rho u) + \nabla_x\!\cdot(\rho u\otimes u) + \nabla_x(\rho\theta) &= 0\,,\\ \partial_t(\rho(\tfrac12|u|^2 + \tfrac{D}{2}\theta)) + \nabla_x\!\cdot(\rho u(\tfrac12|u|^2 + \tfrac{D+2}{2}\theta)) &= 0\,. \end{aligned}$$

Here the pressure satisfies the ideal gas law ($p = \rho\theta$) while the specific internal energy satisfies the polytropic γ-law with $\gamma = \frac{D+2}{D}$ ($\varepsilon = \frac{D}{2}\theta$). Euler systems that govern more general ideal gases can be derived from more complicated molecular models.

When the local Maxwellian approximation is placed into the local entropy dissipation law one obtains the relation

$$\partial_t\left(\rho\log\left(\frac{\rho}{\theta^{\frac{D}{2}}}\right)\right) + \nabla_x\!\cdot\left(\rho u\log\left(\frac{\rho}{\theta^{\frac{D}{2}}}\right)\right) \leq 0\,.$$

This is the local entropy dissipation (production) law one expects from a local version of the second law of thermodynamics.

3.2. Compressible Navier-Stokes System. Corrections to the Euler system can be derived by considering a family F_ϵ of solutions to the scaled Boltzmann equation. The Hilbert or Chapman-Enskog expansions seeks formal solutions of the form

$$F_\epsilon = F^{(0)} + \epsilon\,F^{(1)} + \epsilon^2 F^{(2)} + \cdots\,.$$

One finds that $F^{(0)}$ is a local Maxellian with $\rho^{(0)}$, $u^{(0)}$, and $\theta^{(0)}$ governed by the compressible Euler system [**Br3, Cer**].

Rather than present a derivation of the Navier-Stokes system based on either the Hilbert or the Chapman-Enskog expansion, here we present a simple balance argument similar in spirit to that used by Maxwell [**Max2**] in 1867. Decompose F as

$$F(v,x,t) = \mathcal{M}\big(v\,;\,\rho(x,t),u(x,t),\theta(x,t)\big) + \widetilde{F}(v,x,t)\,,$$

where $\mathcal{M}(v;\rho,u,\theta)$ is the local Maxwellian with (ρ,u,θ) determined by

$$\langle F\rangle = \rho\,, \qquad \langle v\,F\rangle = \rho u\,, \qquad \langle\tfrac12|v-u|^2F\rangle = \tfrac{D}{2}\rho\,\theta\,,$$

and $\widetilde{F}$ is the deviation of F from $\mathcal{M}$. One sees that $\widetilde{F}$ satisfies

$$\langle\widetilde{F}\rangle = 0\,, \qquad \langle v\,\widetilde{F}\rangle = 0\,, \qquad \langle\tfrac12|v-u|^2\widetilde{F}\rangle = 0\,.$$

Placing this decomposition into the local conservation laws yields

$$\partial_t \rho + \nabla_x \cdot (\rho u) = 0\,,$$

$$\partial_t(\rho u) + \nabla_x \cdot (\rho u \otimes u) + \nabla_x(\rho\theta) + \nabla_x \cdot \widetilde{S} = 0\,,$$

$$\partial_t(\rho(\tfrac{1}{2}|u|^2 + \tfrac{D}{2}\theta)) + \nabla_x \cdot (\rho u(\tfrac{1}{2}|u|^2 + \tfrac{D+2}{2}\theta)) + \nabla_x \cdot (\widetilde{S}u + \tilde{q}) = 0\,,$$

where $\widetilde{S}$ and $\tilde{q}$ are defined by

$$\widetilde{S} = \langle (v-u) \otimes (v-u)\, \widetilde{F} \rangle\,, \qquad \tilde{q} = \langle (v-u)\tfrac{1}{2}|v-u|^2 \widetilde{F} \rangle\,.$$

These are the stress and heat flux respectively. They are the only terms that arise from $\widetilde{F}$. Fluid dynamical systems are obtained by making approximations for $\widetilde{F}$. The compressible Euler system is recovered by setting $\widetilde{F} = 0$.

In order to obtain a better fluid dynamical system, one must find a better approximation for the deviation $\widetilde{F}$. One can show that $\widetilde{F}$ satisfies the so-called deviation equation

$$\partial_t \widetilde{F} + \widetilde{\mathcal{P}} v \cdot \nabla_x \widetilde{F} + \widetilde{\mathcal{P}} v \cdot \nabla_x \mathcal{M} = \mathcal{B}(\mathcal{M} + \widetilde{F}, \mathcal{M} + \widetilde{F})\,,$$

where $\widetilde{\mathcal{P}} = \mathcal{I} - \mathcal{P}$ and $\mathcal{P}$ is the operator given by

$$\mathcal{P}G = \mathcal{M}\left[\frac{\langle G \rangle}{\rho} + \frac{(v-u)\cdot\langle (v-u)\, G \rangle}{\theta} + \left(\frac{|v-u|^2}{2\theta} - \frac{D}{2}\right)\left\langle \left(\frac{|v-u|^2}{D\,\theta} - 1\right) G \right\rangle\right].$$

One can show $\mathcal{P}$ is a projection (i.e. $\mathcal{P}^2 = \mathcal{P}$). Because $\mathcal{P}\widetilde{F} = 0$, one sees $\widetilde{\mathcal{P}} = \mathcal{I} - \mathcal{P}$ is a projection onto the deviations from local equilibria. These are orthogonal projections in $L^2(\mathrm{d}v/\mathcal{M})$.

The Navier-Stokes approximation is obtained by arguing that $\widetilde{F}$ is much smaller than $\mathcal{M}$ and taking the dominant term on each side of the deviation equation. More specifically, we make the approximations

$$\partial_t \widetilde{F} + \widetilde{\mathcal{P}} v \cdot \nabla_x \widetilde{F} + \widetilde{\mathcal{P}} v \cdot \nabla_x \mathcal{M} \approx \widetilde{\mathcal{P}} v \cdot \nabla_x \mathcal{M}\,,$$

$$\mathcal{B}(\mathcal{M} + \widetilde{F}, \mathcal{M} + \widetilde{F}) = 2\mathcal{B}(\mathcal{M}, \widetilde{F}) + \mathcal{B}(\widetilde{F}, \widetilde{F}) \approx 2\mathcal{B}(\mathcal{M}, \widetilde{F})\,.$$

We thereby argue that $\widetilde{F} \approx \widetilde{F}_{NS}$ where $\widetilde{F}_{NS}$ satisfies

$$\widetilde{\mathcal{P}} v \cdot \nabla_x \mathcal{M} = 2\mathcal{B}(\mathcal{M}, \widetilde{F}_{NS})\,.$$

This is the Navier-Stokes balance relation. It leads to the compressible Navier-Stokes system, although at this stage it may not be obvious how.

The left-hand side of the Navier-Stokes balance relation has the form

$$\widetilde{\mathcal{P}} v \cdot \nabla_x \mathcal{M} = \mathcal{M}\left(A\left(\frac{v-u}{\theta^{\frac{1}{2}}}\right) : \nabla_x u + B\left(\frac{v-u}{\theta^{\frac{1}{2}}}\right) \cdot \frac{\nabla_x \theta}{\theta^{\frac{1}{2}}} \right),$$

where the non-dimensional functions $A(v)$ and $B(v)$ are defined by

$$A(v) = v \otimes v - \tfrac{1}{D}|v|^2 I\,, \qquad B(v) = \tfrac{1}{2}|v|^2 v - \tfrac{D+2}{2} v\,.$$

Notice that $A(v)$ is a traceless, symmetric matrix, while $B(v)$ is a vector. The solution of the Navier-Stokes balance relation is

$$\widetilde{F}_{NS} = -\frac{\mathcal{M}}{\rho}\left(\widehat{A}(v-u;\theta):\nabla_x u + \widehat{B}(v-u;\theta)\cdot\frac{\nabla_x\theta}{\theta^{\frac{1}{2}}}\right),$$

provided that $\widehat{A}(v;\theta)$ and $\widehat{B}(v;\theta)$ satisfy

$$-\frac{2}{M}\,\mathcal{B}(M,M\widehat{A}) = A(v/\theta^{\frac{1}{2}}), \qquad M\widehat{A} \perp 1, v, |v|^2,$$
$$-\frac{2}{M}\,\mathcal{B}(M,M\widehat{B}) = B(v/\theta^{\frac{1}{2}}), \qquad M\widehat{B} \perp 1, v, |v|^2,$$

with $M = \mathcal{M}(v;1,0,\theta)$. The linear operator on the left-hand side above is symmetric and nonnegative definite in $L^2(M\mathrm{d}v)$.

Maxwell (1866) found explicit solutions to these equations for the case of so-called Maxwell molecules, $\beta = 0$ [**Max2**]. Hilbert (1912) showed that these equations have solutions for the hard-sphere case [**Hil**]. More generally, he showed the linear operator above satisfies a Fredholm alternative. This kind of result has been extended to all classical collision kernels such that $\beta > -D$ and $\hat{b}(\omega\cdot n)$ satisfies the weak cutoff condition [**LS**]. Moreover, one can show that there exist real-valued functions $\tau_A(v;\theta)$ and $\tau_B(v;\theta)$ such that

$$\widehat{A}(v;\theta) = \tau_A(v;\theta)A(v/\theta^{\frac{1}{2}}), \qquad \widehat{B}(v;\theta) = \tau_B(v;\theta)B(v/\theta^{\frac{1}{2}}).$$

The stress and heat flux can then be shown to have the form

$$\widetilde{S}_{NS} = -\mu(\theta)\big[\nabla_x u + (\nabla_x u)^T - \tfrac{2}{D}\nabla_x\cdot u\,I\big],$$
$$\widetilde{q}_{NS} = -\kappa(\theta)\nabla_x\theta,$$

where $\mu(\theta)$ and $\kappa(\theta)$ are positive functions of θ given by

$$\mu(\theta) = \tfrac{1}{(D+2)(D-1)}\,\theta\left\langle \mathcal{M}(v;1,0,\theta)\left|A\left(\frac{v}{\theta^{\frac{1}{2}}}\right)\right|^2\tau_A(v;\theta)\right\rangle,$$
$$\kappa(\theta) = \tfrac{1}{D}\,\theta\left\langle \mathcal{M}(v;1,0,\theta)\left|B\left(\frac{v}{\theta^{\frac{1}{2}}}\right)\right|^2\tau_B(v;\theta)\right\rangle.$$

One immediately sees that $\mu(\theta)$ and $\kappa(\theta)$ are the viscosity and thermal conductivity coefficients respectively. These functions of θ are the only things in the Navier-Stokes system that depend on the collision kernel b, and therefore the only things that depend on details of the microsopic dynamics. The fact that they are independent of ρ was an important early prediction of kinetic theory that was subsequently confirmed by experiment.

3.3. Linear and Weakly Nonlinear Scalings. We consider fluid dynamical regimes in which F is close to a spatially homogeneous Maxwellian $M = M(v)$. By an appropriate choice of a Galilean frame and of mass and velocity units, it can be assumed that this so-called absolute Maxwellian M has the form

$$M(v) \equiv \mathcal{M}(v;1,0,1) = \frac{1}{(2\pi)^{D/2}}\exp(-\tfrac{1}{2}|v|^2).$$

This corresponds to the spatially homogeneous fluid state with mass density and temperature equal to 1 and bulk velocity equal to 0.

We then seek solutions of the form

$$F_\epsilon(v,x,t) = M(v)\big(1+\delta_\epsilon\, g_\epsilon(v,x,t)\big),$$

where δ_ϵ satisfies $\delta_\epsilon \to 0$ as $\epsilon \to 0$. Assuming that g_ϵ converges to g as $\epsilon \to 0$, one can show that

$$g_\epsilon \to g = \rho + v \cdot u + (\tfrac{1}{2}|v|^2 - \tfrac{D}{2})\theta ,$$

where $\rho(x,t)$, $u(x,t)$, and $\theta(x,t)$ are fluctuations of the mass density, bulk velocity, and temperature about their equilibrium values: 1, 0, and 1. This limiting form is a so-called *infinitesimal Maxwellian* because

$$\mathcal{M}(v; 1 + \delta_\epsilon \rho, \delta_\epsilon u, 1 + \delta_\epsilon \theta) = M(v)\big(1 + \delta_\epsilon g + O(\delta_\epsilon^2)\big) .$$

It is the limiting form shared by the leading orders of all linear and weakly nonlinear fluid dynamical approximations.

Moreover, one finds that the fluctuations ρ, u, and θ satisfy

$$\begin{aligned} \partial_t \rho + \nabla_x \cdot u &= 0 , \\ \partial_t u + \nabla_x(\rho + \theta) &= 0 , \\ \tfrac{D}{2} \partial_t \theta + \nabla_x \cdot u &= 0 . \end{aligned}$$

This is the acoustic system. It is the linearization about the homogeneous state of the compressible Euler system. It is one of the simplest systems of fluid dynamical equations imaginable, being essentially the wave equation.

3.4. Derivation of Incompressible Systems. It is easily seen that when ρ, u, and θ satisfy

$$\nabla_x \cdot u = 0 , \qquad \nabla_x(\rho + \theta) = 0 ,$$

they are a stationary solution of the acoustic system which will generally vary in space. On the other hand, it can be shown that absolute Maxwellians are the only stationary solutions of the Boltzmann equation.

It is clear that the time scale at which the acoustic system was derived was not long enough to see the evolution of these solutions. By considering the Boltzmann equation over a longer time scale one can give formal moment derivations of three incompressible fluid dynamical systems, depending on the limiting behavior of the ratio δ_ϵ/ϵ as $\epsilon \to 0$. One finds the following:

- When $\delta_\epsilon/\epsilon \to 0$, one considers time scales of order $1/\epsilon$, and an incompressible Stokes system is derived.
- When $\delta_\epsilon/\epsilon \to 1$ (or any other nonzero number), one considers time scales of order $1/\epsilon$, and an incompressible Navier-Stokes system is derived.
- When $\delta_\epsilon/\epsilon \to \infty$, one considers time scales of order $1/\delta_\epsilon$, and an incompressible Euler system is derived.

What underlies this result is the fact that δ_ϵ/ϵ is the Reynolds number.

The incompressible Stokes system derived by this argument is

$$\begin{aligned} \nabla_x \cdot u &= 0 , & \rho + \theta &= 0 . \\ \partial_t u + \nabla_x p &= \nu \Delta_x u , & u(x,0) &= u^{in}(x) , \\ \tfrac{D+2}{2} \partial_t \theta &= \kappa \Delta_x \theta , & \theta(x,0) &= \theta^{in}(x) . \end{aligned}$$

Here $\nu > 0$ is the kinematic viscosity and $\kappa > 0$ is the thermal diffusivity. Like the acoustic system, the Stokes system is also one of the simplest systems of fluid dynamical equations imaginable, being essentially a system of linear heat equations.

The incompressible Navier-Stokes system one obtains is

$$\nabla_x \cdot u = 0 , \qquad \rho + \theta = 0 .$$

$$\partial_t u + u\cdot\nabla_x u + \nabla_x p = \nu\Delta_x u\,, \qquad u(x,0) = u^{in}(x)\,,$$
$$\tfrac{D+2}{2}\big(\partial_t\theta + u\cdot\nabla_x\theta\big) = \kappa\Delta_x\theta\,, \qquad \theta(x,0) = \theta^{in}(x)\,.$$

Here the kinematic viscosity ν and the thermal diffusivity have the same values as in the Stokes system. Unlike the Stokes system however, the Navier-Stokes system is nonlinear. While this fact does not complicate its formal derivation, it makes the mathematical establishment of its validity much harder.

The incompressible Euler System obtained is

$$\nabla_x\cdot u = 0\,, \qquad \rho + \theta = 0\,.$$

$$\partial_t u + u\cdot\nabla_x u + \nabla_x p = 0\,, \qquad u(x,0) = u^{in}(x)\,,$$
$$\tfrac{D+2}{2}\big(\partial_t\theta + u\cdot\nabla_x\theta\big) = 0\,, \qquad \theta(x,0) = \theta^{in}(x)\,,$$

Like the Navier-Stokes system, the Euler system is nonlinear. The full mathematical establishment of its validity is also an open problem.

As was the case for the acoustic system, the Euler system has stationary solutions that vary in space. It is clear that the time scale at which the Euler system was derived was not long enough to see the evolution of these solutions. Even at a formal level it is unclear how this long-time evolution should be governed.

It should be pointed out that the above systems are not the only incompressible Stokes, Navier-Stokes, and Euler systems that may be derived as fluid dynamical limits of the Boltzmann equation. More refined asymptotic balances lead to incompressible Stokes, Navier-Stokes, and Euler systems that differ from those above in (1) the form of the heat equation and (2) the Boussinesq relation is replaced by $p = \rho + \theta$. These also have moment-based derivations. One should therefore be be careful about referring to "the incompressible Stokes system" (for example) until it is clear to which Stokes system you are referring [**BLP, BL**]. In addition, there are other systems that have moment-based derivations, including some that are "beyond Navier-Stokes" [**Bob, Son1, Son2**]. The goals of the BGL program (1989) are (1) to identify those fluid dynamical systems that can be so derived, and (2) to give a full mathematical justification of those formal derivations.

4. Global Solutions

We now make precise: (1) the notion of solution for the Boltzmann equation, and (2) the notion of solution for the fluid dynamical systems. Ideally, these solutions should be global while the bounds should be physically natural. We therefore work in the setting of DiPerna-Lions renormalized solutions for the Boltzmann equation, and in the setting of Leray solutions for the Navier-Stokes system. These theories have the virtues of considering physically natural classes of initial data, and consequently, of yielding global solutions.

One of the main goals of the BGL program is to connect the DiPerna-Lions theory of renormalized solutions of the Boltzmann equation to the Leray theory of weak solutions of the incompressible Navier-Stokes system.

4.1. DiPerna-Lions Theory. The DiPerna-Lions theory [**DiPL, PLL1**] gives the existence of a global weak solution to a class of formally equivalent initial-value problems that are obtained by multiplying the Boltzmann equation by $\Gamma'(F)$, where

Γ' is the derivative of an admissible function Γ:

$$\big(\tau_\epsilon\, \partial_t + v\cdot\nabla_x\big)\Gamma(F) = \frac{1}{\epsilon}\,\Gamma'(F)\mathcal{B}(F,F)\,,$$
$$F(v,x,0) = F^{in}(v,x) \geq 0\,.$$

This is the so-called renormalized Boltzmann equation. A differentiable function $\Gamma : [0,\infty) \to \mathbb{R}$ is called *admissible* if for some constant $C_\Gamma < \infty$ it satisfies

$$\big|\Gamma'(Z)\big| \leq \frac{C_\Gamma}{\sqrt{1+Z}} \quad \text{for every } Z \geq 0\,.$$

The solutions lie in $C([0,\infty); w\text{-}L^1(M\mathrm{d}v\,\mathrm{d}x))$, where the prefix "$w$-" on a space indicates that the space is endowed with its weak topology.

THEOREM 4.1. (DiPerna-Lions Renormalized Solutions) *Let b satisfy*

$$\lim_{|v|\to\infty} \frac{1}{1+|v|^2} \int_{\mathbb{S}^{D-1}\times K} b(\omega, v_1 - v)\,\mathrm{d}\omega\,\mathrm{d}v_1 = 0$$
$$\text{for every compact } K \subset \mathbb{R}^D\,.$$

Given any initial data F^{in} in the entropy class

$$E(M\mathrm{d}v\,\mathrm{d}x) = \big\{F^{in} \geq 0 \,:\, H(F^{in}) < \infty\big\}\,,$$

there exists at least one $F \geq 0$ in $C([0,\infty); w\text{-}L^1(M\mathrm{d}v\,\mathrm{d}x))$ that for every admissible function Γ is a weak solution of renormalized Boltzmann equation.

This solution satisfies a weak form of the local conservation law of mass

$$\tau_\epsilon\, \partial_t\langle F\rangle + \nabla_x\cdot\,\langle v\,F\rangle = 0\,.$$

Moreover, there exists a martix-valued distribution W such that $W\,\mathrm{d}x$ is nonnegative definite measure and G and W satisfy a weak form of the local conservation law of momentum

$$\tau_\epsilon\, \partial_t\langle v\,F\rangle + \nabla_x\cdot\,\langle v\otimes v\,F\rangle + \nabla_x\cdot\,W = 0\,,$$

and for every $t > 0$, the global energy equality

$$\int_{\mathbb{T}^D} \langle \tfrac{1}{2}|v|^2 F(t)\rangle\,\mathrm{d}x + \int_{\mathbb{T}^D} \tfrac{1}{2}\,\mathrm{tr}(W(t))\,\mathrm{d}x = \int_{\mathbb{T}^D} \langle \tfrac{1}{2}|v|^2 F^{in}\rangle\,\mathrm{d}x\,,$$

and the global entropy inequality

$$H(F(t)) + \int_{\mathbb{T}^D} \tfrac{1}{2}\,\mathrm{tr}(W(t))\,\mathrm{d}x + \frac{1}{\tau_\epsilon \epsilon}\int_0^t R(F(s))\,\mathrm{d}s \leq H(F^{in})\,.$$

Remarks: DiPerna-Lions renormalized solutions are very weak — much weaker than standard weak solutions. They are not known to satisfy many properties that one would formally expect to be satisfied by solutions of the Boltzmann equation. In particular, the theory does not assert either the local conservation of momentum, the global conservation of energy, the global entropy equality, or even a local entropy inequality; nor does it assert the uniqueness of the solution.

4.2. Leray Theory. The DiPerna-Lions theory has many similarities with the Leray theory of global weak solutions of the initial-value problem for Navier-Stokes type systems [**Lry, CF**]. For the Navier-Stokes system with mean zero initial data, we set the Leray theory in the following Hilbert spaces of vector- and scalar-valued functions:

$$\mathbb{H}_v = \left\{ w \in L^2(\mathrm{d}x;\mathbb{R}^D) \,:\, \nabla_x \cdot w = 0\,, \int w \, \mathrm{d}x = 0 \right\},$$

$$\mathbb{H}_s = \left\{ \chi \in L^2(\mathrm{d}x;\mathbb{R}) \,:\, \int \chi \, \mathrm{d}x = 0 \right\},$$

$$\mathbb{V}_v = \left\{ w \in \mathbb{H}_v \,:\, \int |\nabla_x w|^2 \, \mathrm{d}x < \infty \right\},$$

$$\mathbb{V}_s = \left\{ \chi \in \mathbb{H}_s \,:\, \int |\nabla_x \chi|^2 \, \mathrm{d}x < \infty \right\}.$$

Let $\mathbb{H} = \mathbb{H}_v \oplus \mathbb{H}_s$ and $\mathbb{V} = \mathbb{V}_v \oplus \mathbb{V}_s$.

THEOREM 4.2. (Leray Solutions) *Given any initial data $(u^{in}, \theta^{in}) \in \mathbb{H}$, there exists at least one $(u, \theta) \in C([0,\infty); w\text{-}\mathbb{H}) \cap L^2(dt; \mathbb{V})$ that is a weak solution of the Navier-Stokes system. Moreover, for every $t > 0$, (u, θ) satisfies the dissipation inequalities*

$$\int \tfrac{1}{2}|u(t)|^2 \mathrm{d}x + \int_0^t \int \nu |\nabla_x u|^2 \mathrm{d}x \, \mathrm{d}s \le \int \tfrac{1}{2}|u^{in}|^2 dx\,,$$

$$\int \tfrac{D+2}{4}|\theta(t)|^2 \mathrm{d}x + \int_0^t \int \kappa |\nabla_x \theta|^2 \mathrm{d}x \, \mathrm{d}s \le \int \tfrac{D+2}{4}|\theta^{in}|^2 \mathrm{d}x\,.$$

Remarks: By arguing formally from the Navier-Stokes system, one would expect these inequalities to be equalities. However, that is not asserted by the Leray theory. Also, as was the case for the DiPerna-Lions theory, the Leray theory does not assert uniqueness of the solution.

Remark: Because the role of the above dissipation inequalities is to provide a-priori estimates, the existence theory also works if they are replaced by the single dissipation inequality

$$\int \tfrac{1}{2}|u(t)|^2 + \tfrac{D+2}{4}|\theta(t)|^2 dx + \int_0^t \int \nu|\nabla_x u|^2 + \kappa|\nabla_x \theta|^2 dx \, ds$$
$$\le \int \tfrac{1}{2}|u^{in}|^2 + \tfrac{D+2}{4}|\theta^{in}|^2 dx\,.$$

It is this version of the Leray theory that we will obtain in the limit.

5. Survey of Recent Results

The goals of the BGL program are (1) to identify those fluid dynamical systems that can be so derived, and (2) to give a full mathematical justification of those formal derivations. A detail review of its early results can be found in [**Vil**].

The main result of [**BGL3**] for the Navier-Stokes limit is to recover the motion equation for a discrete-time version of the Boltzmann equation assuming the DiPerna-Lions solutions satisfy the local conservation of momentum and with the aid of a mild compactness assumption. This result fell short of the goal in five respects.

- First, the heat equation was not treated because the v^3 terms in the heat flux could not be controlled.
- Second, local momentum conservation was assumed because DiPerna-Lions solutions are not known to satisfy the local conservation law of momentum (or energy) that one would formally expect.
- Third, unnatural technical assumptions were made on the Boltzmann collision kernel.
- Fourth, the discrete-time case was treated in order to avoid having to control the time regularity of the acoustic modes.
- Finally, a mild compactness assumption was required to pass to the limit in certain nonlinear terms.

In recent works all of these shortcomings have been overcome.

Consider the scaled Boltzmann equation

$$\tau_\epsilon \partial_t F_\epsilon + v \cdot \nabla_x F_\epsilon = \frac{1}{\epsilon} \mathcal{B}(F_\epsilon, F_\epsilon) ,$$
$$F_\epsilon = M\big(1 + \delta_\epsilon g_\epsilon\big) .$$

One derives the acoustic system when $\tau_\epsilon = 1$ and

$$\delta_\epsilon \to 0 .$$

One derives the incompressible Stokes, the incompressible Navier-Stokes, and the incompressible Euler system when $\tau_\epsilon = \max\{\epsilon, \delta_\epsilon\}$, $\delta_\epsilon \to 0$ and respectively

$$\frac{\delta_\epsilon}{\epsilon} \to 0 , \qquad \frac{\delta_\epsilon}{\epsilon} \to 1 , \qquad \frac{\delta_\epsilon}{\epsilon} \to \infty .$$

The scaling of the fluctuations is controlled be assuming that

$$\int \left\langle F_\epsilon^{in} \log\left(\frac{F_\epsilon^{in}}{M}\right) - F_\epsilon^{in} + M \right\rangle \mathrm{d}x < C^{in} \delta_\epsilon^2 .$$

The entropy inequality then implies that

$$F_\epsilon = M\big(1 + \delta_\epsilon g_\epsilon\big) ,$$

where g_ϵ is compact in w-L^1. Moreover, every limit point must have the form of an infinitesimal Maxwellian

$$g_\epsilon \to \rho + v \cdot u + (\tfrac{1}{2}|v|^2 - \tfrac{D}{2})\theta ,$$

where ρ, u, θ are in L^2.

Bardos, Golse, and Levermore [**BGL4, BGL5**] recover the acoustic and the Stokes limits for the Boltzmann equation for cutoff collision kernels that arise from Maxwell potentials. In doing so, they control the energy flux and *establish the local conservation laws of momentum and energy in the limit.* The scaling they used was not optimal, essentially requiring

$$\frac{\delta_\epsilon}{\epsilon} \to 0 \quad \text{rather than} \quad \delta_\epsilon \to 0 \quad \text{for the acoustic limit} ,$$
$$\frac{\delta_\epsilon}{\epsilon^2} \to 0 \quad \text{rather than} \quad \frac{\delta_\epsilon}{\epsilon} \to 0 \quad \text{for the Stokes limit} .$$

Lions and Masmoudi [**LM2, LM3**] recover the Navier-Stokes motion equation with the aid of only the local conservation of momentum assumption and the nonlinear compactness assumption that where made in [**BGL3**]. However, they do not recover the heat equation and they retain the same unnatural technical assumptions made in [**BGL3**] on the collision kernel. There were two key new ingredients in their work. First, they were able to control the time regularity of the acoustic modes adapting techniques developed in [**LM1**]. Second, they were able to prove that the contribution of the acoustic modes to the limiting motion equation is just an extra gradiant term that can be incorporated into the pressure term.

Lions and Masmoudi [**LM4**] recover the Stokes motion equation without the local conservation of momentum assumption and with essentially optimal scaling. However, they do not recover the heat equation and they retain the same unnatural technical assumptions made in [**BGL3**] on the collision kernel. There are two reasons they do not recover the heat equation. First, it is unknown whether or not DiPerna-Lions solutions satisfy a local energy conservation law. Second, even if local energy conservation were assumed, the techniques they used to control the momentum flux would fail to control the heat flux.

Golse and Levermore [**GL**] recover the acoustic and Stokes systems. They make natural assumptions on the collision kernel that include those classically derived from hard potentials. For the Stokes limit they recover both the motion and heat equations with a near optimal scaling. For the acoustic limit the scaling they used was not optimal, essentially requiring

$$\frac{\delta_\epsilon}{\epsilon^{\frac{1}{2}}} \to 0 \qquad \text{rather than} \qquad \delta_\epsilon \to 0\,.$$

There were two key new ingredients in this work. First, they control the local momentum and energy conservation defects of the DiPerna-Lions solutions with dissipation rate estimates that allowed them to recover these local conservation laws in the limit. Second, they also control the heat flux with dissipation rate estimates. Because they treat the linear Stokes case, they do not face the need either to control the acoustic modes or for a compactness assumption, both of which are used to pass to the limit in the nonlinear terms in [**LM3**].

Without making any nonlinear compactness hypothesis, Saint Raymond [**StR2**] recovers the Navier-Stokes motion equation for the BGK model. This was a fundamental advance, but it took some time to extract the essential ingredients in a way that would impact the Boltzmann erquation. The flavor of her result is that every appropriately scaled family of BGK solutions has fluctuations that are compact and that every limit point of these fluctuations is an infinitesimal Maxwellian governed by the Navier-Stokes motion equation.

Without making any nonlinear compactness hypothesis, Golse-Saint Raymond [**GStR1, GStR3**] recover the Navier-Stokes system for the Boltzmann equation with Grad-cutoff collision kernels that arise from Maxwell potentials. Their major breakthrough was the development of a new L^1 averaging lemma to prove the compactness assumption [**GStR2**]. This was extracted from Saint Raymond [**StR2**] where she recovered the Navier-Stokes limit for the BGK model. Their proof also employs key elements from [**LM3**] and [**GL**].

Recent work of Levermore and Masmoudi extends the work of Golse and Saint Raymond [**LM**]. It recovers the Navier-Stokes system for the Boltzmann equation with weakly cutoff collision kernels that arise from a wide range of hard and soft potentials. Using the L^1 averaging lemma of Golse-Saint Raymond, they show that this nonlinear compactness hypothesis is satisfied for soft potentials. New estimates allow one to extend the analysis beyond Grad cutoff collision kernels.

These new estimates also allow one to carry out the acoustic and Stokes limits for soft potentials.

Saint Raymond has also established the incompressible Euler limit for so long as the classical solution of the incompressible Euler system exists [**StR4**], thereby completing a partial result in [**LM4**].

A fluid dynamical system that formally includes both the acoustic and the Stokes limits is the so-called compressible Stokes system

$$\begin{aligned} \partial_t \rho_\epsilon + \nabla_x \cdot u_\epsilon &= 0\,, \\ \partial_t u_\epsilon + \nabla_x(\rho_\epsilon + \theta_\epsilon) &= \epsilon\,\nu \nabla_x \cdot \left[\nabla_x u_\epsilon + (\nabla_x u_\epsilon)^T - \tfrac{2}{D}\nabla_x \cdot u_\epsilon I\right], \\ \tfrac{D}{2}\partial_t \theta_\epsilon + \nabla_x \cdot u_\epsilon &= \epsilon\,\kappa \Delta_x \theta_\epsilon\,. \end{aligned}$$

A relative entropy method was used in [**JL1**] to show (assuming a local energy conservation law) that over time scales on the order of $1/\epsilon$ one has

$$g_\epsilon \sim \rho_\epsilon + v \cdot u_\epsilon + \left(\tfrac{1}{2}|v|^2 - \tfrac{D}{2}\right)\theta_\epsilon\,,$$

where ρ_ϵ, u_ϵ, and θ_ϵ solve the weakly compressible Stokes system. Recall that ϵ appears in that system. The key point here is that the convergence is strong. Earlier works on the incompressible Stokes scaling obtained strong convergence only for "well-prepared" initial data — that is, for initial data with no acoustic modes in the limit.

A fluid dynamical system that formally includes both the acoustic and the Navier-Stokes limits is much harder to write down. Such a system can be derived from the Boltzmann equation. The dynamics decomposes into a part that satisfies the incompressible Navier-Stokes system, plus an acoustic part that satisfies a non-local quadratic equation that couples to the incompressible component. In [**JL2**] it is shown that this weakly compressible Navier-Stokes system has a global weak solution in L^2. This result includes the Leray theory, so cannot be improved easily. As with the Leray theory, the key to this result is an "energy" dissipation estimate. Indeed, this global existence result is very general. The acoustic part is unique for a given incompressible component.

The general setting for their global existence result is as follows. Let $U \mapsto H(U)$ be a strictly convex entropy for the system

$$\partial_t U + \nabla_x \cdot F(U) = \epsilon \nabla_x \cdot \left[D(U)\nabla_x H_U(U)\right].$$

This means that there exist $J(U)$ such that (Friedrichs-Lax)

$$H_U(U) F_U(U) = J_U(U)\,,$$

and that

$$\nabla_x H_U(U) \cdot D(U) \nabla_x H_U(U) \geq 0\,.$$

Hence, one has the local dissipation law

$$\begin{aligned} \partial_t H(U) + \nabla_x \cdot J(U) = {}& \epsilon\,\nabla_x \cdot \left[H_U(U)\,D(U)\nabla_x H_U(U)\right] \\ & - \epsilon\,\nabla_x H_U(U) \cdot D(U) \nabla_x H_U(U)\,. \end{aligned}$$

The weakly nonlinear approximation of the solution U_ϵ to this system near a constant solution U_o is $U_\epsilon = U_o + \epsilon \widetilde{U}_\epsilon$ where $\widetilde{U}_\epsilon$ satisfies

$$\partial_t \widetilde{U}_\epsilon + A \cdot \nabla_x \widetilde{U}_\epsilon + \epsilon \nabla_x \cdot \overline{Q}(\widetilde{U}_\epsilon, \widetilde{U}_\epsilon) = \epsilon \nabla_x \cdot \left[\overline{D} \nabla_x \widetilde{U}_\epsilon\right],$$

where $A = F_U(U_o)$,

$$\overline{Q}(V,V) = \lim_{T\to\infty} \frac{1}{T} \int_0^T e^{tA\cdot\nabla_x} F_{UU}(U_o)(e^{-tA\cdot\nabla_x}V, e^{-tA\cdot\nabla_x}V)\,\mathrm{d}t$$
$$\overline{D} = \lim_{T\to\infty} \frac{1}{T} \int_0^T e^{tA\cdot\nabla_x} D(U_o) e^{-tA\cdot\nabla_x}\,\mathrm{d}t\,.$$

This has a quadratic entropy dissipation, and that under mild assumptions (satisfied by the weakly compressible Navier-Stokes approximation), has global solutions [**JL2**].

6. Some Open Problems

Much remains to be done in this program. I close by listing just a few open problems.

- The acoustic limit with optimal scaling ($\delta_\epsilon \to 0$): This problem is an obstruction to establishing the compressible Euler limit, yet might be easier than that limit because the acoustic system is a much tamer target.
- Uniform in time results: Can one remove the local conservation of energy assumption made by Jiang-L for the weakly compressible Stokes approximations? Can one establish the weakly compressible Navier-Stokes approximation?
- Any result for initial data with finite mass, energy, and moment of inertia over the whole space: Formally, every such initial data has a self-similar spreading Maxwellian associated with it. It seems reasonable to work out theories when the kinetic density is close to the associated spreading Maxwellian in the sense of relative entropy. The original DiPerna-Lions theory covers this case.
- Any limit for non-cutoff collision kernels: It is known that the initial-value problem for the spatially homogeneous Boltzmann equation has better regularity for non-cutoff kernels than for cutoff kernels. However, this better regularity has not yet led to an extension of DiPerna-Lions theory for non-cutoff kernels.
- Dominant-balance Stokes, Navier-Stokes, and Euler limits: These limits seem harder than the ones established so far.
- Any results for "beyond Navier-Stokes" approximations.

References

[BGL1] C. Bardos, F. Golse, and D. Levermore, *Sur les limites asymptotiques de la théorie cinétique conduisant à la dynamique des fluides incompressibles*, C.R. Acad. Sci. Paris Sr. I Math. **309** (1989), 727–732.

[BGL2] C. Bardos, F. Golse, and D. Levermore, *Fluid Dynamic Limits of Kinetic Equations I: Formal Derivations*, J. Stat. Phys. **63** (1991), 323–344.

[BGL3] C. Bardos, F. Golse, and C.D. Levermore, *Fluid Dynamic Limits of Kinetic Equations II: Convergence Proofs for the Boltzmann Equation*, Commun. Pure & Appl. Math. **46** (1993), 667–753.

[BGL4] C. Bardos, F. Golse, and C.D. Levermore, *Acoustic and Stokes Limits for the Boltzmann Equation*, C.R. Acad. Sci. Paris **327** (1999), 323–328.

[BGL5] C. Bardos, F. Golse, and C.D. Levermore, *The Acoustic Limit for the Boltzmann Equation*, Archive Rat. Mech. & Anal. **153** (2000), 177–204.

[BL] C. Bardos and C.D. Levermore, *Kinetic Equations and an Incompressible Fluid Dynamical Limit that Recovers Viscous Heating*, (in preparation 2007).

[BU] C. Bardos and S. Ukai, *The Classical Incompressible Navier-Stokes Limit of the Boltzmann Equation*, Math. Models & Meth Appl. Sci. **1** (1991), 235–257.

[BLP] B.J. Bayly, C.D. Levermore, and T. Passot, *Density Variations in Weakly Compressible Fluid Flow*, Phys. Fluids A **4** (1992), 945–954.

[Bob] A.V. Bobylev, *Quasistationary Hydrodynamics for the Boltzmann Equation*, J. Stat. Phys. **80** (1995), 1063–1083.

[Bolt] L. Boltzmann, *Weitere Studien über das Wärmegleichgewicht unter Gasmolekülen*, Sitzungs. Akad. Wiss. Wein **66** (1872), 275–370; English: *Further Studies on the Thermal Equilibrium of Gas Molecules*, in *Kinetic Theory* **2**, S.G. Brush (ed.), Pergamon Press, London, 1966, 88–174.

[Br1] S.G. Brush, *Kinetic Theory* **1**: *The Nature of Gases and of Heat*, Pergamon Press, London, 1965.

[Br2] S.G. Brush, *Kinetic Theory* **2**: *Irreversible Processes*, Pergamon Press, London, 1966.

[Br3] S.G. Brush, *Kinetic Theory* **3**: *The Chapmann-Enskog Solution of the Transport Equation for Moderately Dense Gases*, Pergamon Press, London, 1972.

[Caf1] R. Caflisch, *The Boltzmann Equation with a Soft Potential I: Linear, Spatially Homogeneous*, Commun. Math. Phys. **74** (1980), 71–95.

[Caf2] R.E. Caflisch, *The Fluid Dynamic Limit of the Nonlinear Boltzmann Equation*, Commun. Pure & Appl. Math. **33** (1980), 651–666.

[Cer] C. Cercignani, *The Boltzmann Equation and its Applications*, Springer-Verlag, New York, 1988.

[CIP] C. Cercignani, R. Illner, and M. Pulvirenti, *The Mathematical Theory of Dilute Gases*, Appl. Math. Sci. **106**, Springer-Verlag, New York, 1994.

[CF] P. Constantin and C. Foias, *Navier-Stokes Equations*, Chicago Lectures in Mathematics, The University of Chicago Press, Chicago, 1988.

[DMEL] A. DeMasi, R. Esposito, and J. Lebowitz, *Incompressible Navier-Stokes and Euler Limits of the Boltzmann Equation*, Commun. Pure & Appl. Math. **42** (1990), 1189–1214.

[DiPL] R.J. DiPerna and P.-L. Lions, *On the Cauchy Problem for the Boltzmann Equation: Global Existence and Weak Stability Results*, Annals of Math. **130** (1990), 321–366.

[Ensk] D. Enskog, *Kinetische Theorie der Vorgänge in mässig verdünnten Gasen*, I. Allgemeiner Teil, Almqvist & Wiksell, Uppsala, 1917; English: *Kinetic Theory of Processes in Dilute Gases*, in *Kinetic Theory* **3**; S.G. Brush ed., Pergamon Press, Oxford, 1972, 125–225.

[Eul] L. Euler, *Principes généraux du mouvement des fluides*, Mémoires de l'académie des sciences de Berlin **11** (1757), 274–315. Reprinted in *Erleri Opera Omnia II* **12** *Commentations Mechanicae*, C.A. Trusedell ed., Societatis Scientiarum Naturalium Helveticae, Lausanne 1954, 54–91.

[Gla] R. Glassey, *The Cauchy Problem in Kinetic Theory*, Society for Industrial and Applied Mathematics (SIAM), Philadelphia, 1996.

[Gol] F. Golse, *From Kinetic to Macroscopic Models*, in *Kinetic Equations and Asymptotic Theory*, B. Perthame and L. Desvillettes eds., Series in Applied Mathematics **4**, Gauthier-Villars, Paris, 2000, 41–126.

[GL] F. Golse and C.D. Levermore, *Stokes-Fourier and Acoustic Limits for the Boltsmann Equation: Convergence Proofs*, Commun. Pure & Appl. Math. (submitted 2001).

[GLPS] F. Golse, P.-L. Lions, B. Perthame, R. Sentis, *Regularity of the Moments of the Solution of a Transport Equation*, J. Funct. Anal. **76** (1988), 110–125.

[GPS] F. Golse, B. Perthame, R. Sentis, *Un résultat de compacité pour les équations de transport et application au calcul de la limite de la valeur propre principale de l'opérateur de transport*, C.R. Acad. Sci. Paris Sr. I Math. **301** (1985), 341–344.

[GP] F. Golse and F. Poupaud, *Un rsultat de compacit pour l'quation de Boltzmann avec potentiel mou. Application au problme de demi-espace*, C. R. Acad. Sci. Paris Sr. I Math. **303** (1986), 583–586.

[GStR1] F. Golse and L. Saint-Raymond, *The Navier-Stokes Limit for the Boltzmann Equation*, C. R. Acad. Sci. Paris Sr. I Math. **333** (2001), 897–902.

[GStR2] F. Golse and L. Saint-Raymond, *Velocity Averaging in L^1 for the Transport Equation*, C. R. Acad. Sci. Paris Sr. I Math. **334** (2002), 557–562.

[GStR3] F. Golse and L. Saint-Raymond, *The Navier-Stokes Limit of the Boltzmann Equation for Bounded Collision Kernels*, Invent. Math. **155** (2004), 81–161.

[GStR4] F. Golse and L. Saint-Raymond, *The Navier-Stokes Limit of the Boltzmann Equation for Hard Sphere Collision Kernels*, (talk, April 2004).

[Grad] H. Grad, *Principles of the Kinetic Theory of Gases*, in *Handbuch der Physik* **12**, S. Flügge ed., Springer-Verlag, Berlin, 1958, 205–294.

[Grad2] H. Grad, *Asymptotic Equivalence of the Navier-Stokes and Nonlinear Boltzmann Equations*, Proc. Amer. Math. Symp. on Applications of Partial Differential Equations. (1964), 154–183.

[Guo] Y. Guo, *Boltzmann Diffusive Limit Beyond the Navier-Stokes Approximation*, Cummun. Pure & Appl. Math. **58** (2005), 1–62.

[Hil] D. Hilbert, *Begründung der kinetischen Gastheorie*, Math. Annalen **72** (1912), 562–577; English: *Foundations of the Kinetic Theory of Gases*, in *Kinetic Theory* **3**; S.G. Brush (ed.), Pergamon Press, Oxford, 1972, 89–101.

[JL1] N. Jiang and C.D. Levermore, *Weakly Compressible Stokes Dynamics of the Boltzmann Equation* Prepint 2007.

[JL2] N. Jiang and C.D. Levermore, *Global Weak Solutions to the Weakly Nonlinear Approximation of Hyperbolic-Parabolic Systems with Entropy* Preprint 2007.

[KMN] S. Kawashima, A. Matsumura, and T. Nishida, *On the Fluid Dynamical Approximation to the Boltzmann Equation at the Level of the Navier-Stokes Equation*, Commun. Math. Phys. **70** (1979), 97–124.

[Lan] O.E. Landford, *Time Evolution of Large Classical Systems*, Lecture Notes in Physics **38**, J. Moser ed., Springer-Verlag, Berlin, 1975, 1–111.

[Lry] J. Leray, *Sur le mouvement d'un fluide visqueux emplissant l'espace*, Acta Math. **63** (1934), 193–248.

[LM] C.D. Levermore and N. Masmousdi, *From the Boltzmann Equation to an Incompressible Navier-Stokes-Fourier System*, Preprint 2007.

[PLL1] P.-L. Lions, *Compactness in Boltzmann's Equation via Fourier Integral Operators and Applications, I, II, & III*. J. Math. Kyoto Univ. **34** (1994), 391–427, 429–461, 539–584.

[LM1] P.-L. Lions and N. Masmoudi, *Incompressible limit for a viscous compressible fluid*, J. Math. Pures Appl. **77** (1998), 585–627.

[LM2] P.-L. Lions and N. Masmoudi, *Une approche locale de la limite incompressible*. C. R. Acad. Sci. Paris Sr. I Math. **329** (1999), 387–392.

[LM3] P.-L. Lions and N. Masmoudi, *From the Boltzmann Equations to the Equations of Incompressible Fluid Mechanics, I*, Archive Rat. Mech. & Anal. **158** (2001), 173–193.

[LM4] P.-L. Lions and N. Masmoudi, *From the Boltzmann Equations to the Equations of Incompressible Fluid Mechanics, II*, Archive Rat. Mech. & Anal. **158** (2001), 195–211.

[Max1] J.C. Maxwell, *Illustrations of the Dynamical Theory of Gases*, Phil. Mag. **19** (1860), 19–32, **20** (1860) 21–37; Reprinted in *The Scientific Letters and Papers of James Clerk Maxwell*, Vol. 1, Dover, New York, 1965, 377–409.

[Max2] J.C. Maxwell, *On the Dynamical Theory of Gases*, Philos. Trans. Roy. Soc. London Ser. A **157** (1867), 49–88; Reprinted in *The Scientific Letters and Papers of James Clerk Maxwell*, Vol. 2, Dover, New York, 1965, 26–78.

[Nsh] T. Nishida, *Fluid Dynamical Limit of the Nonlinear Boltzmann Equation to the Level of the Incompressible Euler Equation*, Commun. Math. Phys., **61** (1978), 119–148.

[StR2] Saint-Raymond, *Du modèl BGK de l'équation de Boltzmann aux équations d' Euler des fluides incompressibles*, Bull. Sci. Math. **126** (2002), 493–506.

[StR3] Saint-Raymond, *From the Boltzmann BGK Equation to the Navier-Stokes System*, Ann. Scient. École Norm. Sup. **36** (2003), 271–317.

[StR4] Saint-Raymond, *Convergence of Solutions to the Boltzmann Equation in the Incompressible Euler Limit*, Arch. Rat. Mech, & Anal. **166** (2003), 47–80.

[Son1] Y. Sone, *Kinetic Theory and Fluid Dynamics*, Birkhäuser, Boston, 2002.

[Son2] Y. Sone, *Molecular Gas Dynamics: Theory, Techniques, and Applications*, Birkhäuser, Boston 2007.

[LS] C.D. Levermore and W. Sun, *Fredholm Alternatives for Linearized Boltzmann Collision Operators with Weakly Cutoff Kernels*, (preprint 2007).

[Uk] S. Ukai, *The Incompressible Limit and the Initial Layer of the Compressible Euler Equation*, J. Math. Kyoto Univ. **26** (1986) 323–331.

[Vil] C. Villani, *Limites hydrodynamiques de l'équation de Boltzmann [d'après C. Bardos, F. Golse, D. Levermore, P.-L. Lions, N. Masmoudi, L. Saint-Raymond]*, Séminaire Bourbaki **2000-2001**, exposition 893.

(C.D. Levermore) Department of Mathematics, *and* Institute for Physical Science and Technology, University of Maryland, College Park, MD 20742-4015, USA
E-mail address: `lvrmr@math.umd.edu`

Proceedings of Symposia in Applied Mathematics
Volume **65**, 2007

From the Boltzmann equation to the incompressible Navier-Stokes equations

François Golse

Dedicated to P. Lax and L. Nirenberg

ABSTRACT. This paper surveys the derivation of Leray solutions of the incompressible Navier-Stokes equations from the DiPerna-Lions theory of renormalized solutions of the Boltzmann equation, following the detailed proof given in [F. Golse & L. Saint-Raymond, Invent. Math. **155** (2004), 81–161] in the case of cutoff Maxwell molecules.

1. Introduction

A classical theme in mathematical physics is the derivation of models describing the state of matter at the macroscopic scale from information on the interaction of its elementary constituents at the microscopic scale. For instance, the viscosity of a fluid can be measured experimentally; on the other hand, in the case where the fluid is a gas, its viscosity can also be computed by the methods of statistical mechanics.

In addition to the fundamental issue recalled above, understanding the transition from the microscopic to the macroscopic description of matter can be of considerable help in designing hierarchies of models in the context of engineering and technology. In the case of gas dynamics, the interaction of an immersed body with the surrounding fluid leads to remarkable effects, some of which are explained by the kinetic theory of gases. A famous example is that of thermal transpiration: consider a pipe filled with gas, with a temperature gradient along its surface. Then, a gas flow is induced in the direction of the gradient. Thermal transpiration was explained by Maxwell in 1879 in his last paper on the kinetic theory of gases, with further contributions by Knudsen in 1910. Modern applications of such phenomena are numerous: see for instance the discussion about Knudsen compressors — how to make a pumping system without mechanical part — in the two recent books by Y. Sone [**55**] and [**56**].

1991 *Mathematics Subject Classification.* Primary 35Q35, 35Q30; Secondary 82C40.
Key words and phrases. Kinetic theory of gases, Statistical Mechanics, Fluid mechanics, Boltzmann equation, Navier-Stokes equations.

In the present paper, we survey recent progress on the derivation of the Navier-Stokes equations for incompressible flows from the Boltzmann equation.

Deriving the macroscopic equations of fluid mechanics from the kinetic theory of gases is an old problem, which already appears somehow in Maxwell's 1866 paper on the kinetic theory of gases. Perhaps the first mathematician to consider this problem as one of "pure" mathematics was D. Hilbert — see the statement of his 6th problem in [**35**], as well as his own contribution to a solution in [**36**].

Hilbert's ideas in [**36**] have deeply influenced the field, and are still in use today. In [**15**], R. Caflisch completed Hilbert's argument in [**36**] and proved that the Euler system of gas dynamics can be derived from the Boltzmann equation on short time intervals, before shock waves appear. A major drawback in Hilbert's method is that it applies only to smooth solutions of both the Boltzmann equation and its hydrodynamic limits. In most fluid models, smooth solutions are not known to exist for all times and initial data of arbitrary size.

The scope of the present paper is different: in order to derive fluid models from the Boltzmann equation for all positive times and without restriction on the size of the initial data, we give up the smoothness requirements in Hilbert's method and use instead compactness arguments.

So far, this approach has been successfully applied to the derivation of the acoustic system ([**7**], [**26**]), of the time-dependent Stokes ([**46**]) and Stokes-Fourier equations ([**26**]), and of the Navier-Stokes-(Fourier) system ([**30**], [**31**]) from the Boltzmann equation.

This last contribution connects the Leray theory of "turbulent solutions" of the incompressible Navier-Stokes equations in 3 space dimensions to the DiPerna-Lions theory of renormalized solutions of the Boltzmann equation. To this date, both theories are the only ones giving global existence of solutions of either the Navier-Stokes or the Boltzmann equation without limitation on the size of the initial data.

Unfortunately, the discussion in [**30**] and [**31**] involves some amount of technique. The present paper surveys the results and methods in those works, while trying to avoid as much as possible the technicalities in the complete proof.

Beyond their essential contributions to some of the topics considered here, P. Lax and L. Nirenberg are at the origin of several beautiful ideas that have shaped modern analysis. They have been a source of inspiration, not only to their own students and collaborators, but also to many mathematicians around the world. I am offering them this modest contribution with great pleasure.

2. The Boltzmann equation: formal structure

The Maxwell-Boltzmann kinetic theory of gases is a statistical theory in the single-molecule phase space. It is a description of matter that is intermediate between molecular dynamics, i.e. the N-body problem of classical mechanics for the system of all the molecules in the amount of gas considered, and the macroscopic theory of gas dynamics, following the general equations for fluids established first by Euler, and later by Navier and Stokes.

Strictly speaking, the kinetic theory of gases is not a first principle of physics, but rather some subtle limit of molecular dynamics, whose precise formulation became clear only after the fundamental work of H. Grad [**32**], and has been known ever since as *the Boltzmann-Grad limit.* (Grad's argument was made even more clear in the case of hard spheres, by C. Cercignani, who proposed in [**18**] a lucid, formal derivation of the Boltzmann equation from molecular dynamics.) The validity of the Boltzmann-Grad limit was rigorously established for the first time by O. Lanford [**41**] in 1975, on very short time intervals and in the case of a hard sphere gas. Lanford's argument was later proved to hold for all times in the case of a very rarefied gas of hard spheres expanding in the vacuum, by R. Illner and M. Pulvirenti [**38**].

In the Maxwell-Boltzmann kinetic theory of gases, the state of the gas at time t is described by the single-molecule phase-space number density $F \equiv F(t,x,v) \geq 0$. In other words, $F(t,x,v)dxdv$ is the average number of gas molecules to be found at time t in any infinitesimal phase-space volume $dxdv$ around the point (x,v), where $x \in \mathbf{R}^3$ designates the position and $v \in \mathbf{R}^3$ the velocity of such molecules.

In the absence of external forces (such as the electromagnetic force in the case of ionized gases, or gravity) the number density F satisfies the Boltzmann equation

$$\partial_t F + v \cdot \nabla_x F = \mathcal{C}(F) ,$$

where $\mathcal{C}(F)$ is a rather intricate nonlocal expression, usually called *the Boltzmann collision integral.*

Before giving the detailed expression of the collision integral, we want to stress some important features thereof. First, all collisions other than binary are neglected in Boltzmann's equation; besides, these collisions are viewed as instantaneous and purely local events. This results from the Grad's scaling assumption, according to which the molecular radius r vanishes in the limit leading to the Boltzmann equation. Specifically, Grad's scaling consists in considering a gas of a large number N of molecules viewed as hard spheres of radius r, with N and r such that

$$N \to +\infty \text{ and } r \to 0 , \qquad \text{with } Nr^2 \to \text{finite, positive quantity.}$$

Since $\mathcal{C}(F)$ only accounts for purely local and instanteneous binary collisions, one anticipates that

$$\mathcal{C} \text{ is a bilinear operator acting only on the } v \text{ variable in } F .$$

Hence we define the collision integral for all rapidly decaying functions $f \equiv f(v)$, by

$$\mathcal{C}(f)(v) = \iint_{\mathbf{R}^3 \times \mathbf{S}^2} (f(v')f(v'_*) - f(v)f(v_*))|v - v_*| dv_* d\sigma ,$$

where the velocities v' and v'_* are defined in terms of $v, v_* \in \mathbf{R}^3$ and $\sigma \in \mathbf{S}^2$ by the formulas

$$v' \equiv v'(v, v_*, \sigma) = \tfrac{1}{2}(v + v_*) + \tfrac{1}{2}|v - v_*|\sigma ,$$
$$v'_* \equiv v'_*(v, v_*, \sigma) = \tfrac{1}{2}(v + v_*) - \tfrac{1}{2}|v - v_*|\sigma .$$

As σ runs through the unit sphere $\mathbf{S}^2$, the velocity pair (v', v'_*) runs through the set of all possible pre-collision velocities for a pair of point particles with equal masses whose post-collision velocities are (v, v_*). The geometrical meaning of the unit vector σ is particularly obvious in the center of mass reference frame: σ is the direction of the pre-collision relative velocity of the pair of point particles (see figure 1).

For a space-time dependent number density $F \equiv F(t,x,v)$, the collision integral $\mathcal{C}(F)$ designates the function

$$\mathcal{C}(F)(t,x,v) := \mathcal{C}(F(t,x,\cdot))(v) .$$

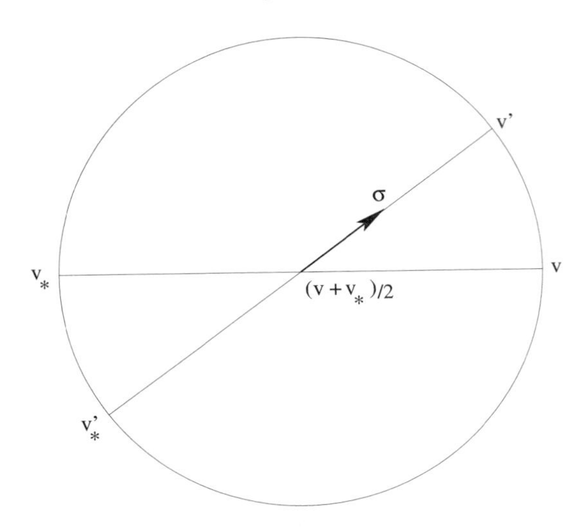

FIGURE 1. Geometry of collisions in the center of mass reference frame

The mathematical properties of the Boltzmann collision integral have been extensively studied in the literature (most notably the spectral properties of its linearization at local Maxwellian equilibria, see below). However, not much of this material is needed our discussion of the Navier-Stokes limit. Instead, we mostly use the following fundamental features of this collision integral, which are consequences of the symmetries in the collision integral.

Local conservation laws. The first such feature is the set of space-time conservation laws that are consequences of the symmetries in the Boltzmann collision integral.

THEOREM 2.1 (Local conservation laws). *Assume that $f \equiv f(v)$ is a rapidly decaying measurable function; then*

$$\int_{\mathbf{R}^3} \mathcal{C}(F) \begin{pmatrix} 1 \\ v_k \\ |v|^2 \end{pmatrix} dv = 0\,, \qquad k = 1, 2, 3\,.$$

In particular, any weak solution F of the Boltzmann equation that is rapidly decaying in v satisfies the following local conservation laws

$$\begin{aligned}
&\partial_t \int_{\mathbf{R}^3} F dv + \operatorname{div}_x \int_{\mathbf{R}^3} F v dv = 0\,, && \text{(mass)} \\
&\partial_t \int_{\mathbf{R}^3} F v dv + \operatorname{div}_x \int_{\mathbf{R}^3} F v \otimes v dv = 0\,, && \text{(momentum)} \\
&\partial_t \int_{\mathbf{R}^3} F \tfrac{1}{2}|v|^2 dv + \operatorname{div}_x \int_{\mathbf{R}^3} F v \tfrac{1}{2}|v|^2 dv = 0\,, && \text{(energy)}
\end{aligned} \tag{2.1}$$

These relations are obviously similar to Euler's system of space-time conservation laws for compressible fluids, with however one considerable difference. Indeed, the relations above hold for *any* classical solution of the Boltzmann equation, so that one cannot expect in general that the flux terms

$$\int_{\mathbf{R}^3} Fv\otimes v dv\,, \quad \text{(momentum flux)} \quad \int_{\mathbf{R}^3} Fv\tfrac{1}{2}|v|^2 dv\,, \qquad \text{(energy flux)}$$

be expressed as functions of the conserved densities

$$\int_{\mathbf{R}^3} F dv\,, \quad \text{(mass density)} \quad \int_{\mathbf{R}^3} Fv dv\,, \quad \text{(momentum density)}$$

$$\int_{\mathbf{R}^3} F\tfrac{1}{2}|v|^2 dv\,, \quad \text{(kinetic energy density)}$$

In other words, the system of space-time conservation laws above is not in closed form. This should not be surprising since these conservation laws hold for each solution of the Boltzmann that is rapidly decaying at infinity. Since the same system of conservation laws above holds in all possible flow regimes for the Boltzmann equation that can be described with the help of fluid mechanics models, one cannot expect that this system of conservation laws exactly coincide with one particular choice of these models. However, this system of conservation laws, even though not in closed form, is one of the most important features of the Boltzmann collision integral, especially in the context of hydrodynamic limits.

In any case, the discussion above suggests to consider the notion of *macroscopic observable*. These are quantities defined as averages of the number density in the velocity variable, of the form

$$\rho_\phi(t,x) := \int_{\mathbf{R}^3} \phi(v)F(t,x,v)dv\,.$$

Typically, in order to compute the local density of some mechanical quantity at the hydrodynamic level — say, for instance, the energy — one sets $\phi(v)$ to be that quantity for a single molecule — in the case of the kinetic energy for a molecule of mass 1, one has $\phi(v) = \frac{1}{2}|v|^2$, so that the macroscopic energy density is indeed

$$\int_{\mathbf{R}^3} \tfrac{1}{2}|v|^2 F(t,x,v)dv\,.$$

All the conserved quantities considered above, or fluxes thereof are examples of macroscopic observables.

That the quantities considered in all hydrodynamic limits of kinetic models are macroscopic observables plays an important role in the mathematical justification of these limits.

The entropy and Boltzmann's H theorem. Although the Boltzmann equation is a description of matter in terms of molecules, it is not equivalent to molecular dynamics which is a mechanically reversible[1] system. By contrast, the Boltzmann equation is irreversible in the sense that it possess a Lyapunov function that is precisely minus the entropy of the gas. This Lyapunov function is usually called *Boltzmann's H-function*, and its evolution is governed by Boltzmann's famous H theorem recalled below.

[1]The precise meaning of mechanical reversibility being as follows: starting with any initial configuration of a system of N molecules — i.e. the string of positions and velocities of the system of molecules at time $t = 0$ — let the system evolve until times T. At time T, change the velocity of each molecule into its opposite while leaving its position unchanged, and let the system evolve as before; then, at time $2T$, all the molecules are back at their original positions.

THEOREM 2.2 (Boltzmann's H theorem). *Assume that $f \equiv f(v) > 0$ a.e. is a rapidly decaying measurable function such that $\ln f$ has polynomial growth at infinity. Then, the local entropy production rate*

$$R(f) = -\int_{\mathbf{R}^3} \mathcal{C}(f) \ln f dv \geq 0 \,.$$

Besides, the following conditions are equivalent:

$$R(f) = 0 \;\Leftrightarrow\; \mathcal{C}(f) = 0 \text{ a.e. } \;\Leftrightarrow f \text{ is a Maxwellian}$$

i.e. there exists $\rho, \theta > 0$ and $u \in \mathbf{R}^3$ such that

$$f(v) = \mathcal{M}_{(\rho,u,\theta)}(v) := \frac{\rho}{(2\pi\theta)^{3/2}} e^{-\frac{|v-u|^2}{2\theta}} \qquad \text{a.e. in } v \in \mathbf{R}^3 \,.$$

This theorem has the following interesting consequence: assume that F is a solution of the Boltzmann equation such that $F(t,x,\cdot)$ is a measurable, a.e. positive, rapidly decaying function. Then, upon multiplying each side of the Boltzmann equation by $\ln F$, and then integrating in the v variable, one finds that

$$\partial_t \int_{\mathbf{R}^3} F \ln F dv + \operatorname{div}_x \int_{\mathbf{R}^3} vF \ln F dv = -R(F) \leq 0 \,. \tag{2.2}$$

When considered jointly with the system of local conservation laws (2.1), this differential inequality is formally analogous to the notion of "convex extensions" to a system of conservation laws proposed by K. Friedrichs and P. Lax in [**24**] and [**42**]. And indeed, if F is assumed to be a *local Maxwellian*, i.e. if

$$F(t,x,v) = \mathcal{M}_{(\rho(t,x),u(t,x),\theta(t,x))}(v)$$

then the local conservation laws (2.1) coincide with the Euler system of gas dynamics

$$\begin{aligned} \partial_t \rho + \operatorname{div}_x(\rho u) &= 0 \,, \\ \partial_t(\rho u) + \operatorname{div}_x(\rho u \otimes u) + \nabla_x(\rho\theta) &= 0 \,, \\ \partial_t \left(\rho \left(\tfrac{1}{2}|u|^2 + \tfrac{3}{2}\theta\right)\right) + \operatorname{div}_x \left(\rho u \left(\tfrac{1}{2}|u|^2 + \tfrac{5}{2}\theta\right)\right) &= 0 \,, \end{aligned}$$

while the differential inequality (2.2) reduces to

$$\partial_t \left(\rho \ln\left(\frac{\rho}{\theta^{3/2}}\right)\right) + \operatorname{div}_x \left(\rho \ln\left(\frac{\rho}{\theta^{3/2}}\right)\right) \leq 0 \tag{2.3}$$

That this differential inequality is a convex extension of the Euler system of gas dynamics is a well-known fact, and the prototype of this notion of convex extension in the theory of hyperbolic systems of conservation laws. The quantity

$$-\int_{\mathbf{R}^3} F \ln F(t,x,v) dv$$

represents the entropy density at time t and at the position x in the gas. Accordingly, differential inequalities such as (2.2) or (2.3) are referred to as "entropy inequalities" in the context of systems of conservation laws.

The few remarks above already make a connection between the Boltzmann equation and one important model in fluid mechanics, namely the Euler system of gas dynamics. This is however far from satisfying, since this connection is based upon assuming that the number density F is a local Maxwellian. Unfortunately, except in very special cases, local Maxwellians are usually not *exact* solutions of the Boltzmann equation.

Going back to (2.2), one can further integrate in x, assuming that F vanishes as $|x| \to +\infty$ uniformly in v. After integration, the divergence term on the left-hand side of (2.2) disappears and one finds

$$\frac{d}{dt}H(F) = -R(F)$$

where $H(F)$ is Boltzmann's H-function

$$H(F)(t) = \iint_{\mathbf{R}^3 \times \mathbf{R}^3} F \ln F(t,x,v)dv\,.$$

Hence $H(F)(t)$ decreases along solutions of the Boltzmann equation, unless $F(t,\cdot,\cdot)$ is a local Maxwellian distribution, i.e. is of the form

$$\mathcal{M}_{(\rho(t,\cdot),u(t,\cdot),\theta(t,\cdot))}(v)$$

where $\rho > 0$, $u \in \mathbf{R}^3$ and $\theta > 0$ are functions of t and x.

3. The incompressible Navier-Stokes limit of the Boltzmann equation

Weakly nonlinear kinetic regimes. Throughout this paper, we shall be concerned with flows that are weakly nonlinear at the kinetic level — in some sense to be made precise below — while not necessarily so at the hydrodynamic level.

Weakly nonlinear regimes at the kinetic level correspond with states that are perturbations of some uniform, global equilibrium. After rescaling and using the Galilean invariance of the Boltzmann equation, one can choose this global equilibrium to be

$$M = \mathcal{M}_{(1,0,1)} \qquad \text{(the centered, reduced Gaussian distribution)}$$

Next, one needs to measure the size of number density fluctuations around the equilibrium state M with a notion of distance that is well propagated by the dynamics of the Boltzmann equation.

In view of Boltzmann's H Theorem, the most natural way to do so is to use the notion of relative entropy.

Definition 3.1. The relative entropy of a measurable number density $F \equiv F(x,v) \geq 0$ a.e. with respect to some Maxwellian state $\mathcal{M}$ is

$$H(F|\mathcal{M}) = \iint_{\mathbf{R}^3 \times \mathbf{R}^3} \left[F \ln\left(\frac{F}{\mathcal{M}}\right) - F + \mathcal{M}\right] dxdv\,.$$

Notice that the integrand in the definition of the relative entropy above is unconditionally nonnegative, so that $H(F|\mathcal{M})$ is always defined as an element of $[0,+\infty]$.

Global existence theories for the Boltzmann equation. Although mathematicians have been interested in the Boltzmann equation for a long time — see for instance the work of D. Hilbert [**36**] or of T. Carleman [**16**] — the first global existence results on the Cauchy problem for that equation appeared relatively recently.

The first such result is due to S. Ukai [**58**], who showed the global existence and uniqueness of the solution of the Boltzmann equation for any initial data sufficiently close to some uniform Maxwellian equilibrium state.

Another class of global existence results is due to R. Illner and M. Shinbrot [**37**] who proved the global existence and uniqueness of solutions of the Boltzmann equation for any initial data dominated by some Gaussian in phase-space (i.e. in (x, v)) with small enough $L^1 \cap L^\infty$ norm. Physically, this regime corresponds with small perturbations of the vacuum. As such, it is not too useful in the context of hydrodynamic limits, for which perturbations of uniform, global equilibria are more natural.

Lastly, R. DiPerna and P.-L. Lions [**21**] came up with a powerful theory of global, weak solutions of the Boltzmann equation that have the virtue to exist for initial data of arbitrary size — in some sense explained below. It is this notion of solution of the Boltzmann equation that will be used in the present work — the reason for this choice being postponed to the end of the present section.

In our presentation of the DiPerna-Lions theory, we slightly deviate from the original framework in [**21**] and introduce a variant thereof due to P.-L. Lions [**45**].

Consider the Boltzmann equation in $\mathbf{R}^3$ with Maxwellian equilibrium M at infinity:

$$\begin{aligned} \partial_t F + v \cdot \nabla_x F &= \mathcal{C}(F)\,, & (t, x, v) &\in \mathbf{R}_+^* \times \mathbf{R}^3 \times \mathbf{R}^3 \\ F(t, x, v) &\to M\,, & \text{as } |x| &\to +\infty\,, \\ F\big|_{t=0} &= F^{in}\,. \end{aligned}$$

The convergence of F to M at infinity is assumed to hold in a very weak sense: what is meant is only that

$$H(F(t, \cdot, \cdot)|M) < +\infty \qquad \text{for all } t \geq 0\,.$$

Since the collision integral $\mathcal{C}(F)$ is quadratic and acts as a pointwise multiplication in the variables t and x, an L^2 control in the x variable seems to be needed in order to define $\mathcal{C}(F)$ as an L^1_{loc} function. Hence the main difficulty in defining a notion of weak solutions of the Boltzmann equation is that Boltzmann's H Theorem naturally provides only an $L \ln L$ control on the number density F, which is a priori not sufficient to define $\mathcal{C}(F)$ as a bona fide distribution.

In view of this difficulty, R. DiPerna and P.-L. Lions conceived the idea of writing the Boltzmann equation in the sense of distributions not for the number density itself, but for some nonlinear truncation thereof. In other words, while all the terms in the equation

$$(\partial_t + v \cdot \nabla_x) F = \mathcal{C}(F)$$

may not be well defined as distributions in (t, x, v), one can replace the equation above with

$$(\partial_t + v \cdot \nabla_x) \beta(F) = \beta'(F) \mathcal{C}(F)\,,$$

both equations being equivalent for classical C^1 solutions provided that $\beta'(F) \neq 0$ everywhere. DiPerna and Lions' original idea was to use this second form of the Boltzmann equation as the definition of their notion of weak solutions, for nonlinear truncations β such that $\beta'(F)\mathcal{C}(F)$ be locally integrable assuming only an $L \ln L$ bound on F.

More precisely, observe that for each $r > 0$, one has

$$\iint_{|x|+|v| \leq r} \frac{\mathcal{C}(F)}{\sqrt{1+F}} dv dx \leq C_r \int_{|x| \leq r} \left(R(F) + \int_{\mathbf{R}^3} (1 + |v|^2) F dv \right) dx\,.$$

This suggests the following

DEFINITION 3.2. A renormalized solution relative to M of the Boltzmann equation is a nonnegative $F \in C(\mathbf{R}_+, L^1_{loc}(\mathbf{R}^3 \times \mathbf{R}^3))$ such that

$$H(F(t)|M) < +\infty \quad \text{and} \quad \Gamma'\left(\frac{F}{M}\right)\mathcal{C}(F) \in L^1_{loc}(\mathbf{R}_+ \times \mathbf{R}^3 \times \mathbf{R}^3)$$

and that satisfies

$$M(\partial_t + v \cdot \nabla_x)\Gamma\left(\frac{F}{M}\right) = \Gamma'\left(\frac{F}{M}\right)\mathcal{C}(F)$$

in the sense of distributions on $\mathbf{R}_+^* \times \mathbf{R}^3 \times \mathbf{R}^3$, for each $0 \le \Gamma \in C^1(\mathbf{R}_+)$ such that

$$\Gamma'(Z) \le \frac{1}{\sqrt{1+Z}}, \qquad Z \ge 0\,.$$

With this definition, we can state the following global existence result

THEOREM 3.3 (R. DiPerna - P.-L. Lions). *For each measurable $F^{in} \ge 0$ a.e. such that*

$$H(F^{in}|M) < +\infty$$

there is a renormalized solution relative to M of the Boltzmann equation with initial data F^{in}.

This solution satisfies the continuity equation and the local conservation law of momentum up to the divergence of a Radon measure m with values in the cone of nonnegative symmetric matrices, i.e.

$$\begin{aligned}
&\partial_t \int_{\mathbf{R}^3} F dv + \operatorname{div}_x \int_{\mathbf{R}^3} vF dv = 0\,, \\
&\partial_t \int_{\mathbf{R}^3} vF dv + \operatorname{div}_x \int_{\mathbf{R}^3} v \otimes vF dv + \operatorname{div}_x m = 0\,,
\end{aligned} \tag{3.1}$$

together with the following entropy inequality, for each $t > 0$:

$$H(F(t)|M) + \int_{\mathbf{R}^3} \operatorname{trace} m(t) + \int_0^t \int_{\mathbf{R}^3} R(F)(s,x)dxds \le H(F^{in}|M)\,.$$

If one compares the equalities (3.1) with the full system of local conservation laws (2.1) one would obtain in the case of classical solutions of the Boltzmann equation with rapid decay at infinity in the v variable, one notices that (a) the local conservation of energy is not guaranteed, and (b) the local conservation of momentum holds only up to the potentially nontrivial defect term[2] $\operatorname{div}_x m$. In other words, the fundamental laws of the mechanics of continua (Newton's second law and the conservation of energy) are not known to be implied by the DiPerna-Lions theory of renormalized solutions of the Boltzmann equation. This seems to be a rather strong argument against using this notion of solution in deriving hydrodynamic models from the Boltzmann equation. However, we shall later show a way around this seemingly formidable difficulty.

[2]This formulation of the local conservation law of momentum modulo a defect measure is an improvement over the original DiPerna-Lions result in [**21**], which appeared later in [**46**]. R. DiPerna and P.-L. Lions had obtained the entropy inequality above without the defect measure term in [**22**].

The incompressible Navier-Stokes limit. The degree of rarefaction of a gas is usually described in terms of a dimensionless number, the Knudsen number, denoted Kn, defined as the ratio of the mean free path of gas molecules to some characteristic length scale of the flow. (For instance, this characteristic length scale could be either the size of a solid body immersed in the gas, or the typical wavelength in the Fourier decomposition of the number density at time $t = 0$).

All hydrodynamic limits of the Boltzmann equation assume that $\mathrm{Kn} \ll 1$.

The incompressible Navier-Stokes limit assumes in addition that the Mach number Ma is small, while the Reynolds number Re is of order 1. More specifically, in this incompressible Navier-Stokes limit

$$\mathrm{Ma} \simeq \mathrm{Kn} \ll 1$$

while Re converges to a finite, positive number. This last requirement is in fact a consequence of the assumption, since the Knudsen, Mach and Reynolds numbers satisfy von Karman's relation [**39**]

$$\mathrm{Kn} = \mathrm{Const.} \frac{\mathrm{Ma}}{\mathrm{Re}} \,.$$

Obviously, a gas is not an incompressible fluid. However, if the velocity field of a gas flow is small compared to the speed of sound, the acoustic and vortical modes decouple, so that the velocity field of the gas satisfy the same equations as that of an incompressible fluid. This is precisely the type of gas flows, henceforth referred to as "incompressible flows", that we are concerned with in the present work.

For notational convenience, we introduce the small parameter

$$\epsilon = \mathrm{Ma} = \mathrm{Kn} \,.$$

THEOREM 3.4 (Golse & Saint-Raymond [**30**], [**31**]). *Let $u^{in} \in L^2(\mathbf{R}^3)$ and $\theta^{in} \in L^2 \cap L^\infty(\mathbf{R}^3)$ satisfy $\operatorname{div}_x u^{in} = 0$.*

For each small enough $\epsilon > 0$, let F_ϵ be a renormalized solution relative to M of the Boltzmann equation with initial data

$$F_\epsilon^{in}(x,v) = \mathcal{M}_{(1-\epsilon\theta^{in}(\epsilon x), \epsilon u^{in}(\epsilon x), 1+\epsilon\theta^{in}(\epsilon x))}(v) \,.$$

Then, in the limit as $\epsilon \to 0$

$$\frac{1}{\epsilon} \int_{\mathbf{R}^3} \left(F_\epsilon \left(\frac{t}{\epsilon^2}, \frac{x}{\epsilon}, v \right) - M \right) \begin{pmatrix} v \\ (\frac{1}{3}|v|^2 - 1) \end{pmatrix} dv \to \begin{pmatrix} u(t,x) \\ \theta(t,x) \end{pmatrix}$$

weakly in $L^1_{loc}(\mathbf{R}_+ \times \mathbf{R}^3)$ modulo extraction of a subsequence, where (u, θ) is a "Leray solution" of the "Navier-Stokes-Fourier" system

$$\begin{aligned} \partial_t u + \operatorname{div}_x(u \otimes u) + \nabla_x p = \nu \Delta_x u \,, \qquad \operatorname{div}_x u = 0 \,, \\ \tfrac{5}{2}(\partial_t \theta + \operatorname{div}_x(u\theta)) = \kappa \Delta_x \theta \,, \end{aligned}$$

with initial data (u^{in}, θ^{in}).

The viscosity and heat conductivity in the theorem above are given by the formulas

$$\nu = \tfrac{1}{5} \mathcal{D}^*(v \otimes v - \tfrac{1}{3}|v|^2 I) \,, \qquad \kappa = \tfrac{2}{3} \mathcal{D}^*(\tfrac{1}{2}(|v|^2 - 5)v)$$

where $\mathcal{D}$ is the Dirichlet form of the collision integral linearized at M, i.e[3]:

$$\mathcal{D}(\Phi) = \tfrac{1}{2} \iiint_{\mathbf{R}^3\times\mathbf{R}^3\times\mathbf{S}^2} |\Phi + \Phi_* - \Phi' - \Phi'_*|^2 |v - v_*| M M_* dv dv_* d\sigma \,.$$

We recall the formula for the Legendre dual:

$$\mathcal{D}^*(\alpha) = \sup_a (\alpha \cdot a - \mathcal{D}(a))$$

whenever α is a vector fied on the Euclidian space $\mathbf{R}^3$, or

$$\mathcal{D}^*(\alpha) = \sup_a (\alpha : a - \mathcal{D}(a))$$

when α is a matrix field with values in $M_3(\mathbf{R})$, with the usual notation

$$\alpha : a = \mathrm{trace}(\alpha^T a)\,.$$

We also recall the definition of the notion of Leray solution of a solution of the Navier-Stokes equations.

Definition 3.5. A "Leray solution" of the Navier-Stokes-Fourier system is an element (u, θ) of $C(\mathbf{R}_+; w - L^2(\mathbf{R}^3))$ that solves the system in the sense of distributions on $\mathbf{R}_+^* \times \mathbf{R}^3$, satisfies the initial condition at $t = 0$, and verifies the "energy inequality" for each $t > 0$:

$$\begin{aligned} \tfrac{1}{2}\int_{\mathbf{R}^3} (|u|^2 + \tfrac{5}{2}|\theta|^2)(t,x)dx + \int_0^t \int_{\mathbf{R}^3} (\nu|\nabla_x u|^2 + \kappa|\nabla_x \theta|^2) dx ds \\ \le \tfrac{1}{2}\int_{\mathbf{R}^3} (|u^{in}|^2 + \tfrac{5}{2}|\theta^{in}|^2)(x)dx\,. \end{aligned}$$

There is an obvious formal similarity between the Leray energy inequality and the entropy inequality in the DiPerna-Lions existence theorem for the Boltzmann equation. Both the DiPerna-Lions and the Leray existence theorems are proved by applying compactness arguments in some appropriate weak topology, and this explains the analogies in both statements. In fact, more is true: our result establishes that the Leray energy inequality is the limiting form of the DiPerna-Lions entropy inequality.

This observation clarifies some subtleties about the interpretation of the temperature equation in the Navier-Stokes-Fourier system above (see also [**56**], section 3.2.4 on pp. 107–108.)

First, observe that, for an ideal monatomic gas, the internal energy is proportional to the temperature. In the present case, since θ is not the temperature itself but the fluctuation thereof, the total energy of the fluid at time t is of the form

$$\int_{\mathbf{R}^3} \left(\tfrac{1}{2}\epsilon^2|u|^2 + \tfrac{3}{2}(1+\epsilon\theta)\right) dx$$

(assuming that the density of the gas is 1). In other words, the term

$$\tfrac{1}{2}\int_{\mathbf{R}^3} (|u|^2 + \tfrac{5}{2}|\theta|^2)(t,x)dx$$

[3]When Φ is a vector field in the Euclidian space $\mathbf{R}^3$, the notation $|\Phi|$ designates its Euclidian norm. When Φ is a matrix field with values in $M_3(\mathbf{R})$, $|\Phi|$ designates the Frobenius norm of Φ, i.e. $|\Phi| = \mathrm{trace}(\Phi^T\Phi)^{1/2}$.

that appears in the "Leray energy inequality" is not really an energy (or more appropriately a fluctuation thereof at the scale ϵ^2), but the fluctuation of entropy at the scale ϵ^2.

There is another indication that the temperature equation in the Navier-Stokes-Fourier system above differs from that which appears in classical incompressible fluid models. Indeed, the temperature equation in the compressible Navier-Stokes system is

$$\rho\left(\partial_t \tfrac{3}{2}T + u\cdot\nabla_x \tfrac{3}{2}T\right) + p\operatorname{div}_x u = \operatorname{div}_x(\chi(T)\nabla_x\theta) + \mu(T)\tfrac{1}{2}D(u):D(u)$$

where ρ is the macroscopic density, T the temperature and $\mu(T)$ and $\chi(T)$ designate respectively the viscosity and heat conduction, while $D(u)$ is the symmetric, traceless part of $\nabla_x u$ and p is the pressure.

First, we assume incompressibility, i.e. both $\rho =$ Const. and $\operatorname{div}_x u = 0$. Besides, since the temperature fluctuations and the velocity field are both of order ϵ, the viscous heating term

$$\mu(T)\tfrac{1}{2}D(u):D(u)$$

disappears in the limit considered. This leaves us with the equation

$$\partial_t \tfrac{3}{2}T + u\cdot\nabla_x \tfrac{3}{2}T = \operatorname{div}_x\left(\rho^{-1}\chi(T)\nabla_x\theta\right)$$

where

$$\chi(T) = \tfrac{2}{3}T\mathcal{D}_T^*(\tfrac{1}{2}v(|v|^2-5))$$

where $\mathcal{D}_T$ is the same Dirichlet form as above, where M is replaced with $\mathcal{M}_{(1,0,T)}$.

Setting the constant density to be $\rho = 1$ and the temperature $T = 1+\epsilon\theta$, one has, in the limit as $\epsilon \to 0$, one arrives at the following equation for the temperature fluctuation

$$\tfrac{3}{2}\left(\partial_t\theta + u\cdot\nabla_x\theta\right) = \operatorname{div}_x\left(\chi(1)\nabla_x\theta\right)\,.$$

Clearly, one has $\chi(1) = \kappa$, so that the diffusion coefficient in this equation is $\frac{2}{3}\kappa$, instead of $\frac{2}{5}\kappa$ as in the temperature equation obtained in the theorem above.

The reason for this difference is to be found in the fact that the gas is not exactly incompressible, in the sense that its density is not exactly a constant, as assumed when deriving the above temperature equation from the compressible Navier-Stokes equations. More precisely, in the incompressible Navier-Stokes scaling on the Boltzmann equation, the leading order of the momentum equation implies that the density fluctuations compensate the temperature fluctuations.

To summarize, the Navier-Stokes-Fourier system in the theorem above does not coincide with the system obtained from the compressible Navier-Stokes system when setting the density to be a constant. Notice however that the motion equation in the theorem above coincides with the momentum equation of the compressible Navier-Stokes system when assuming that the density is a $O(\epsilon)$ perturbation of a constant. The difference between both systems appears on the temperature equation only. It should be mentionned that other scaling assumptions than the one considered here lead to a different temperature equation, as explained in a very interesting paper by C. Bardos and C.D. Levermore [**8**].

Let us conclude this section with a discussion of the literature on the proof of the incompressible Navier-Stokes limit of the Boltzmann equation. In [**19**], Caflisch's method [**15**] using a truncated Hilbert expansion of the solution of the Boltzmann equation in powers of ϵ was applied to the incompressible Navier-Stokes limit. As in the work of Caflisch, this derivation assumes that the solution of the Navier-Stokes equations obtained in the limit is smooth.

An obvious shortcoming of this result is that a smooth solution of the Navier-Stokes equations in 3 space dimensions is not known to exist for all positive times and without restriction on the size of the initial data. The best regularity result on the Navier-Stokes equations in 3 space dimensions known to this date is due to L. Caffarelli, R. Kohn, and L. Nirenberg [**13**]: roughly speaking, it says that the set of singularities for some appropriate notion of weak solutions of the Navier-Stokes equations cannot be larger than a curve in space-time. Unfortunately, the argument in [**19**] requires global (instead of partial) regularity on the Navier-Stokes solution, so that the result in [**19**] is limited to either small times or small initial data.

A complete proof of the Navier-Stokes limit for small data and all positive times was given by C. Bardos and S. Ukai [**9**] by a different approach.

The same problem for large data was systematically investigated by C. Bardos, F. Golse and C.D. Levermore in [**6**]. This paper defined a program, whose steps were subsequently completed in a series of articles by various authors: see [**46**], [**7**], [**26**], [**52**] and [**53**]. In the sequel, we shall briefly recall the input of each of these contributions to the final statement above. The reference [**30**] treated only the case of bounded collision kernels (such as in the case of cutoff Maxwell molecules) while [**31**] extends the method of [**30**] to all hard cutoff potentials in the sense of Grad, which includes the case of hard spheres discussed here.

4. Main ideas in the proof of the Navier-Stokes limit

Formal argument. Before sketching the crucial ideas in the proof of Theorem 3.4, we first present a formal argument for the incompressible Navier-Stokes limit.

Although the problem of deriving incompressible hydrodynamic models from the Boltzmann equation is a fairly natural — and fundamental — issue, it seems that the first author to have addressed it is Y. Sone in [**54**], in a time-independent setting.

The formal argument for the time-dependent problem was subsequently given in [**4**], [**5**].

Introduce the relative number density fluctuation g_ϵ:

$$g_\epsilon(t,x,v) = \frac{F_\epsilon\left(\frac{t}{\epsilon^2},\frac{x}{\epsilon},v\right) - M(v)}{\epsilon M(v)}, \qquad \text{where } M(v) = \tfrac{1}{(2\pi)^{3/2}}e^{-\frac{|v|^2}{2}}.$$

In terms of g_ϵ, the Boltzmann equation becomes

$$\epsilon\partial_t g_\epsilon + v\cdot\nabla_x g_\epsilon + \frac{1}{\epsilon}\mathcal{L}g_\epsilon = \mathcal{Q}(g_\epsilon, g_\epsilon)$$

where the linearized collision operator $\mathcal{L}$ and $\mathcal{Q}$ are defined by

$$\mathcal{L}g = -M^{-1}D\mathcal{C}[M](Mg)\,, \qquad \mathcal{Q}(g,g) = \tfrac{1}{2}M^{-1}D^2\mathcal{C}[M](Mg,Mg)$$

Observe that, in the above equation for g_ϵ, the highest order term is $\frac{1}{\epsilon}\mathcal{L}g_\epsilon$. Therefore, it is a natural to investigate the nullspace of $\mathcal{L}$. In the case of a gas of hard spheres considered in the present paper, the linearized collision operator has been studied in great detail by Hilbert.

LEMMA 4.1 (Hilbert [**36**]). *The operator $\mathcal{L}$ is a self-adjoint, nonnegative, Fredholm, unbounded operator on $L^2(\mathbf{R}^3; Mdv)$ with* $\ker\mathcal{L} = \mathrm{span}\{1, v_1, v_2, v_3, |v|^2\}$.

The formal derivation of the incompressible Navier-Stokes limit of the Boltzmann equation involves two steps:

(a) identifying the limiting form of g_ϵ as $\epsilon \to 0$, and

(b) taking limits in the local conservation laws of mass, momentum and energy.

(a) Identifying the limiting number density fluctuation. Multiplying the Boltzmann equation by ϵ and letting $\epsilon \to 0$ suggests that

$$g_\epsilon \to g \quad \text{with } \mathcal{L}g = 0\,.$$

By Hilbert's lemma, g is an *infinitesimal Maxwellian*, i.e. is of the form

$$g(t,x,v) = \rho(t,x) + u(t,x)\cdot v + \tfrac{1}{2}\theta(t,x)(|v|^2-3)\,.$$

Notice that g is parametrized by its own macroscopic observables, since

$$\rho = \langle g\rangle\,, \quad u = \langle vg\rangle\,, \quad \text{ and } \theta = \langle(\tfrac{1}{3}|v|^2-1)g\rangle$$

with the single bracket notation for the Gaussian average:

$$\langle\phi\rangle = \int_{\mathbf{R}^3}\phi(v)M(v)dv\,.$$

(b) Passing to the limit in the local conservation laws. Starting from the local conservation of mass

$$\epsilon\partial_t\langle g_\epsilon\rangle + \operatorname{div}_x\langle vg_\epsilon\rangle = \frac{1}{\epsilon^2}\int_{\mathbf{R}^3}\mathcal{C}(F_\epsilon)dv = 0\,,$$

we pass to the limit in the sense of distributions as $\epsilon \to 0$ and obtain the relation

$$\operatorname{div}_x\langle vg\rangle = \operatorname{div}_x u = 0\,.$$

which is the incompressibility condition in the Navier-Stokes equations.

Likewise, starting from the local conservation of momentum

$$\epsilon\partial_t\langle vg_\epsilon\rangle + \operatorname{div}_x\langle v\otimes vg_\epsilon\rangle = \frac{1}{\epsilon^2}\int_{\mathbf{R}^3}v\mathcal{C}(F_\epsilon)dv = 0\,,$$

we infer that

$$\operatorname{div}_x\langle v\otimes vg\rangle = \nabla_x(\rho+\theta) = 0\,.$$

Assuming some decay for g at infinity in some very weak sense — typically, that $g \in L^\infty_t(L^2(\mathbf{R}^3\times\mathbf{R}^3; Mdvdx))$, we conclude that

$$\rho + \theta = 0\,.$$

Going back to the formula for g, we conclude that

$$g(t,x,v) = u(t,x)\cdot v + \theta(t,x)\tfrac{1}{2}(|v|^2-5)\,.$$

Next we explain how we derive the Navier-Stokes momentum equation. Starting from the local conservation of momentum, we separate the tensor $\langle v\otimes vg_\epsilon\rangle$ into its traceless part and a scalar field:

$$\langle v\otimes vg_\epsilon\rangle = \langle A(v)g_\epsilon\rangle + \tfrac{1}{3}\langle|v|^2g_\epsilon\rangle I\,,$$

where

$$A(v) = v\otimes v - \tfrac{1}{3}|v|^2 I\,.$$

Hence

$$\partial_t\langle vg_\epsilon\rangle + \operatorname{div}_x\frac{1}{\epsilon}\langle A(v)g_\epsilon\rangle + \nabla_x\frac{1}{\epsilon}\langle\tfrac{1}{3}|v|^2g_\epsilon\rangle = 0\,.$$

Observe that, componentwise

$$A(v)\perp\ker\mathcal{L} \text{ in } L^2(Mdv)\,.$$

Hence, by the Fredholm alternative in Hilbert's lemma above, we infer the existence of a unique tensor field $\hat{A}(v) \in L^2(Mdv)$ such that

$$\mathcal{L}\hat{A} = A \text{ componentwise, and } \hat{A} \perp \ker \mathcal{L}\,.$$

Since $\mathcal{L}$ is self-adjoint on $L^2(Mdv)$, one has

$$\begin{aligned}\frac{1}{\epsilon}\langle A(v)g_\epsilon\rangle = \left\langle \hat{A}(v)\frac{1}{\epsilon}\mathcal{L}g_\epsilon\right\rangle &= \langle \hat{A}\mathcal{Q}(g_\epsilon, g_\epsilon)\rangle - \langle \hat{A}(\epsilon\partial_t + v\cdot\nabla_x)g_\epsilon\rangle \\ &\to \langle \hat{A}\mathcal{Q}(g,g)\rangle - \langle \hat{A}v\cdot\nabla_x g\rangle\end{aligned}$$

in the limit as $\epsilon \to 0$. Recalling the classical formula

$$\mathcal{Q}(g,g) = \tfrac{1}{2}\mathcal{L}(g^2)$$

we see that

$$\langle \hat{A}\mathcal{Q}(g,g)\rangle = \tfrac{1}{2}\langle \hat{A}\mathcal{L}(g^2)\rangle = \tfrac{1}{2}\langle Ag^2\rangle\,.$$

Summarizing, we have obtained

$$\frac{1}{\epsilon}\langle A(v)g_\epsilon\rangle \to \tfrac{1}{2}\langle Ag^2\rangle - \langle \hat{A}v\cdot\nabla_x g\rangle\,.$$

With the formula for g given above,

$$\tfrac{1}{2}\langle Ag^2\rangle = A(\langle vg\rangle) = A(u)$$

and

$$\langle \hat{A}v\cdot\nabla_x g\rangle = \nu(\nabla_x\langle vg\rangle + (\nabla_x\langle vg\rangle)^T - \tfrac{2}{3}\operatorname{div}_x\langle vg\rangle) = \nu(\nabla_x u + (\nabla_x u)^T)$$

with

$$\nu = \tfrac{1}{10}\langle \hat{A} : v\otimes v\rangle = \tfrac{1}{10}\langle \hat{A} : A\rangle = \tfrac{1}{10}\langle \hat{A} : \mathcal{L}\hat{A}\rangle > 0\,.$$

Hence, passing to the limit in the local conservation of momentum gives

$$\partial_t u + \operatorname{div}_x(u\otimes u) - \nu\Delta_x u = 0 \text{ modulo gradient fields.}$$

Notice that we do not seek the limit of the gradient field

$$\nabla_x \frac{1}{\epsilon}\langle \tfrac{1}{3}|v|^2 g_\epsilon\rangle\,.$$

Indeed, following Leray [**43**], we pass to the limit in the local conservation of momentum by integrating against divergence-free test vector fields, so that all gradient fields disappear before passing to the limit.

Hence we have derived the incompressibility condition, and the Navier-Stokes motion equation.

In order to derive the temperature equation, we combine the local conservation laws of mass and energy to obtain

$$\partial_t\langle \tfrac{1}{2}(|v|^2 - 5)g_\epsilon\rangle + \frac{1}{\epsilon}\operatorname{div}_x\langle \tfrac{1}{2}v(|v|^2-5)g_\epsilon\rangle = 0\,.$$

We proceed as in the case of the momentum equation, observing that the vector field

$$B(v) = \tfrac{1}{2}v(|v|^2 - 5) \perp \ker \mathcal{L}$$

so that there exists a unique $\hat{B} \equiv \hat{B}(v) \in L^2(Mdv)$ satisfying

$$\mathcal{L}\hat{B} = B \text{ componentwise and } \hat{B} \perp \ker \mathcal{L}\,.$$

We find that

$$\frac{1}{\epsilon}\langle Bg_\epsilon\rangle \to \tfrac{1}{2}\langle Bg^2\rangle - \langle \hat{B}v\cdot\nabla_x\rangle$$
$$= \tfrac{5}{2}u\theta - \kappa\nabla_x\theta$$

where

$$\kappa = \tfrac{1}{3}\langle \hat{B}\cdot B\rangle = \tfrac{1}{3}\langle \hat{B}\cdot\mathcal{L}\hat{B}\rangle > 0\,.$$

Substituting this into the linear combination of the local conservation laws of mass and energy considered above, we find that

$$\tfrac{5}{2}(\partial_t u + \operatorname{div}_x(u\theta)) - \kappa\Delta_x\theta = 0\,.$$

That the macroscopic density and temperature fluctuations equilibrate, i.e. formula

$$\rho + \theta = 0$$

obviously plays a crucial role in the derivation of the temperature equation, and explains some of the remarks above (at the end of section 2) on the meaning of the temperature equation in the Navier-Stokes-Fourier system considered here.

Conservation defects. As already mentionned above, a first, major difficulty in justifying the formal arguments above for renormalized solutions of the Boltzmann equation is that these solutions are not known to satisfy the local conservation laws of momentum and energy.

Therefore, we give up the *exact* local conservation laws and use instead conservation laws modulo defect terms, satisfied by renormalized solutions of the Boltzmann equation. These conservation laws modulo defect terms take the form

$$\partial_t\int_{\mathbf{R}^3}\Gamma\left(\frac{F_\epsilon}{M}\right)\begin{pmatrix}1\\ v\\ \frac{1}{2}|v|^2\end{pmatrix}M\,dv + \operatorname{div}_x\int_{\mathbf{R}^3}\Gamma\left(\frac{F_\epsilon}{M}\right)v\otimes\begin{pmatrix}1\\ v\\ \frac{1}{2}|v|^2\end{pmatrix}M\,dv$$
$$= \int_{\mathbf{R}^3}\Gamma'\left(\frac{F_\epsilon}{M}\right)\mathcal{C}(F_\epsilon)\begin{pmatrix}1\\ v\\ \frac{1}{2}|v|^2\end{pmatrix}dv\,.$$

In the weakly nonlinear hydrodynamic limits considered here, $F_\epsilon/M \to 1$ at a speed that is controlled by the entropy production rate and the size of the number density fluctuations: hence one expects that the usual symmetries of the collision integral that imply the local conservation laws are recovered in the rhs. of the above equalities as the $\epsilon \to 0$.

The idea of recovering local conservation laws at the hydrodynamic level only, instead of postulating them for the renormalized solutions of the Boltzmann equation considered before passing to the limit came up in [**7**]; at the same time, P.-L. Lions and N. Masmoudi [**46**] found another way of achieving the same result for the motion equation and in the Stokes limit only; Finally, a rather systematic method for these conservation defects was proposed in [**26**].

Compactness arguments. The DiPerna-Lions entropy inequality gives a priori bounds on the number density fluctuations that are uniform in ϵ: specifically, one has

$$H(F_\epsilon(t,\cdot,\cdot)|M) \le H(F_\epsilon(0,\cdot,\cdot)|M) \text{ for each } t,\epsilon > 0$$

and since

$$F_\epsilon\big|_{t=0} = \mathcal{M}_{(1-\epsilon\theta^{in},\epsilon u^{in},1+\epsilon\theta^{in})}$$

with $\theta^{in} \in L^2 \cap L^\infty(\mathbf{R}^N)$ and $u^{in} \in L^2(\mathbf{R}^N)$, one has

$$H(M(1+\epsilon g_\epsilon)(t,\cdot,\cdot)|M) = \iint_{\mathbf{R}^3\times\mathbf{R}^3} h(\epsilon g_\epsilon(t,\cdot,\cdot))M dv dx \le C\epsilon^2 \text{ for each } t,\epsilon > 0\,,$$

where h is the nonlinear integrand in the definition of the relative entropy:

$$h(z) = (1+z)\ln(1+z) - z\,, \quad z > -1\,.$$

Since $h(z) = \frac{1}{2}z^2$ as $z \to 0$, for each measurable $g \equiv g(x,v)$ such that $1+\epsilon g_\epsilon(x,v) \ge 0$ a.e.,

$$\text{if } \iint_{\mathbf{R}^3\times\mathbf{R}^3} h(\epsilon g)M dv dx = O(\epsilon^2) \text{ then } g \in L^2(M dv dx)$$

and one has

$$\lim_{\epsilon\to 0^+} \frac{1}{\epsilon^2}\iint_{\mathbf{R}^3\times\mathbf{R}^3} h(\epsilon g)M dv dx = \tfrac{1}{2}\iint_{\mathbf{R}^3\times\mathbf{R}^3} g^2 M dv dx\,.$$

Thus, for a single g, the relative entropy $\frac{1}{\epsilon^2}H(M(1+\epsilon g)|M)$ is asymptotically equivalent to the $L^2(Mdvdx)$-norm of g.

Things are quite different for a family of g's: in the first place, the bound

$$\iint_{\mathbf{R}^3\times\mathbf{R}^3} h(\epsilon g_\epsilon(t,\cdot,\cdot))M dv dx \le C\epsilon^2$$

does not guarantee that $g_\epsilon(t,\cdot,\cdot) = O(1)$ in $L^2(Mdvdx)$.

However, the relative entropy bound guarantees that

$$(1+|v|^2)g_\epsilon \text{ is relatively compact in weak-}L^1_{loc}(\mathbf{R}_+; L^1(\mathbf{R}^3\times\mathbf{R}^3))$$

Modulo extracting subsequences, for each $\phi = O(|v|^2)$ at infinity

$$\phi g_\epsilon \to \phi g \text{ weakly in } L^1_{loc}(\mathbf{R}_+; L^1(\mathbf{R}^3\times\mathbf{R}^3))$$

and this justifies passing to the limit in expressions that are linear in g_ϵ.

It remains to pass to the limit in the nonlinear term, i.e. to justify that

$$\langle v g_\epsilon\rangle \otimes \langle v g_\epsilon\rangle \to \langle v g\rangle \otimes \langle v g\rangle \text{ as } \epsilon \to 0$$

and this requires a.e. pointwise, instead of weak convergence.

In fact, it would also require some equiintegrability control on g_ϵ — which we do not know how to verify. Fortunately, when dealing with renormalized solutions of the Boltzmann equation, one has to replace the term $\langle v g_\epsilon\rangle$ with $\langle v g_\epsilon \gamma(\epsilon g_\epsilon)$, where γ is a nonlinear truncation that eliminates the large values of ϵg_ϵ. With such a truncation, the relative entropy bound is enough to provide the equiintegrability needed to pass to the limit assuming a.e. convergence.

Both this equiintegragility control and the a.e. convergence of averages such as $\langle v g_\epsilon\rangle$ (or $\langle v g_\epsilon \gamma(\epsilon g_\epsilon)\rangle$) are obtained by using a "velocity averaging" lemma, a typical example of which (in a time-independent situation) is as follows:

LEMMA 4.2 (Golse-Saint-Raymond [**29**]). *Let $f_n \equiv f(x,v)$, a bounded sequence in $L^1(\mathbf{R}^D_x; L^p(\mathbf{R}^D_v))$ for some $p > 1$ be such that the sequence $v\cdot\nabla_x f_n$ is bounded in $L^1(\mathbf{R}^D\times\mathbf{R}^D)$. Then*
(a) the sequence f_n is weakly relatively compact in $L^1_{loc}(\mathbf{R}^D\times\mathbf{R}^D)$; and
(b) for each $\phi \in C_c(\mathbf{R}^D)$, the sequence of moments

$$\int_{\mathbf{R}^D} f_n(x,v)\phi(v)dv \textit{ is strongly relatively compact in } L^1_{loc}(\mathbf{R}^D)\,.$$

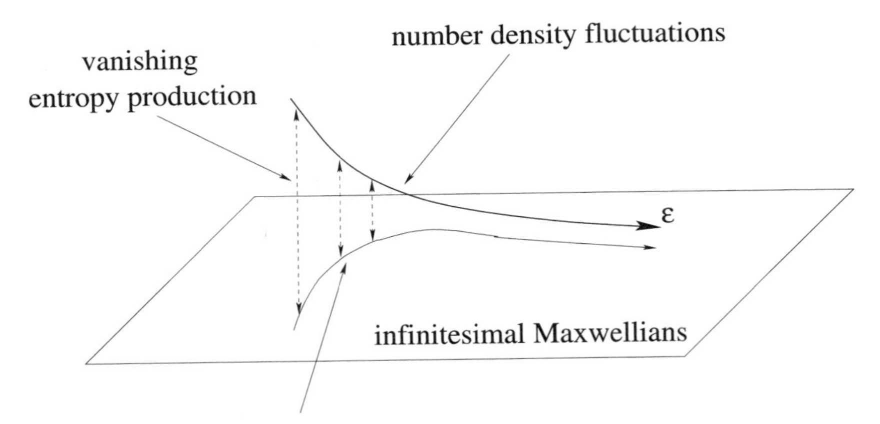

FIGURE 2. Strategy for compactness

The idea of velocity averaging was published for the first time independently by V. Agoshkov [**1**] and F. Golse, B. Perthame and R. Sentis in [**28**]. There, an L^2-variant of this result is proved with Fourier techniques, controlling in particular the small divisors coming from the symbol of $v \cdot \nabla_x$. In a later reference [**27**], the same authors and P.-L. Lions extended the method to give

(a) the optimal regularity of averages in the Sobolev spaces H^s,

(b) some regularization effect[4] for averages in the Sobolev space $W^{s,p}$ with $p \neq 2$,

(c) a compactness result in L^1 (somewhat weaker than the one stated above), and

(d) a discussion on the optimality of the results in (a-c) based on counterexamples.

For the $L^1_x(L^p_v)$ result above, a significant part of the proof is done in *physical* instead of Fourier space.

One important idea in that proof is that the group generated by $v \cdot \nabla_x$ exchanges x- and v- regularity for $t \neq 0$, as can be seen easily from the formula

$$e^{tv\cdot\nabla_x}\phi(x,v) = \phi(x+tv,v)\,.$$

This observation entails dispersion estimates "à la Strichartz", due to C. Bardos and P. Degond [**3**], and F. Castella and B. Perthame [**17**].

The L^1 velocity averaging lemma above uses their ideas together with some interpolation argument along the lines of [**44**], where the time $t > 0$ is viewed as the interpolation parameter. A precursor of the lemma above was found earlier by L. Saint-Raymond in 2000 in her PhD thesis — and reported in [**52**].

With this result, L. Saint-Raymond subsequently produced a complete proof of the Navier-Stokes limit for the BGK model in [**53**]. An important feature of that proof is the entropy production term, which controls the distance from the solution

[4]Regularity of velocity averages in L^p has been recently improved with the help of more refined harmonic analysis methods: see [**23**], [**10**], [**51**], [**20**], [**57**], or [**11**] and [**50**] for a survey of these results, together with applications thereof.

to the manifold of local Maxwellian equilibria in a way that is quite obvious for the BGK model. The analogous question for the Boltzmann equation is known to be much more delicate, and was handled first in a rather intricate manner in [**30**]. This argument is simplified to some extent in the more recent result [**31**], which uses normalizing nonlinearities that are better suited to the quadratic nature of the Boltzmann collision integral.

Going back to velocity averaging and dispersion estimates, it should be noted that these are essentially the only nontrivial regularity statements known about the free transport operator $\partial_t + v \cdot \nabla_x$. None of these results — except maybe their limiting variants involving L^1 or Hardy spaces — requires using sophisticated techniques. Therefore, it may seem somewhat surprising that such results came out only relatively recently.

Perhaps the only serious difficulty about velocity averaging was to realize that the natural quantities on which to look for smoothing by the transport operator are macroscopic observables (i.e. velocity averages).

This may explain why velocity averaging results appeared for the first time in connection with the problem of hydrodynamic (or macroscopic) limits for kinetic models.

5. Perspectives, open problems

We have described a derivation of the incompressible Navier-Stokes equations from the Boltzmann equation, in the framework of Leray "turbulent solutions" and DiPerna-Lions renormalized solutions. As emphasized above, perhaps the most remarkable feature in this derivation is the fact that the Leray energy inequality is the limiting form of Boltzmann's H Theorem. Our presentation is limited to the case of hard spheres; however, the same method allows dealing with any hard cutoff power law potential in Grad's sense: we refer to [**33**] and [**14**] for a description of the cutoff assumption, and to [**31**] for a complete proof of that limit.

So far, we have discussed only the case of flows that are not driven by external forces. It would be good to derive also the Navier-Stokes equations with some external force density on its right-hand side from the Boltzmann equation — only the case of conservative forces defined by a potential was investigated in [**4**]; however, a similar formal argument also applies to the case of a solenoidal (i.e. divergence-free) force field. A complete proof, however, would be most welcome, as it might require some modification in the velocity averaging part of the argument.

Likewise, steady Navier-Stokes equations — with a solenoidal force field on the right-hand side — are known to have at least one strong solution. (This can be proved by a Leray-Schauder argument, see [**40**].)

On the contrary, the steady Boltzmann equation seems even less well understood as the corresponding evolution problem, at least as regards the issue of finding solutions in the large (see however [**2**], [**34**]). Some new understanding on this difficult problem would be most welcome. Considering the steady Boltzmann equation with some external, solenoidal force field in "nearly Navier-Stokes" regimes might be helpful in this direction.

But the greatest challenge at present about hydrodynamic limits of the Boltzmann equation is undoubtedly the compressible Euler limit.

As recalled in the introduction, the Euler system of gas dynamics is known to describe some limit of the Boltzmann equation for a short time, before shock waves appear in the solution of the Euler system.

The first result in that direction is due to Nishida [**49**], who used a variant of L. Nirenberg's abstract Cauchy-Kovalewski theorem [**48**]. Later on, R. Caflisch obtained an analogous result in [**15**] following Hilbert's method [**36**]. One advantage in Caflisch's method is that the Euler limit of the Boltzmann equation holds at least up to the first time when a shock wave develops in the Euler solution.

On the other hand, in the one-dimensional case, global BV solutions of the compressible Euler system have been constructed by T.-P. Liu [**47**] following Glimm's celebrated method [**25**] — for initial data with small total variation. Whether such solutions are limits of solutions of the Boltzmann equation remains an outstanding open problem. Perhaps Bressan's ideas [**12**] on the stability of Glimm's solutions could be used in this context.

References

[1] V. Agoshkov, *Space of functions with differential difference characteristics and smoothness of solutions of the transport equation*, Dokl. Akad. Nauk SSSR **276** (1984), 1289–1293.

[2] L. Arkeryd, A. Nouri, *The stationary Boltzmann equation in* $\mathbf{R}^n$ *with given indata*, Ann. Sc. Norm. Super. Pisa Cl. Sci. (5) **1** (2002), 359–385.

[3] C. Bardos, P. Degond, *Global existence for the Vlasov-Poisson equation in 3 space variables with small initial data*, Ann. Inst. H. Poincaré Anal. Non Linéaire **2** (1985), 101–118.

[4] C. Bardos, F. Golse, C.D. Levermore, *Sur les limites asymptotiques de la théorie cinétique conduisant à la dynamique des fluides incompressibles* C.R. Acad. Sci. **309** (1989), 727–732.

[5] C. Bardos, F. Golse, C.D. Levermore, *Fluid Dynamic Limits of the Boltzmann Equation I*, J. Stat, Phys. **63** (1991), 323–344.

[6] C. Bardos, F. Golse, C.D. Levermore, *Fluid Dynamic Limits of Kinetic Equations II: Convergence Proofs for the Boltzmann Equation*, Comm. Pure & Appl. Math **46** (1993), 667–753.

[7] C. Bardos, F. Golse, C.D. Levermore, *The Acoustic Limit for the Boltzmann Equation*, Archive for Rational Mech. Anal. **153** (2000), 177-204.

[8] C. Bardos, C.D. Levermore: *Kinetic equations and an incompressible limit that recovers viscous heating*, preprint.

[9] C. Bardos, S. Ukai, *The classical incompressible Navier-Stokes limit of the Boltzmann equation*, Math. Models and Methods in the Appl. Sci. **1** (1991), 235–257.

[10] M. Bézard, *Régularité* L^p *précisée des moyennes dans les équations de transport*, Bull. Soc. Math. France **122** (1994), 29–76.

[11] F. Bouchut, F. Golse, M. Pulvirenti, *Kinetic Equations and Asymptotic Theory*, L. Desvillettes & B. Perthame ed., Editions scientifiques et médicales Elsevier, Paris, 2000.

[12] A. Bressan, *Hyperbolic systems of conservation laws. The one-dimensional Cauchy problem*, Oxford University Press, Oxford, 2000.

[13] L. Caffarelli, R.V. Kohn, L. Nirenberg, *Partial regularity of suitable weak solutions of the Navier-Stokes equations*, Comm. Pure Appl. Math. **35** (1982), 771–831.

[14] R. E. Caflisch, *The Boltzmann equation with a soft potential. I. Linear, spatially homogeneous*, Commun. Math. Phys. **74** (1980), 71–95.

[15] R. E. Caflisch, *The fluid dynamic limit of the nonlinear Boltzmann equation*, Comm. on Pure and Appl. Math. **33** (1980), 651–666.

[16] T. Carleman, *Sur la théorie de l'équation intégro-différentielle de Boltzmann*, Acta Math. **60** (1933), 91–146.

[17] F. Castella, B. Perthame, *Estimations de Strichartz pour les équations de transport cinétiques*, C.R. Acad. Sci. Sér. I **322** (1996), 535–540.

[18] C. Cercignani, *On the Boltzmann equation for rigid spheres*, Transp. Theory and Stat. Phys. (1972), 211–225.

[19] A. DeMasi, R. Esposito, J. Lebowitz, *Incompressible Navier-Stokes and Euler limit of the Boltzmann equations*, Comm. on Pure and Appl. Math. **42** (1989), 1189–1214.

[20] R. DeVore, G. Petrova, *The averaging lemma*, J. Amer. Math. Soc. **14** (2001), 279–296.

[21] R.J. DiPerna, P.-L. Lions, *On the Cauchy problem for the Boltzmann equation: global existence and weak stability results*, Ann. of Math. **130** (1990), 321–366.

[22] R.J. DiPerna, P.-L. Lions, *Global solutions of Boltzmann's equation and the entropy inequality*, Arch. Rational Mech. Anal. **114** (1991), 47–55.

[23] R.J. DiPerna, P.-L. Lions, Y. Meyer, *L^p regularity of velocity averages*, Ann. Inst. Henri Poincaré, Anal. Non-linéaire **8** (1991), 271–287.

[24] K.O. Friedrichs, P.D. Lax, *Systems of conservation equations with a convex extension*, Proc. Nat. Acad. Sci. U.S.A. **68** (1971), 1686–1688.

[25] J. Glimm: *Solutions in the large for nonlinear hyperbolic systems of equations*, Comm. Pure Appl. Math. **18** (1965), 697–715.

[26] F. Golse, C.D. Levermore: *The Stokes-Fourier and Acoustic Limits for the Boltzmann Equation*, Comm. on Pure and Appl. Math. **55** (2002), 336–393.

[27] F. Golse, P.-L. Lions, B. Perthame, R. Sentis, *Regularity of the moments of the solution of a transport equation*, J. Funct. Anal. **76** (1988), 110–125.

[28] F. Golse, B. Perthame, R. Sentis, *Un résultat de compacité pour les équations de transport et application au calcul de la limite de la valeur propre principale de l'opérateur de transport*, C.R. Acad. Sci. **301** (1985), 341–344.

[29] F. Golse, L. Saint-Raymond, *Velocity averaging in L^1 for the transport equation*, C. R. Acad. Sci. **334** (2002), 557–562.

[30] F. Golse, L. Saint-Raymond, *The Navier-Stokes limit of the Boltzmann equation for bounded collision kernels*, Invent. Math. 155 (2004), no. 1, 81–161.

[31] F. Golse, L. Saint-Raymond, *The Incompressible Navier-Stokes limit of the Boltzmann equation for hard cutoff potentials*, preprint.

[32] H. Grad, *On the kinetic theory of rarefied gases*, Comm. on Pure and Appl. Math. **2** (1949), 331–407.

[33] H. Grad, *Asymptotic theory of the Boltzmann equation II*, in *Rarefied Gas Dynamics. Proc. 3rd Internat. Sympos., Palais de l'Unesco* **1** (1963), Paris.

[34] J.-P. Guiraud, *Problème aux limites intérieures pour l'équation de Boltzmann*, (French) in "Actes du Congrès International des Mathématiciens" (Nice, 1970), vol. 3, pp. 115–122. Gauthier-Villars, Paris, 1971.

[35] D. Hilbert, *Mathematical Problems*, International Congress of Mathematicians, Paris 1900, translated and reprinted in Bull. Amer. Math. Soc. **37** (2000), 407-436.

[36] D. Hilbert, *Begründung der kinetischen Gastheorie* Math. Ann. **72** (1912), 562–577.

[37] R. Illner, M. Shinbrot, *The Boltzmann equation: global existence for a rare gas in an infinite vacuum*, Commun. Math. Phys. **95** (1984), 217–226.

[38] R. Illner, M. Pulvirenti, *Global validity of the Boltzmann equation for a two-dimensional rare gas in a vacuum*, Commun. Math. Phys. **105** (1986), 189–203.

[39] T. von Karman, *From Low-Speed Aerodynamics to Astronautics*, Pergamon Press, Oxford, 1963.

[40] O.A. Ladyzhenskaya: *The mathematical theory of viscous incompressible flows*, Gordon and Breach, Science Publishers, New York-London-Paris 1969.

[41] O.E. Lanford III, *Time evolution of large classical systems*, in "Dynamical systems, theory and applications" (Rencontres, Battelle Res. Inst., Seattle, Wash., 1974), pp. 1–111. Lecture Notes in Phys., Vol. 38, Springer, Berlin, 1975.

[42] P.D. Lax, *Shock waves and entropy*, in *Contributions to nonlinear functional analysis* (Proc. Sympos., Math. Res. Center, Univ. Wisconsin, Madison, Wis., 1971), pp. 603–634. Academic Press, New York, 1971.

[43] J. Leray, *Sur le mouvement d'un liquide visqueux emplissant l'espace*, Acta Math. **63** (1934), 193-248.

[44] J.-L. Lions, *Théorèmes de trace et d'interpolation I, II*, Ann. Scuola Norm. di Pisa **13** (1959), pp. 389–403, & **14** (1960), pp. 317–331.

[45] P.-L. Lions, *Conditions at infinity for Boltzmann's equation* Comm. in Partial Differential Equations **19** (1994), 335–367.

[46] P.-L. Lions, N. Masmoudi *From Boltzmann Equation to the Navier-Stokes and Euler Equations I & II*, Archive Rat. Mech. & Anal. **158** (2001), 173–193 & 195–211.

[47] T.-P. Liu, *Solutions in the large for the equations of nonisentropic gas dynamics*, Indiana Univ. Math. J. **26** (1977), 147–177.

[48] L. Nirenberg, *An abstract form of the nonlinear Cauchy-Kowalewski theorem*, J. Differential Geometry **6** (1972), 561–576.

[49] T. Nishida: *Fluid dynamical limit of the nonlinear Boltzmann equation to the level of the compressible Euler equation*, Comm. Math. Phys. **61** (1978), 119–148.

[50] B. Perthame, *Kinetic formulation of conservation laws*, Oxford University Press, Oxford, 2002.

[51] B. Perthame, P. Souganidis, *A limiting case for velocity averaging*, Ann. Scient. Ecole Norm. Sup. (4) **31** (1998), 591–598.

[52] L. Saint-Raymond, *Discrete time Navier-Stokes limit for the BGK Boltzmann equation*, Comm. Partial Diff. Eq. **27** (2002), 149–184.

[53] L. Saint-Raymond, *From the BGK model to the Navier-Stokes equations*, Ann. Sci. Ecole Norm. Sup. (4) **36** (2003), 271–317.

[54] Y. Sone, *Asymptotic theory of flow of a rarefied gas over a smooth boundary II*, in Rarefied Gas Dynamics, Vol. II, 737–749, ed. by D. Dini, Pisa, Editrice Tecnico Scientifica (1971).

[55] Y. Sone, *Kinetic theory and fluid dynamics*, Birkhäuser, Boston (2002).

[56] Y. Sone, *Molecular Gas Dynamics: Theory, Techniques, and Applications*, Birkhäuser, Boston (2007).

[57] E. Tadmor, T. Tao, *Velocity averaging, kinetic formulations and regularizing effetcs in quasi-linear PDEs*, Comm. on Pure and Appl. Math., to appear.

[58] S. Ukai, *On the existence of global solutions of a mixed problem for the nonlinear Boltzman equation*, Proc. of the Japan Acad. **50** (1974), 179–184.

Laboratoire J.-L. Lions, Université Paris 7 -Denis Diderot, BP187, 4 place Jussieu, 75252 Paris Cedex 05, France

Current address: Centre de Mathématiques Laurent Schwartz, Ecole Polytechnique, 91128 Palaiseau Cedex, France

E-mail address: golse@math.polytechnique.fr

Proceedings of Symposia in Applied Mathematics
Volume **65**, 2007

Hyperbolic Conservation Laws with Involutions and Contingent Entropies

C.M. Dafermos[1]

In honor of Peter Lax and Louis Nirenberg, on their eightieth birthday, with admiration and affection.

ABSTRACT. The equations of elastodynamics and the Born-Infeld version of Maxwell's equations are examples of hyperbolic systems of conservation laws in which the principal entropy fails to be convex. It is shown that this is compensated by the presence of "involutions" and supplementary "contingent entropies," so that the Cauchy problem is locally well-posed for classical solutions, which in turn are unique and stable within the broader class of admissible weak solutions.

1. Introduction

The term "hyperbolic conservation laws" was coined by Peter Lax, in his pioneering paper [16], to designate the class of quasilinear first order hyperbolic systems of partial differential equations in divergence form. The framework of the general theory was designed by distilling the properties of the Euler equations in gas dynamics, which still serve as the prototypical example of such a system. The reader may find accounts of the current state of the subject in the books [2,6,10,14,19,22].

Hyperbolic systems of conservation laws typically express the balance laws of continuum physics for media with "elastic" response. In particular, any system of conservation laws originating in physics is endowed with an "entropy" - a term introduced in that context by Lax [17] - which induces an additional conservation law satisfied identically by smooth solutions. This natural entropy is akin to physical free energy and should not be confused with the physical entropy, even though these two notions are intimately related. Indeed, in the case of physically admissible discontinuous weak solutions of the system, the extra conservation law induced by the natural entropy is converted into an inequality that expresses the Second Law of thermodynamics.

The natural entropy function for the Euler equations is convex. This is fortunate, because any system of conservation laws endowed with a convex entropy is symmetrizable hyperbolic [13]. In turn, the Cauchy problem for symmetric

[1]1991 Mathematics Subject classification. 35L65, 74B20, 76L05, 78A25.
Supported by the NSF under grants DMS-0202888 and DMS-0244295.

hyperbolic systems is locally well-posed in the classical sense [2,15,20,24]; moreover, classical solutions are unique and L^2-stable within the broader class of admissible weak solutions [7,8,10,12].

The aim of this work is to investigate the existence and stability of classical solutions to the Cauchy problem for a class of hyperbolic systems of conservation laws that are not symmetrizable in the usual sense, as their natural entropy fails to be convex. The system of nonlinear elastodynamics and the Born-Infeld version of Maxwell's equations will be discussed here as representative members of that class. Physically relevant solutions of these systems must satisfy certain conditions of invariance, called involutions. The notion of entropy function should be accordingly extended to encompass entropies that are relevant only contingently upon the involutions. It will be shown that the involutions in conjunction with supplementary contingent entropies may compensate for the lack of convexity of the natural entropy.

The results of this paper, together with a sketch of the proofs, have already been published in [10]. Detailed proofs, however, are presented here for the first time.

2. Hyperbolic systems of Conservation Laws with Involutions

Throughout the paper, we shall be employing the notation of [10]. The canonical form of a system of n conservation laws in m spatial variables is

$$\partial_t U(x,t) + \sum_{\alpha=1}^{m} \partial_\alpha G_\alpha(U(x,t)) = 0. \tag{2.1}$$

The *state vector* U, with values in some open subset $\mathcal{O}$ of $\mathbb{R}^n$, is a function of the spatial variable $x \in \mathbb{R}^m$ and the temporal variable t. For $\alpha = 1, \cdots, m$, the *fluxes* G_α are given smooth functions from $\mathcal{O}$ to $\mathbb{R}^n$. The symbol ∂_α denotes partial derivative with respect to the variable x_α and ∂_t stands for the partial derivative with respect to t.

The system (2.1) is *hyperbolic* if for any $U \in \mathcal{O}$ and $\xi \in S^{m-1}$, the $n \times n$ matrix

$$\Lambda(\xi; U) = \sum_{\alpha=1}^{m} \xi_\alpha \mathrm{D}G_\alpha(U) \tag{2.2}$$

has real eigenvalues and n linearly independent eigenvectors. We will be dealing here with special hyperbolic systems having the following structure: For some positive integer k and $\alpha = 1, \cdots, m$ there are $k \times n$ matrices M_α such that

$$M_\alpha G_\beta(U) + M_\beta G_\alpha(U) = 0, \qquad \alpha, \beta = 1, \cdots, m, \tag{2.3}$$

for any $U \in \mathcal{O}$. A direct consequence of (2.3) is that, classical or even weak, solutions of the Cauchy problem for (2.1) satisfy identically the equation

$$\sum_{\alpha=1}^{m} M_\alpha \partial_\alpha U(x,t) = 0, \tag{2.4}$$

provided that the initial data do so. We then call (2.4) an *involution* for (2.1) [3,9].

Involutions are connected with the zero eigenvalues of the matrix $\Lambda(\xi;U)$ defined by (2.2). Indeed, for $\xi \in S^{m-1}$, setting

$$N(\xi) = \sum_{\alpha=1}^{m} \xi_\alpha M_\alpha, \tag{2.5}$$

(2.3) implies

$$N(\xi)\Lambda(\xi;U) = 0. \tag{2.6}$$

The following two systems, originating in physics, exemplify hyperbolic conservation laws endowed with involutions.

In Lagrangian coordinates, isentropic motion of (nonlinear) elastic media is governed by the equations [10]

$$\begin{cases} \partial_t F_{i\alpha} - \partial_\alpha v_i = 0, & i = 1,2,3;\ \alpha = 1,2,3 \\ \partial_t v_i - \sum_{\alpha=1}^{3} \partial_\alpha S_{i\alpha}(F) = 0, & i = 1,2,3. \end{cases} \tag{2.7}$$

Here F stands for the (3×3 matrix-valued) *deformation gradient* while v is the (3-vector-valued) *velocity*. The (3×3 matrix-valued) $S(F)$ denotes the *Piola-Kirchhoff stress tensor*, which is the gradient of the (scalar-valued) *strain energy* $\varepsilon(F)$:

$$S_{i\alpha}(F) = \frac{\partial \varepsilon(F)}{\partial F_{i\alpha}}, \qquad i = 1,2,3;\ \alpha = 1,2,3. \tag{2.8}$$

$\mathcal{O}$ consists of the set of (F, v) with $\det F > 0$, and typically $\varepsilon(F) \to \infty$ as $\det F \to 0$. The first nine equations in (2.7) express the kinematic compatibility between deformation gradient and velocity, while the remaining three equations express conservation of linear momentum. The system (2.7) is hyperbolic when $\varepsilon(F)$ is *rank-one convex*, i.e.,

$$\sum_{i,j=1}^{3} \sum_{\alpha,\beta=1}^{3} \frac{\partial^2 \varepsilon(F)}{\partial F_{i\alpha} \partial F_{j\beta}} \zeta_i \zeta_j \xi_\alpha \xi_\beta > 0 \tag{2.9}$$

holds for all ζ and ξ in S^2. Furthermore, (2.7) is endowed with the involutions

$$\partial_\beta F_{i\alpha} - \partial_\alpha F_{i\beta} = 0, \qquad i = 1,2,3;\ \ \alpha, \beta = 1,2,3, \tag{2.10}$$

which express the symmetry of second spatial derivatives of the motion and should thus hold for any physically meaningful solution.

The second example is provided by the Born-Infeld version of Maxwell's equations [4]. These are the classical Maxwell's equations

$$\begin{cases} \partial_t B + \operatorname{curl} E = 0 \\ \partial_t D - \operatorname{curl} H = 0, \end{cases} \tag{2.11}$$

where, however, the (3-vector-valued) *electric field* E and the (3-vector-valued) *magnetic field* H are now related to the (3-vector-valued) *magnetic induction* B and the (3-vector-valued) *electric displacement* D by nonlinear constitutive relations

$$E = \frac{\partial \eta}{\partial D} = \frac{1}{\eta}[D + B \wedge Q], \qquad H = \frac{\partial \eta}{\partial B} = \frac{1}{\eta}[B - D \wedge Q], \tag{2.12}$$

with

$$\eta = \sqrt{1 + |B|^2 + |D|^2 + |Q|^2}\ , \qquad Q = D \wedge B\,. \tag{2.13}$$

Under the above assumptions, (2.11) is hyperbolic. Furthermore, this system is endowed with the involutions

$$\operatorname{div} B = 0, \qquad \operatorname{div} D = 0, \tag{2.14}$$

which must hold for all physically meaningful solutions.

3. Entropies and Contingent Entropies

A (scalar-valued) function $\eta(U)$ is an *entropy* for the system (2.1), with associated (m-vector-valued) *entropy flux* $Q(U)$ if

$$\mathrm{D}Q_\alpha(U) = \mathrm{D}\eta(U)\mathrm{D}G_\alpha(U), \qquad \alpha = 1, \cdots, m, \tag{3.1}$$

holds for all $U \in \mathcal{O}$. This is equivalent to the statement that any classical solution U of (2.1) satisfies identically the additional conservation law

$$\partial_t \eta(U(x,t)) + \sum_{\alpha=1}^{m} \partial_\alpha Q_\alpha(U(x,t)) = 0. \tag{3.2}$$

The components $U^1, \cdots, U^n$ of the state vector U may be regarded (and will be employed below) as entropies-albeit trivial ones. In systems originating in physics, the fluxes G_α are designed in such a way that the formally overdetermined system (3.1) generates a natural, nonlinear entropy-entropy flux pair (η, Q) with the property that the inequality

$$\partial_t \eta(U(x,t)) + \sum_{\alpha=1}^{m} \partial_\alpha Q_\alpha(U(x,t)) \le 0, \tag{3.3}$$

in the sense of distributions, expresses the compatibility of bounded measurable weak solutions U of (2.1) with the Second Law of thermodynamics. This special entropy, which is akin to physical free energy, will henceforth be called *principal*. Whenever the principal entropy is convex, the Cauchy problem for the system is locally well-posed, in the classical sense [2,15,20,24]. Moreover, classical solutions are L^2-stable within the broader class of L^∞ weak solutions that satisfy the physical admissibility condition (3.3) [8,10,12]. Convexity of the principal entropy is common but not ubiquitous in physics; in particular, it fails in the two physical examples considered in Section 2.

The principal entropy-entropy flux pair for the system (2.7) of elastodynamics is

$$\eta(F,v) = \frac{1}{2}v^2 + \varepsilon(F), \qquad Q_\alpha(F,v) = -\sum_{i=1}^{3} v_i S_{i\alpha}(F). \tag{3.4}$$

The principle of material frame indifference in continuum mechanics does not allow the strain energy function $\varepsilon(F)$ to be convex, and thus $\eta(F,v)$ cannot be convex. Similarly, in the Born-Infeld theory the principal entropy-entropy flux pair (η, Q) is given by (2.13), with $\eta(B,D)$ nonconvex.

In elastodynamics, rank-one convexity (2.9) of the strain energy function, in conjunction with the involutions (2.10), partially compensates for the lack of convexity of $\eta(F,v)$ [7,9]. However, in order to get full compensation and also treat the Born-Infeld equations, one must look more closely into the structure of these systems, in order to identify supplementary entropies.

With regard to the system (2.7) of elastodynamics, it has been shown [10,11,21] that any L^∞ weak solution (F,v) that satisfies the involutions (2.10) also satisfies the kinematic conservation laws

$$\partial_t(\det F) - \sum_{i=1}^{3}\sum_{\alpha=1}^{3} \partial_\alpha(v_i F^*_{\alpha i}) = 0, \tag{3.5}$$

$$\partial_t F^*_{\gamma k} - \sum_{i,j=1}^{3}\sum_{\alpha,\beta=1}^{3} \partial_\alpha(\epsilon_{ijk}\epsilon_{\alpha\beta\gamma} v_i F_{j\beta}) = 0, \qquad k = 1,2,3;\ \gamma = 1,2,3, \tag{3.6}$$

where $F^* = (\det F)F^{-1}$ is the matrix of minors of F and $\epsilon_{ijk}, \epsilon_{\alpha\beta\gamma}$ are the usual permutation symbols. It should be emphasized that the validity of (3.5) and (3.6) is contingent upon F satisfying the involutions (2.10). Thus $\det F$ and $F^*_{\gamma k}$ are not entropies for (2.7) in the sense of (3.1), above.

A similar situation obtains in the Born-Infeld version of Maxwell's equations [5,23]: Under the constitutive assumptions (2.12), (2.13), classical, or even BV weak, solutions (B,D) of (2.11) that satisfy the involutions (2.14) also satisfy the additional conservation law

$$\partial_t Q - \operatorname{div}\left[\eta^{-1}(I + B\otimes B + D\otimes D - Q\otimes Q)\right] = 0. \tag{3.7}$$

Here, too, the components of Q are not entropies for (2.11), in the sense of (3.1), because the validity of (3.7) is contingent upon (B,D) satisfying the involutions (2.14).

The above two examples suggest that, for systems with involutions, the notion of entropy should be broadened as follows [23]. For a system (2.1), equipped with the involution (2.4), a (scalar-valued) function $\eta(U)$ is a *contingent entropy*, associated with the (m-vector-valued) *contingent entropy flux* Q if, for any $U \in \mathcal{O}$,

$$\mathrm{D}Q_\alpha(U) = \mathrm{D}\eta(U)\mathrm{D}G_\alpha(U) + \Xi(U)^\top M_\alpha, \qquad \alpha = 1, \cdots, m, \tag{3.8}$$

where $\Xi(U)$ is some (k-vector-valued) function defined on $\mathcal{O}$ and playing the role of a Lagrange multiplier. In particular, any entropy is a contingent entropy with $\Xi(U) = 0$.

Clearly, (3.8) implies that any classical solution U of (2.1) that satisfies the involution (2.4) must also satisfy the conservation law (3.2). In our examples, $\det F$ and the nine entries of F^* are contingent entropies for the system (2.7), and the three components of Q are contingent entropies for the system (2.11).

After this preparation, let us consider a system (2.1) of n conservation laws, endowed with (a) k involutions (2.4), (b) a principal, generally contingent, entropy-entropy flux pair (η, Q), and (c) ℓ supplementary contingent entropy-entropy flux pairs.

We set $N = n + \ell$ and compose the N-vector-valued function $\Phi(U)$ whose first n components are the components $U^1, \cdots, U^n$ of the state vector U and the remaining ℓ components are the supplementary contingent entropies. Thus, for classical solutions U of (2.1) satisfying the involutions (2.4),

$$\partial_t \Phi(U(x,t)) + \sum_{\alpha=1}^{m} \partial_\alpha \Psi_\alpha(U(x,t)) = 0, \tag{3.9}$$

where Ψ_α is the α-th column vector of the $N \times m$ matrix-valued function $\Psi(U)$ whose I-th row vector Ψ^I is the flux associated with the contingent entropy Φ^I. In particular, for $I = 1, \cdots, n$, $\Psi^I_\alpha = G^I_\alpha$.

The principal contingent entropy-entropy flux pair (η, Q) satisfies (3.8), for some Lagrange multiplier Ξ. Similarly,

$$\mathrm{D}\Psi_\alpha(U) = \mathrm{D}\Phi(U)\mathrm{D}G_\alpha(U) + \Omega(U)^\top M_\alpha, \qquad \alpha = 1, \cdots, m, \tag{3.10}$$

where Ω is the $k \times N$ matrix-valued function whose I-th column vector Ω^I is the Lagrange multiplier associated with the contingent entropy-entropy flux pair (Φ^I, Ψ^I). In particular, for $I = 1, \cdots, n$, $\Omega^I = 0$.

The main point of this paper is to demonstrate that in the above setting the requirement of convexity on the principal entropy may be relaxed into the following weaker condition:

DEFINITION 3.1. The principal contingent entropy $\eta(U)$ is called *polyconvex*, relative to the contingent entropies $\Phi(U)$, if it admits a representation

$$\eta(U) = \theta(\Phi(U)), \qquad U \in \mathcal{O}, \tag{3.11}$$

where θ is a smooth function defined on an open neighborhood $\mathcal{F}$ of $\Phi(\mathcal{O})$ in $\mathbb{R}^N$, whose Hessian matrix is positive definite at every $\Phi \in \mathcal{F}$.

In the example of elastodynamics, $\Phi = (F, v, F^*, \det F)$, arranged as a 22-vector. The entropy η, defined by (3.4), will be polyconvex when the strain energy $\varepsilon(F)$ admits a representation

$$\varepsilon(F) = \phi(F, F^*, \det F), \tag{3.12}$$

where $\phi(F, H, \delta)$ is a smooth function with positive definite Hessian on an open neighborhood of the manifold $\{(F, H, \delta) : \det F > 0, H = F^*, \delta = \det F\}$, embedded in $\mathbb{R}^{19}$. This is a physically reasonable assumption which has been discussed thoroughly in the literature, especially in the context of elastostatics [1]. In particular, the strain energy for any elastic fluid (gas) is polyconvex, as it is of the form $\varepsilon(F) = \phi(\det F)$, with ϕ convex on $(0, \infty)$. Similarly, for the Born-Infeld theory, where $\Phi = (B, D, Q)$, arranged as a 9-vector, the entropy η, defined by (2.13), is polyconvex.

One may attempt to treat systems (2.1) endowed with a polyconvex entropy by seeking N-vector-valued functions $\Pi_\alpha(\Phi)$ and scalar-valued functions $H_\alpha(\Phi), \alpha = 1, \cdots, m$, defined on $\mathcal{F}$ and satisfying the following conditions: $\Pi_\alpha(\Phi(U)) = \Psi_\alpha(U), H_\alpha(\Phi(U)) = Q_\alpha(U)$, for $U \in \mathcal{O}$, and (θ, H) is an entropy-entropy flux pair for the system

$$\partial_t \Phi(x,t) + \sum_{\alpha=1}^{m} \partial_\alpha \Pi_\alpha(\Phi(x,t)) = 0. \tag{3.13}$$

Indeed, whenever such a construction can be carried out, the system (2.1) becomes embedded in the extended system (3.13) which is endowed with a convex entropy and may thus be treated by the standard theory. This program has been implemented successfully for the system of elastodynamics [11], for the Born-Infeld version of Maxwell's equations [5,23], as well as for systems (2.1) endowed with a polyconvex entropy [18]. It is currently unknown, however, whether such an approach is also applicable to general systems (2.1) endowed with involutions and a contingent polyconvex entropy. Here we are pursuing an alternative avenue, working directly with the original system (2.1).

The following implications of polyconvexity will serve to prepare the ground for the analysis in the next two sections. Notice first that (3.8) and (3.10) yield the symmetry relations

$$\mathrm{D}^2\eta(U)\mathrm{D}G_\alpha(U) + \mathrm{D}\Xi(U)^\top M_\alpha = \mathrm{D}G_\alpha(U)^\top \mathrm{D}^2\eta(U) + M_\alpha^\top \mathrm{D}\Xi(U), \tag{3.14}$$

$$\mathrm{D}^2\Phi^I(U)\mathrm{D}G_\alpha(U) + \mathrm{D}\Omega^I(U)^\top M_\alpha = \mathrm{D}G_\alpha(U)^\top \mathrm{D}^2\Phi^I(U) + M_\alpha^\top \mathrm{D}\Omega^I(U), \tag{3.15}$$

for $I = 1, \cdots, N$.

In the sequel, $\theta_\Phi(\Phi)$ will denote the differential of the function $\theta(\Phi)$, as a row N-vector with components $\theta_{\Phi^I}(\Phi) = \partial\theta(\Phi)/\partial\Phi^I, I = 1, \cdots, N$; and $\theta_{\Phi\Phi}(\Phi)$ will stand for the $N \times N$ Hessian matrix of $\theta(\Phi)$.

For any $U \in \mathcal{O}$, we define the symmetric $n \times n$ matrix

$$A(U) = \mathrm{D}^2\eta(U) - \sum_{I=1}^{N} \theta_{\Phi^I}(\Phi(U))\mathrm{D}^2\Phi^I(U). \tag{3.16}$$

Using (3.11),

$$A(U) = \mathrm{D}\Phi(U)^{\top}\theta_{\Phi\Phi}(\Phi(U))\mathrm{D}\Phi(U), \tag{3.17}$$

so that $A(U)$ is positive definite when $\eta(U)$ is polyconvex. Furthermore, by virtue of (3.14) and (3.15), the $n \times n$ matrices $R_\alpha(U)$ defined by

$$R_\alpha(U) = A(U)\mathrm{D}G_\alpha(U) + \Gamma(U)^{\top} M_\alpha, \qquad \alpha = 1, \cdots, m, \tag{3.18}$$

where

$$\Gamma(U) = \mathrm{D}\Xi(U) - \sum_{I=1}^{N} \theta_{\Phi^I}(\Phi(U))\mathrm{D}\Omega^I(U), \tag{3.19}$$

are symmetric:

$$R_\alpha(U)^{\top} = R_\alpha(U), \qquad \alpha = 1, \cdots, m. \tag{3.20}$$

4. Classical Solutions to the Cauchy Problem

A *classical solution* of the initial-value problem

$$\partial_t U(x,t) + \sum_{\alpha=1}^{m} \partial_\alpha G_\alpha(U(x,t)) = 0, \qquad x \in \mathbb{R}^m, \ t > 0, \tag{4.1}$$

$$U(x,0) = U_0(x), \qquad x \in \mathbb{R}^m, \tag{4.2}$$

on the time interval $[0, T)$, with $0 < T \leq \infty$, is a locally Lipschitz function U, defined on $\mathbb{R}^m \times [0, T)$ and taking values in $\mathcal{O}$, which satisfies (4.1) almost everywhere and (4.2) for all x in $\mathbb{R}^m$. The main result of this section extends the standard theorem for the local existence of classical solutions from systems endowed with a convex entropy to systems that are equipped only with a polyconvex entropy contingent upon involutions.

We consider initial data $U_0(x)$ that decay, as $|x| \to \infty$, to a constant state $\bar{U} \in \mathcal{O}$. For convenience, we shift the origin of the state coordinate system so that $\bar{U} = 0$.

Throughout this section, $H^r, r = 0, 1, 2, \cdots$, stands for the Sobolev space $W^{r,2}(\mathbb{R}^m; \mathbb{R}^n)$, and its norm is denoted by $\|\cdot\|_r$. The symbol ν will stand for the typical multi-index, $\nu = (\nu_1, \cdots, \nu_m)$, of order $|\nu| = \nu_1 + \cdots + \nu_m$, and ∂^ν will denote the differential operator $\partial^{|\nu|}/\partial x_1^{\nu_1} \cdots \partial x_m^{\nu_m}$. We will employ the notation $\partial^\nu U = U_\nu$ and $\partial^\nu U_0 = U_{\nu 0}$.

THEOREM 4.1. *Assume* (4.1) *is a hyperbolic system of conservation laws endowed with the involution* (2.4), *the principal contingent entropy-entropy flux pair* (η, Q), *and a collection of supplementary contingent entropy-entropy flux pairs. Suppose, further, that* η *is polyconvex, in the sense of Definition* 3.1. *Given initial data* $U_0 \in H^s$, *for some* $s > \frac{m}{2} + 1$, *taking values in a compact* *subset of $\mathcal{O}$ and satisfying the involution, there exists a continuously differentiable classical solution U of the Cauchy problem* (4.1), (4.2) *on a time interval* $[0, T_\infty)$, *and*

$$U(\cdot, t) \in \bigcap_{r=0}^{s} C^r([0, T_\infty); H^{s-r}). \tag{4.3}$$

The interval $[0, T_\infty)$ is maximal in the sense that if $T_\infty < \infty$, then

$$\int_0^{T_\infty} \|\nabla U(\cdot, t)\|_{L^\infty(\mathbb{R}^m)} dt = \infty \tag{4.4}$$

and/or the range of $U(\cdot, t)$ escapes from every compact subset of $\mathcal{O}$ as $t \uparrow T_\infty$.

In the traditional situation where $\eta(U)$ is a convex entropy, rather than merely a polyconvex contingent entropy as is the case here, the solution to (4.1), (4.2) is constructed [2,10,20] as the fixed point of the map that carries a function V to the solution U of the Cauchy problem for the linear symmetrizable hyperbolic system

$$\partial_t U(x,t) + \sum_{\alpha=1}^{m} \mathrm{D}G_\alpha(V(x,t))\partial_\alpha U(x,t) = 0. \tag{4.5}$$

This approach is not applicable here, because the involution (2.4) is not generally preserved by solutions U to (4.5), (4.2), even though it may hold for V and U_0.[2] The suggestion by Serre [23] to employ the alternative linearization

$$\partial_t U(x,t) + \sum_{\alpha=1}^{m} \partial_\alpha[G_\alpha(V(x,t)) + \mathrm{D}G_\alpha(V(x,t))(U(x,t) - V(x,t))] = 0, \tag{4.6}$$

in the place of (4.5), resolves the above difficulty but creates a new one, because, as U and V are now entering at the same order, the map $V \mapsto U$ incurs loss of regularity. We will overcome these obstacles by applying the method of *vanishing viscosity*, which yields solutions to (4.1) as the $\epsilon \downarrow 0$ limit of solutions of the parabolic system

$$\partial_t U(x,t) + \sum_{\alpha=1}^{m} \partial_\alpha G_\alpha(U(x,t)) = \epsilon \Delta U(x,t). \tag{4.7}$$

The first step is to establish existence for the system (4.7), for fixed $\epsilon > 0$.

LEMMA 4.1. *Assume U_0 is in H^s, for some $s > \frac{m}{2} + 1$, and takes values in a compact subset of an open bounded set $\mathcal{B}$ of $\mathbb{R}^n$ whose closure $\overline{\mathcal{B}}$ is contained in $\mathcal{O}$. For any fixed $\omega > \|U_0\|_s$, there exists a (generally weak) solution U of* (4.7),

[2] The system of elastodynamics is exceptional in this respect, as it can be treated through (4.5) [10].

(4.2) *on a time interval* $[0,T_\epsilon), 0 < T_\epsilon \leq \infty$, *such that* $U(\cdot,t)$ *takes values in some compact subset of* $\mathcal{B}$, *for any fixed* $t \in [0,T_\epsilon)$, *and*

$$U(\cdot,t) \in C^0([0,T_\epsilon); H^s) \cap L^2([0,T_\epsilon); H^{s+1}), \tag{4.8}$$

with

$$\|U(\cdot,t)\|_s < \omega, \tag{4.9}$$

for all $t \in [0,T_\epsilon)$. *The interval* $[0,T_\epsilon)$ *is maximal, in the sense that if* $T_\epsilon < \infty$ *then, as* $t \to T_\epsilon$, $\|U(\cdot,t)\|_s \to \omega$ *and/or the range of* $U(\cdot,t)$ *escapes from every compact subset of* $\mathcal{B}$.

PROOF. With the fixed ω and with T to be selected later, we associate the class $\mathcal{V}$ of Lipschitz functions V, defined on $\mathbb{R}^m \times [0,T]$, taking values in $\overline{\mathcal{B}}$, and

$$V(\cdot,t) \in L^\infty([0,T]; H^s), \qquad \sup_{[0,T]} \|V(\cdot,t)\|_s \leq \omega. \tag{4.10}$$

By standard weak lower semicontinuity of norms, $\mathcal{V}$ is a complete metric space under the metric

$$\rho(V,\bar{V}) = \sup_{[0,T]} \|V(\cdot,t) - \bar{V}(\cdot,t)\|_0. \tag{4.11}$$

For given $V \in \mathcal{V}$, we construct the solution U on $\mathbb{R}^m \times [0,T]$ of the linear parabolic system

$$\partial_t U(x,t) - \epsilon\Delta U(x,t) = -\sum_{\alpha=1}^m \partial_\alpha G_\alpha(V(x,t)), \tag{4.12}$$

with initial condition (4.2). Thus

$$(4\pi\epsilon)^{\frac{m}{2}} U(x,t) = \int_{\mathbb{R}^m} t^{-\frac{m}{2}} \exp\left[-\frac{|x-y|^2}{4\epsilon t}\right] U_0(y)dy \tag{4.13}$$
$$- \int_0^t \int_{\mathbb{R}^m} (t-\tau)^{-\frac{m}{2}} \exp\left[-\frac{|x-y|^2}{4\epsilon(t-\tau)}\right] \sum_{\alpha=1}^m \partial_\alpha G_\alpha(V(y,\tau))dyd\tau\ .$$

We proceed to show that if T is sufficiently small, then $U \in \mathcal{V}$ and the map that carries V to U is a contraction. The unique fixed point of that map will then be the desired solution of (4.7), (4.2) on $[0,T]$.

In what follows, c will stand for a generic constant that may depend solely on s and on bounds of the G_α and their derivatives on $\overline{\mathcal{B}}$.

From (4.13) and (4.10) we deduce

$$\sup_{[0,T]} \|U(\cdot,t) - U_0(\cdot)\|_{L^\infty} \leq c\sqrt{\epsilon T}\|\nabla U_0\|_{L^\infty} + cT\|\sum_{\alpha=1}^m \partial_\alpha G_\alpha(V)\|_{L^\infty}, \tag{4.14}$$

which shows that when T is sufficiently small then, for any fixed $t \in [0,T]$, $U(\cdot,t)$ takes values in a compact subset of $\mathcal{B}$.

We now fix any multi-index ν of order $|\nu| \le s$ and apply ∂^ν to (4.12), which yields

$$\partial_t U_\nu(x,t) - \epsilon \Delta U_\nu(x,t) = -\sum_{\alpha=1}^{m} \partial_\alpha \partial^\nu G_\alpha(V(x,t)). \tag{4.15}$$

By virtue of (4.10) and standard Moser-type inequalities [20],

$$\sup_{[0,T]} \|\partial^\nu G_\alpha(V(\cdot,t))\|_0 \le c\omega. \tag{4.16}$$

By standard theory of the heat equation, the solution U_ν of (4.15) lies in $C^0([0,T];H^0)$, whence $U \in C^0([0,T];H^s)$. Furthermore, the "energy" estimate, obtained formally by multiplying (4.15) by $2U_\nu^\top$, integrating over $\mathbb{R}^m \times [0,t]$, for $t \in [0,T]$, and integrating by parts, is here valid:

$$\begin{aligned} &\int_{\mathbb{R}^m} |U_\nu(x,t)|^2 dx + 2\epsilon \int_0^t \int_{\mathbb{R}^m} |\nabla U_\nu|^2 dx d\tau \\ &= \int_{\mathbb{R}^m} |U_{\nu 0}(x)|^2 dx + 2 \int_0^t \int_{\mathbb{R}^m} \sum_{\alpha=1}^{m} \partial_\alpha U_\nu^\top \partial^\nu G_\alpha(V) dx d\tau \\ &\le \int_{\mathbb{R}^m} |U_{\nu 0}(x)|^2 dx + \epsilon \int_0^t \int_{\mathbb{R}^m} |\nabla U_\nu|^2 dx d\tau + \frac{cT\omega^2}{\epsilon}. \end{aligned} \tag{4.17}$$

Therefore, upon summing over all multi-indices ν of order $|\nu| \le s$,

$$\sup_{[0,T]} \|U(\cdot,t)\|_s^2 \le \|U_0(\cdot)\|_s^2 + \frac{cT\omega^2}{\epsilon}. \tag{4.18}$$

Since $\|U_0\|_s < \omega$, for T sufficiently small (4.9) holds for all $t \in [0,T]$ and thus $U \in \mathcal{V}$.

We now fix V and $\bar{V}$ in $\mathcal{V}$ and consider the solutions U and $\bar{U}$ of (4.12), (4.2) induced by them. Then

$$\partial_t(U - \bar{U}) - \epsilon \Delta (U - \bar{U}) = -\sum_{\alpha=1}^{m} \partial_\alpha [G_\alpha(V) - G_\alpha(\bar{V})]. \tag{4.19}$$

Multiplying (4.19) by $2(U-\bar{U})^\top$, integrating over $\mathbb{R}^m \times [0,t]$, and integrating by parts yields

$$(4.20)\quad \int_{\mathbb{R}^m} |U(x,t)-\bar{U}(x,t)|^2 dx + 2\epsilon \int_0^t \int_{\mathbb{R}^m} |\nabla(U-\bar{U})|^2 dx d\tau$$

$$= 2\int_0^t \int_{\mathbb{R}^m} \sum_{\alpha=1}^m \partial_\alpha (U-\bar{U})^\top [G_\alpha(V)-G_\alpha(\bar{V})] dx d\tau$$

$$\leq \epsilon \int_0^t \int_{\mathbb{R}^m} |\nabla(U-\bar{U})|^2 dx d\tau + \frac{cT}{\epsilon} \sup_{[0,T]} \int_{\mathbb{R}^m} |V(x,\tau)-\bar{V}(x,\tau)|^2 dx.$$

Recalling (4.11), we conclude that when $cT/\epsilon = \mu^2 < 1$, we have contraction, $\rho(U,\bar{U}) \leq \mu\rho(V,\bar{V})$, which establishes the existence of a single fixed point and thereby the existence of a unique solution to (4.7), (4.2) on $[0,T]$.

Since $\|U(\cdot,T)\|_s < \omega$ and $U(\cdot,T)$ takes values in a compact subset of $\mathcal{B}$, we may repeat the above construction and extend U to a larger interval $[0,T']$. Continuing this process, we end up with a solution U defined on a maximal interval $[0,T_\epsilon)$, such that if $T_\epsilon < \infty$ then, as $t \to T_\epsilon$, $\|U(\cdot,t)\|_s \to \omega$ and/or the range of $U(\cdot,t)$ escapes from every compact subset of $\mathcal{B}$. The proof is complete.

Next we construct solutions to the hyperbolic system (4.1):

LEMMA 4.2. *Let U_0 and $\mathcal{B}$ be as in Lemma* 4.1. *There is a positive constant c_0, depending solely on s and $\mathcal{B}$, such that with any fixed $\omega > c_0\|U_0\|_s$ is associated $T > 0$ and a unique classical solution U of* (4.1), (4.2) *on* $[0,T]$, *having the following properties: For each $t \in [0,T], U(\cdot,t)$ takes values in $\overline{\mathcal{B}}$, $U(\cdot,t) \in H^s$ and $\|U(\cdot,t)\|_s \leq \omega$.*

PROOF. Assume, temporarily, that $U_0 \in H^{s+2}$. Fix $\epsilon \in (0,1)$ and consider the solution U of (4.7), (4.2), defined on the maximal interval $[0,T_\epsilon)$, as established by Lemma 4.1. The aim is to show that T_ϵ is bounded below by some $T > 0$, uniformly in ϵ, and establish a priori bounds for U on $[0,T]$, independent of ϵ, which will allow us to construct the solution to (4.1), (4.2) by passing to the limit $\epsilon \downarrow 0$.

By virtue of (2.3) and (4.7), the function $X = \sum M_\alpha \partial_\alpha U$ solves the parabolic system $\partial_t X = \epsilon\Delta X$. Since $X(\cdot,0) = 0$, X must vanish identically on $\mathbb{R}^m \times [0,T_\epsilon)$ and thus U satisfies the involution (2.4).

For the remainder of this section, c will stand for a generic constant that depends solely on s and on bounds of the functions G_α, R_α, A and their derivatives, on $\overline{\mathcal{B}}$.

We fix any multi-index ν of order $|\nu| \leq s$ and apply ∂^ν to (4.7), which yields

$$(4.21)\quad \partial_t U_\nu(x,t) + \sum_{\alpha=1}^m \partial^\nu \partial_\alpha G_\alpha(U(x,t)) = \epsilon\Delta U_\nu(x,t).$$

Since $U_0 \in H^{s+2}, \partial_t U_\nu \in L^\infty([0,T_\epsilon); H^0)$. Multiplying (4.21) by $\partial_t U_\nu^\top$, integrating over $\mathbb{R}^m \times [0,t], 0 < t < T_\epsilon$, and integrating by parts, we obtain

$$\int_0^t \int_{\mathbb{R}^m} |\partial_t U_\nu|^2 dx d\tau + \frac{\epsilon}{2} \int_{\mathbb{R}^m} |\nabla U_\nu(x,t)|^2 dx \tag{4.22}$$

$$= \frac{\epsilon}{2} \int_{\mathbb{R}^m} |\nabla U_{\nu 0}(x)|^2 dx - \int_0^t \int_{\mathbb{R}^m} \partial_t U_\nu^\top \sum_{\alpha=1}^m \partial^\nu \partial_\alpha G_\alpha(U) dx d\tau.$$

By Moser-type estimates, for any ν of order $|\nu| \le s-1$,

$$\|\partial^\nu \partial_\alpha G_\alpha(U(\cdot,\tau))\|_0 \le c\|U(\cdot,\tau)\|_s. \tag{4.23}$$

Hence, applying the Cauchy-Schwarz inequality to (4.22) and then summing over all multi-indices ν of order $|\nu| \le s-1$, we conclude

$$\int_0^t \|\partial_t U(\cdot,\tau)\|_{s-1}^2 d\tau \le \epsilon \|U_0(\cdot)\|_s^2 + c\int_0^t \|U(\cdot,\tau)\|_s^2 d\tau. \tag{4.24}$$

Next we rewrite (4.21) as

$$\partial_t U_\nu + \sum_{\alpha=1}^m \mathrm{D}G_\alpha(U)\partial_\alpha U_\nu \tag{4.25}$$

$$= \sum_{\alpha=1}^m \{\mathrm{D}G_\alpha(U)\partial_\alpha U_\nu - \partial^\nu[\mathrm{D}G_\alpha(U)\partial_\alpha U]\} + \epsilon\Delta U_\nu,$$

we multiply by $2U_\nu^\top A(U)$, where $A(U)$ is the symmetric positive definite matrix (3.17), and integrate over $\mathbb{R}^m \times [0,t]$. Notice that

$$2U_\nu^\top A(U)\partial_t U_\nu = \partial_t[U_\nu^\top A(U)U_\nu] - U_\nu^\top \partial_t A(U) U_\nu, \tag{4.26}$$

$$U_\nu^\top A(U)\Delta U_\nu = \sum_{\alpha=1}^m \partial_\alpha[U_\nu^\top A(U)\partial_\alpha U_\nu] - \sum_{\alpha=1}^m \partial_\alpha U_\nu^\top A(U)\partial_\alpha U_\nu \tag{4.27}$$

$$- \sum_{\alpha=1}^m U_\nu^\top \partial_\alpha A(U)\partial_\alpha U_\nu.$$

By the Cauchy inequality, since $A(U)$ is positive definite,

$$- \sum_{\alpha=1}^m U_\nu^\top \partial_\alpha A(U)\partial_\alpha U_\nu \le \sum_{\alpha=1}^m \partial_\alpha U_\nu^\top A(U)\partial_\alpha U_\nu + c|\nabla U|^2|U_\nu|^2. \tag{4.28}$$

Furthermore, recalling that U satisfies the involution (2.4), and using (3.18) and (3.20), we deduce

$$\sum_{\alpha=1}^{m} 2U_\nu^\top A(U)\mathrm{D}G_\alpha(U)\partial_\alpha U_\nu \tag{4.29}$$

$$= \sum_{\alpha=1}^{m} 2U_\nu^\top [A(U)\mathrm{D}G_\alpha(U) + \Gamma(U)^\top M_\alpha]\partial_\alpha U_\nu = \sum_{\alpha=1}^{m} 2U_\nu^\top R_\alpha(U)\partial_\alpha U_\nu$$

$$= \sum_{\alpha=1}^{m} \partial_\alpha [U_\nu^\top R_\alpha(U) U_\nu] - \sum_{\alpha=1}^{m} U_\nu^\top \partial_\alpha R_\alpha(U) U_\nu.$$

We thus obtain

$$\int_{\mathbb{R}^m} U_\nu^\top(x,t)A(U(x,t))U_\nu(x,t)dx \le \int_{\mathbb{R}^m} U_{\nu 0}^\top(x)A(U_0(x))U_{\nu 0}(x)dx \tag{4.30}$$

$$+ \int_0^t \int_{\mathbb{R}^m} 2U_\nu^\top A(U) \sum_{\alpha=1}^{m} \{\mathrm{D}G_\alpha(U)\partial_\alpha U_\nu - \partial^\nu[\mathrm{D}G_\alpha(U)\partial_\alpha U]\} dx d\tau$$

$$+ \int_0^t \int_{\mathbb{R}^m} U_\nu^\top \{\partial_t A(U) + \sum_{\alpha=1}^{m} \partial_\alpha R_\alpha(U)\} U_\nu dx d\tau + c\epsilon \int_0^t \int_{\mathbb{R}^m} |\nabla U|^2 |U_\nu|^2 dx d\tau.$$

Notice the Moser-type estimate [20]

$$\|\mathrm{D}G_\alpha(U(\cdot,\tau))\partial_\alpha U_\nu(\cdot,\tau) - \partial^\nu[\mathrm{D}G_\alpha(U(\cdot,\tau))\partial_\alpha U(\cdot,\tau)]\|_0 \tag{4.31}$$

$$\le c\|\nabla U(\cdot,\tau)\|_{L^\infty} \|U(\cdot,\tau)\|_s.$$

Therefore, upon summing (4.29) over all multi-indices ν of order $|\nu| \le s$, we obtain an estimate

$$\|U(\cdot,t)\|_s^2 \le c\|U_0(\cdot)\|_s^2 + \int_0^t g(\tau)\|U(\cdot,\tau)\|_s^2 d\tau, \tag{4.32}$$

where

$$0 \le g(\tau) \le c\{\|\nabla U(\cdot,\tau)\|_{L^\infty} + \|\partial_t U(\cdot,\tau)\|_{L^\infty} + \epsilon\|\nabla U(\cdot,\tau)\|_{L^\infty}^2\}. \tag{4.33}$$

By Gronwall's inequality,

$$\|U(\cdot,t)\|_s^2 \le c\|U_0(\cdot)\|_s^2 \exp \int_0^t g(\tau)d\tau. \tag{4.34}$$

On the other hand, by virtue of (4.33), (4.24), Schwarz's inequality and the Sobolev lemma,

$$\left\{\int_0^t g(\tau)d\tau\right\}^2 \le ct\left\{\epsilon\|U_0(\cdot)\|_s^2 + \int_0^t \|U(\cdot,\tau)\|_s^2 d\tau\right\} + c\epsilon^2 \left\{\int_0^t \|U(\cdot,\tau)\|_s^2 d\tau\right\}^2. \tag{4.35}$$

It is now clear that there is c_0 such that if one fixes $\omega > c_0\|U_0\|_s$ then, for T sufficiently small, (4.14) implies that $U(\cdot,t)$ takes values in $\mathcal{B}$ and (4.34), (4.35) together imply that $\|U(\cdot,t)\|_s < \omega$, for any $t \in [0,T]$. Moreover, by (4.24),

$$\int_0^T \|\partial_t U(\cdot,t)\|_{s-1}^2 dt \le c(T+1)\omega^2. \tag{4.36}$$

It should be emphasized that some T with the above specifications may be selected independently of ϵ and it is thus a common lower bound of T_ϵ, for all ϵ.

Suppose now $U_0 \in H^s$. We fix a sequence $\{U_{0\ell}\}$ in H^{s+2} and a sequence $\{\epsilon_\ell\}$ in $(0,1)$, such that $U_{0\ell} \to U_0$ in H^s, and $\epsilon_\ell \to 0$, as $\ell \to \infty$. Let U_ℓ denote the solution of (4.7), with $\epsilon = \epsilon_\ell$, and initial data $U_\ell(\cdot,0) = U_{0\ell}(\cdot)$, restricted on the time interval $[0,T]$. By virtue of the above estimates, $\{U_\ell\}$ is contained in a bounded set of $L^\infty([0,T];H^s)$ and $\{\partial_t U_\ell\}$ is contained in a bounded set of $L^2([0,T];H^{s-1})$. By standard embedding theorems, $\{U_\ell\}$ is then uniformly equicontinuous on $\mathbb{R}^m \times [0,T]$. Therefore, we may extract a subsequence, again denoted by $\{U_\ell\}$, that converges to a continuous function U, uniformly on compact subsets of $\mathbb{R}^m \times [0,T]$. In particular, $U(\cdot,t)$ takes values in $\overline{\mathcal{B}}$, for all $t \in [0,T]$. Clearly, U satisfies the system (4.1) on $[0,T]$, in the sense of distributions, as well as the initial condition (4.2). Furthermore, for any fixed $t \in [0,T]$, $\{U_\ell(\cdot,t)\}$ converges to $U(\cdot,t)$, weakly in H^s, and hence $U(\cdot,t) \in L^\infty([0,T];H^s)$, with $\|U(\cdot,t)\|_s \le \omega$. It then follows from (4.1) that $\partial_t U(\cdot,t) \in L^\infty([0,T];H^{s-1})$. In particular, since $s > \frac{m}{2}+1, \nabla U$ and $\partial_t U$ are bounded, $|\nabla U| \le c\omega, |\partial_t U| \le c\omega$, so that U is a classical solution of (4.1), (4.2) on $[0,T]$.

The uniqueness of the above solution will be established, in a far more general setting, in Section 5, and so it will be taken henceforth for granted. The proof is complete.

The next proposition highlights the local dependence property for solutions of the hyperbolic system (4.1). In what follows, B_δ, with $\delta > 0$, will denote the ball of radius δ in $\mathbb{R}^m$, centered at the origin, and $H^s(B_\delta)$ will stand for the Sobolev space $W^{s,2}(B_\delta;\mathbb{R}^n)$.

LEMMA 4.3. *In the setting of Lemma 4.2, there are positive constants a, b, λ depending solely on s and $\mathcal{B}$, such that, for any $t \in [0,T]$ and $\rho > 0$,*

(4.37)

$$\|U(\cdot,t)\|_{H^s(B_\rho)}^2 \le a\|U_0(\cdot)\|_{H^s(B_{\rho+\lambda t})}^2 \exp\left\{ b\int_0^t \|\nabla U(\cdot,\tau)\|_{L^\infty(B_{\rho+\lambda(t-\tau)})} d\tau \right\}.$$

In particular, letting $\rho \to \infty$,

$$\|U(\cdot,t)\|_s^2 \le a\|U_0(\cdot)\|_s^2 \exp\left\{ b\int_0^t \|\nabla U(\cdot,\tau)\|_{L^\infty(\mathbb{R}^m)} d\tau \right\}. \tag{4.38}$$

PROOF. Assume, temporarily, $U_0 \in H^{s+m}$, so that $U(\cdot,t) \in L^\infty([0,T];H^{s+m})$. For any multi-index ν of order $|\nu| \leq s$, we rewrite (4.25), with $\epsilon = 0$, and then multiply this equation by $2U_\nu^\top A(U)$. Recalling (4.26) and (4.29), we deduce

$$\partial_t[U_\nu^\top A(U)U_\nu] + \sum_{\alpha=1}^m \partial_\alpha[U_\nu^\top R_\alpha(U)U_\nu] \tag{4.39}$$

$$= 2U_\nu^\top A(U)\sum_{\alpha=1}^m \{\mathrm{D}G_\alpha(U)\partial_\alpha U_\nu - \partial^\nu[\mathrm{D}G_\alpha(U)\partial_\alpha U]\}$$

$$+U_\nu^\top\{\partial_t A(U) + \sum_{\alpha=1}^m \partial_\alpha R_\alpha(U)\}U_\nu.$$

We fix $\lambda > 0$ so large that the matrices

$$\lambda A(U) + \sum_{\alpha=1}^m \xi_\alpha R_\alpha(U) \tag{4.40}$$

are positive definite, for all $\xi \in S^{m-1}$ and $U \in \overline{\mathcal{B}}$. Then we integrate (4.39) over the frustum $\{(x,\tau) : 0 \leq \tau \leq \sigma, |x| \leq \rho + \lambda(t-\tau)\}$, where $\sigma \in (0,t]$, and apply the divergence theorem. The resulting integral over the lateral surface of the frustum is nonnegative, in consequence of the choice of λ. We now sum over all multi-indices ν of order $|\nu| \leq s$, we use the Moser-type estimate (4.31), albeit for the ball $B_{\rho+\lambda(t-\tau)}$ rather than for the whole of $\mathbb{R}^m$, and set $u(\tau) = \|U(\cdot,\tau)\|^2_{H^s(B_{\rho+\lambda(t-\tau)})}$. We thus arrive at an estimate of the form

$$u(\sigma) \leq au(0) + \int_0^\sigma g(\tau)u(\tau)d\tau, \qquad 0 \leq \sigma \leq t, \tag{4.41}$$

with

$$0 \leq g(\tau) \leq b\|\nabla U(\cdot,\tau)\|_{L^\infty(B_{\rho+\lambda(t-\tau)})}. \tag{4.42}$$

Therefore, (4.37) follows by Gronwall's inequality.

Assume now U is the solution of (4.1), (4.2) with $U_0 \in H^s$. We fix a sequence $\{U_{0\ell}\}$ in H^{s+m} such that $U_{0\ell} \to U_0$ in H^s, as $\ell \to \infty$, and consider the solution U_ℓ of (4.1) with initial value $U_\ell(\cdot,0) = U_{0\ell}(\cdot)$. We write (4.37) for the solutions U_ℓ and then let $\ell \to \infty$. As $\{U_\ell\}$ is contained in a bounded set of $L^\infty([0,T];H^s)$, $\{U_\ell(\cdot,t)\}$ converges to $U(\cdot,t)$, weakly in H^s, for any fixed $t \in [0,T]$. Since $s > \frac{m}{2}+1$, $\nabla U_\ell(\cdot,t) \to \nabla U(\cdot,t)$, uniformly on $B_{\rho+\lambda(t-\tau)}$, for all $\tau \in [0,t]$. Hence (4.37) also holds for solutions with initial values in H^s. The proof is complete.

The remaining ingredient is the following regularity result:

LEMMA 4.4. *Let U be the solution of* (4.1), (4.2) *on* $[0,T]$ *obtained in Lemma* 4.2. *Then $t \mapsto U(\cdot,t)$ is continuous in H^s on* $[0,T]$.

PROOF. Because of the divergence structure of (4.11), $t \mapsto U(\cdot,t)$ is at least weakly continuous in H^s on $[0,T]$. To demonstrate strong continuity, we have to prove that, for any multi-index ν of order $|\nu| \leq s, t \mapsto U_\nu(\cdot,t)$ is (strongly) continuous in H^0 on $[0,T]$. To that end, since (4.1) is invariant under time translations $(x,t) \mapsto (x,t+\tau)$, as well as under reflections $(x,t) \mapsto (-x,-t)$ it will suffice to show that $t \mapsto U_\nu(\cdot,t)$ is right continuous in H^0 at $t=0$.

Let us fix $t \in (0,T)$. Assuming, temporarily, that U is a solution of (4.1), (4.2) with initial data $U_0 \in H^{s+m}$, we integrate (4.39) over the frustum $\{(x,\tau) : 0 \leq \tau \leq t, |x| \leq \rho + \lambda(t-\tau)\}$, where $\rho > 0$ and λ sufficiently large, as in the proof of Lemma 4.3. After an integration by parts and upon using the estimates $\|\nabla U\|_{L^\infty} \leq c\omega$, $\|\partial_t U\|_{L^\infty} \leq c\omega$, $\|U(\cdot,t)\|_s \leq \omega$ and (4.31), we deduce

(4.43)
$$\int_{B_\rho} U_\nu^\top(x,t)A(U(x,t))U_\nu(x,t)dx - \int_{B_{\rho+\lambda t}} U_{\nu 0}^\top(x)A(U_0(x))U_{\nu 0}(x)dx \leq ct\omega^3.$$

Suppose now U is the solution of (4.1), (4.2) with initial data $U_0 \in H^s$. As in the proof of Lemma 4.3, we fix a sequence $\{U_{0\ell}\}$ in H^{s+m} such that $U_{0\ell} \to U_0$ in H^s, as $\ell \to \infty$, and consider the solution U_ℓ of (4.1) with initial value $U_\ell(\cdot,0) = U_{0\ell}(\cdot)$. We write (4.43) for the solution U_ℓ, and let $\ell \to \infty$. In order to pass to the limit, we consider the identity

$$(4.44) \qquad (\bar{V}-V)^\top A(U)(\bar{V}-V) = \bar{V}^\top A(\bar{U})\bar{V} - V^\top A(U)V$$
$$-\bar{V}^\top[A(\bar{U})-A(U)]\bar{V} - 2(\bar{V}-V)^\top A(U)V,$$

with $U = U(x,t), \bar{U} = U_\ell(x,t), V = U_\nu(x,t), \bar{V} = U_{\ell\nu}(x,t)$, integrate it with respect to x over the ball B_ρ, and let $\ell \to \infty$. In the resulting equation, the left-hand side is nonnegative, since $A(U)$ is positive definite, while the last two terms on the right-hand side tend to zero, the first one because $U_\ell(\cdot,t) \to U(\cdot,t)$, uniformly on B_ρ, and the second because $U_{\ell\nu}(\cdot,t) \to U_\nu(\cdot,t)$, weakly in H^0. Thus the lim inf of the difference of the first two terms on the right-hand side is nonnegative, whence (4.43) holds even for solutions U with initial values that are merely in H^s.

The next step is to rewrite (4.44) with $U = U_0(x), \bar{U} = U(x,t), V = U_{\nu 0}(x), \bar{V} = U_\nu(x,t)$, and integrate it with respect to x over $\mathbb{R}^m$. On the right-hand side of the resulting equation: The difference of the first two terms is bounded from above by $ct\omega^3$, on account of (4.43), for $\rho = \infty$. The next term is also bounded by $ct\omega^3$, since $|U(x,t) - U_0(x)| \leq ct\omega$. Finally, the last term tends to zero, as $t \to 0$, because $U_\nu(\cdot,t) \to U_{\nu 0}(\cdot)$, weakly in H^0. Thus the integral of $(U_\nu - U_{\nu 0})^\top A(U_0)(U_\nu - U_{\nu 0})$ over $\mathbb{R}^m$ tends to zero, as $t \to 0$, and this implies that $U_\nu(\cdot,t) \to U_{\nu 0}(\cdot)$, strongly in H^0. The proof is complete.

PROOF OF THEOREM 4.1. By Lemma 4.2, a classical solution U to (4.1), (4.2) exists on an interval $[0,T]$ and $U(\cdot,T)$ is in H^s and takes values in a compact subset of $\mathcal{O}$. We may thus repeat the above construction and extend U to a larger interval $[0,T']$. Continuing this process, we end up with a solution U defined on a maximal interval $[0,T_\infty)$, and if $T_\infty < \infty$ then, as

$t \to T_\infty$, $\|U(\cdot,t)\|_s \to \infty$ and/or the range of U escapes from every compact subset of $\mathcal{O}$. However, (4.38) implies that $\|U(\cdot,t)\|_s$ cannot blow up as $t \to T_\infty$, unless (4.4) holds.

Lemma 4.4 implies $U(\cdot,t) \in C^0([0,T_\infty); H^s)$. Applying to (4.1) the operator ∂_t^r, for $r = 0, \cdots, s-1$, we show by induction that $\partial_t^r U(\cdot,t) \in C^0([0,T_\infty); H^{r-s})$, i.e., (4.3) holds. The proof is complete.

5. Uniqueness and Stability of Classical Solutions

It has been known for a long time [7,8,12] that classical solutions of hyperbolic systems of conservation laws endowed with a convex entropy are unique and stable within the broader class of physically admissible weak solutions. Here we address this issue for systems possessing only polyconvex contingent entropies.

We consider the hyperbolic system (2.1) of n conservation laws in m spatial variables, and assume that it is endowed with the involution (2.4), as well as with a principal (generally) contingent entropy-entropy flux pair (η, Q) together with a collection of ℓ supplementary contingent entropy-entropy flux pairs. As in Section 3, we set $N = n + \ell$ and assemble the N-vector $\Phi(U)$ whose first n components are the components $U^1, \cdots, U^n$ of U and the remaining ℓ components are the supplementary contingent entropies. For $\alpha = 1, \cdots, m$, the α-components of the associated fluxes are similarly assembled into the N-vector $\Psi_\alpha(U)$.

A *weak solution* on the time interval $[0,T)$ is a bounded measurable function U, defined on $\mathbb{R}^m \times [0,T)$ and taking values in $\mathcal{O}$, which satisfies the system (4.1) as well as the involution (2.4), in the sense of distributions. It is known [10] that any weak solution may be normalized so that $t \mapsto U(\cdot,t)$, as a function from $[0,T)$ to $L^\infty(\mathbb{R}^m;\mathbb{R}^n)$, is weak* continuous. In that case, initial conditions (4.2) may be imposed in the classical sense.

Here we shall deal with the restricted class of *mild weak solutions* distinguished by their feature of satisfying as equalities the conservation laws induced by the supplementary contingent entropies. To be precise, a weak solution is mild if

(5.1)
$$\int_0^T \int_{\mathbb{R}^m} [\partial_t V^\top \Phi(U) + \sum_{\alpha=1}^m \partial_\alpha V^\top \Psi_\alpha(U)] dx dt + \int_{\mathbb{R}^m} V^\top(x,0)\Phi(U_0(x)) dx = 0$$

holds for all Lipschitz, N-vector-valued, test functions V, with compact support in $\mathbb{R}^m \times [0,T)$. Note that (5.1) will hold when U satisfies (3.9) in the sense of distributions and $\Phi(U(\cdot,t)) \to \Phi(U_0(\cdot))$ in L^∞ weak*, as $t \to 0$.

Clearly, any classical solution qualifies as a mild weak solution, since (3.9) holds identically for any Lipschitz solution. It is surprising, however, that one often encounters even discontinuous mild weak solutions, in systems arising in physics. For example, any weak solution (F, v) of the system (2.7) of elastodynamics, with involution (2.10), is necessarily mild. Indeed, as we saw in Section 3, any L^∞ solution of (2.7) that satisfies (2.10) also satisfies the supplementary conservation laws (3.5) and (3.6), in the sense of distributions. Moreover, the stricter condition (5.1), which involves the initial data, will also hold, because $t \mapsto (F(\cdot,t), v(\cdot,t))$ is weak* continuous in L^∞ and this, together with the involution (2.10), renders the functions $t \mapsto F^*(\cdot,t)$ and $t \mapsto \det F(\cdot,t)$ weak* continuous in L^∞; see [1]. Similarly, weak solutions (B, D) of bounded variation for the system (2.11), with involutions (2.14), are mild. The reason is that shocks of moderate strength compatible with (2.11), (2.14) are linearly degenerate and hence do not incur entropy production. Thus, (3.7) must hold as an equality.

A weak solution U of (4.1), (4.2) will be dubbed *admissible* if it satisfies the inequality

$$(5.2)\qquad \int_0^T \int_{\mathbb{R}^m} [\partial_t\psi\eta(U) + \sum_{\alpha=1}^m \partial_\alpha\psi Q_\alpha(U)]dxdt + \int_{\mathbb{R}^m} \psi(x,0)\eta(U_0(x))dx \geq 0,$$

for all Lipschitz continuous, nonnegative test functions ψ, with compact support in $\mathbb{R}^m \times [0, T)$. Notice that (5.2) is slightly more stringent than requiring that U satisfy the inequality (3.3), in the sense of distributions. Clearly, all classical solutions are admissible, since they satisfy (3.2), and thereby also (5.2), as an equality.

The following proposition states that when the principal contingent entropy is polyconvex then classical solutions are unique and stable within the broader class of admissible mild weak solutions.

THEOREM 5.1.[3] *Consider the Cauchy problem* (4.1), (4.2), *where the system* (4.1) *is endowed with the involution* (2.4), *the principal contingent entropy-entropy flux pair* (η, Q), *and a collection of supplementary contingent entropy-entropy flux pairs. Assume* η *is polyconvex, in the sense of Definition* 3.1. *Let* $\bar{U}$ *be a classical solution and* U *be an admissible mild weak solution, both defined on* $\mathbb{R}^m \times [0, T)$ *and taking values in a compact subset* $\mathcal{D}$ *of* $\mathcal{O}$, *such that* $\Phi(\mathcal{D})$ *is contained in a convex subset of* $\mathcal{F}$. *The corresponding initial data* $\bar{U}_0$ *and* U_0 *satisfy the involution* (2.4). *Then*

$$(5.3)\qquad \int_{|x|<\rho} |U(x,t) - \bar{U}(x,t)|^2 dx \leq ae^{bt} \int_{|x|<\rho+\lambda t} |U_0(x) - \bar{U}_0(x)|^2 dx$$

holds for all $\rho > 0$ *and* $t \in [0, T)$, *with* a, b *and* λ *positive constants, where* a *and* λ *depend solely on* $\mathcal{D}$ *while* b *also depends on the Lipschitz constant of* $\bar{U}$ *on* $\{(x,\tau) : 0 \leq \tau \leq t, |x| < \rho + \lambda(t-\tau)\}$. *In particular,* $\bar{U}$ *is the unique admissible mild weak solution of* (4.1) *with initial data* $\bar{U}_0$ *and values in* $\mathcal{D}$.

PROOF. For any U and $\bar{U}$ in $\mathcal{D}$, we set

[3] In the statement of this theorem in [10] the list of assumptions is incomplete.

$$h(U,\bar{U}) = \eta(U) - \eta(\bar{U}) - \theta_\Phi(\Phi(\bar{U}))[\Phi(U) - \Phi(\bar{U})], \tag{5.4}$$

$$\begin{aligned} Y_\alpha(U,\bar{U}) = Q_\alpha(U) - Q_\alpha(\bar{U}) - \theta_\Phi(\Phi(\bar{U}))[\Psi_\alpha(U) - \Psi_\alpha(\bar{U})] \\ +[\theta_\Phi(\Phi(\bar{U}))\Omega(\bar{U})^\top - \Xi(\bar{U})^\top]M_\alpha[U - \bar{U}], \end{aligned} \tag{5.5}$$

$$\begin{aligned} Z_\alpha(U,\bar{U}) = \quad & -\mathrm{D}G_\alpha(\bar{U})^\top \mathrm{D}\Phi(\bar{U})^\top \theta_{\Phi\Phi}(\Phi(\bar{U}))[\Phi(U) - \Phi(\bar{U})] \\ & +\mathrm{D}\Phi(\bar{U})^\top \theta_{\Phi\Phi}(\Phi(\bar{U}))[\Psi_\alpha(U) - \Psi_\alpha(\bar{U})] \\ & -\mathrm{D}\Phi(\bar{U})^\top \theta_{\Phi\Phi}(\Phi(\bar{U}))\Omega(\bar{U})^\top M_\alpha[U - \bar{U}] \\ & +\Gamma(\bar{U})^\top M_\alpha[U - \bar{U}], \end{aligned} \tag{5.6}$$

where $\Gamma(U)$ is defined by (3.19).

Recalling Definition 3.1, we see that $h(U,\bar{U})$ is of quadratic order in $U - \bar{U}$, and positive definite. Upon using (3.8), (3.10) and (3.11), we deduce

$$\begin{aligned} \mathrm{D}Y_\alpha(U,\bar{U}) = [\theta_\Phi(\Phi(U)) - \theta_\Phi(\Phi(\bar{U}))]\mathrm{D}\Phi(U)\mathrm{D}G_\alpha(U) \\ +[\Xi(U) - \Xi(\bar{U})]^\top M_\alpha - \theta_\Phi(\Phi(\bar{U}))[\Omega(U) - \Omega(\bar{U})]^\top M_\alpha\,, \end{aligned} \tag{5.7}$$

which vanishes at $U = \bar{U}$, so that $Y(U,\bar{U})$ is also of quadratic order in $U - \bar{U}$. In particular,

$$|Y(U,\bar{U})| \le \lambda h(U,\bar{U}). \tag{5.8}$$

Turning to $Z(U,\bar{U})$, and by virtue of (3.10),

$$\begin{aligned} \mathrm{D}Z_\alpha(U,\bar{U}) = \quad & -\mathrm{D}G_\alpha(\bar{U})^\top \mathrm{D}\Phi(\bar{U})^\top \theta_{\Phi\Phi}(\Phi(\bar{U}))\mathrm{D}\Phi(U) \\ & +\mathrm{D}\Phi(\bar{U})^\top \theta_{\Phi\Phi}(\Phi(\bar{U}))\mathrm{D}\Phi(U)\mathrm{D}G_\alpha(U) \\ & +\mathrm{D}\Phi(\bar{U})^\top \theta_{\Phi\Phi}(\Phi(\bar{U}))[\Omega(U) - \Omega(\bar{U})]^\top M_\alpha \\ & +\Gamma(\bar{U})^\top M_\alpha\,. \end{aligned} \tag{5.9}$$

Recalling (3.17), (3.18) and (3.20), we conclude that

$$\mathrm{D}Z_\alpha(\bar{U},\bar{U}) = -\mathrm{D}G_\alpha(\bar{U})^\top A(\bar{U}) + A(\bar{U})\mathrm{D}G_\alpha(\bar{U}) + \Gamma(\bar{U})^\top M_\alpha = M_\alpha^\top \Gamma(\bar{U}). \tag{5.10}$$

We fix a nonnegative, Lipschitz continuous test function ψ with compact support in $\mathbb{R}^m \times [0,T)$, and evaluate h, Y and Z along the two solutions $U(x,t)$ and $\bar{U}(x,t)$. As an admissible weak solution, U will satisfy the inequality (5.2), while $\bar{U}$, being a classical solution, will satisfy (5.2) as equality. Thus

(5.11)
$$\int_0^T \int_{\mathbb{R}^m} [\partial_t \psi\, h(U,\bar{U}) + \sum_{\alpha=1}^m \partial_\alpha \psi\, Y_\alpha(U,\bar{U})]dxdt + \int_{\mathbb{R}^m} \psi(x,0)\, h(U_0(x),\bar{U}_0(x))\, dx$$
$$\geq - \int_0^T \int_{\mathbb{R}^m} \Big\{ \partial_t \psi\, \theta_\Phi(\Phi(\bar{U}))[\Phi(U) - \Phi(\bar{U})]$$
$$+ \sum_{\alpha=1}^m \partial_\alpha \psi \{\theta_\Phi(\Phi(\bar{U}))[\Psi_\alpha(U) - \Psi_\alpha(\bar{U})] - [\theta_\Phi(\Phi(\bar{U}))\Omega(\bar{U})^\top - \Xi(\bar{U})^\top] M_\alpha [U-\bar{U}]\} \Big\} dxdt$$
$$- \int_{\mathbb{R}^m} \psi(x,0)\theta_\Phi(\Phi(\bar{U}_0(x)))[\Phi(U_0(x)) - \Phi(\bar{U}_0(x))]dx.$$

Next we write (5.1) for both U and $\bar{U}$, with test function $V^\top = \psi\theta_\Phi(\Phi(\bar{U}))$. This yields

$$\int_0^T \int_{\mathbb{R}^m} \Big\{ \partial_t[\psi\theta_\Phi(\Phi(\bar{U}))][\Phi(U) - \Phi(\bar{U})] \tag{5.12}$$
$$+ \sum_{\alpha=1}^m \partial_\alpha[\psi\theta_\Phi(\Phi(\bar{U}))][\Psi_\alpha(U) - \Psi_\alpha(\bar{U})] \Big\} dxdt$$
$$+ \int_{\mathbb{R}^m} \psi(x,0)\theta_\Phi(\Phi(\bar{U}_0(x)))[\Phi(U_0(x)) - \Phi(\bar{U}_0(x))]dx = 0.$$

Furthermore, since both U and $\bar{U}$ satisfy the involution (2.4),

$$\int_0^T \int_{\mathbb{R}^m} \sum_{\alpha=1}^m \partial_\alpha \left\{ \psi[\theta_\Phi(\Phi(\bar{U}))\Omega(\bar{U})^\top - \Xi(\bar{U})^\top] \right\} M_\alpha[U - \bar{U}]dxdt = 0. \tag{5.13}$$

By virtue of (3.10) and $\sum M_\alpha \partial_\alpha \bar{U} = 0$,

$$\partial_t \theta_\Phi(\Phi(\bar{U})) = \partial_t \Phi(\bar{U})^\top \theta_{\Phi\Phi}(\Phi(\bar{U})) \tag{5.14}$$
$$= - \sum_{\alpha=1}^m \partial_\alpha \Psi_\alpha(\bar{U})^\top \theta_{\Phi\Phi}(\Phi(\bar{U}))$$
$$= - \sum_{\alpha=1}^m \partial_\alpha \bar{U}^\top \mathrm{D}\Psi_\alpha(\bar{U})^\top \theta_{\Phi\Phi}(\Phi(\bar{U}))$$
$$= - \sum_{\alpha=1}^m \partial_\alpha \bar{U}^\top [\mathrm{D}\Phi(\bar{U})\mathrm{D}G_\alpha(\bar{U}) + \Omega(\bar{U})^\top M_\alpha]^\top \theta_{\Phi\Phi}(\Phi(\bar{U}))$$
$$= - \sum_{\alpha=1}^m \partial_\alpha \bar{U}^\top \mathrm{D}G_\alpha(\bar{U})^\top \mathrm{D}\Phi(\bar{U})^\top \theta_{\Phi\Phi}(\Phi(\bar{U})).$$

Similarly,

$$\partial_\alpha \theta_\Phi(\Phi(\bar U)) = \partial_\alpha \bar U^\top \mathrm{D}\Phi(\bar U)^\top \theta_{\Phi\Phi}(\Phi(\bar U)), \tag{5.15}$$

$$\partial_\alpha[\theta_\Phi(\Phi(\bar U))\Omega(\bar U)^\top - \Xi(\bar U)^\top] = \partial_\alpha \bar U^\top \mathrm{D}\Phi(\bar U)^\top \theta_{\Phi\Phi}(\Phi(\bar U))\Omega(\bar U)^\top - \partial_\alpha \bar U^\top \Gamma(\bar U)^\top. \tag{5.16}$$

Therefore, recalling (5.6),

$$\begin{aligned}\partial_t\theta_\Phi(\Phi(\bar U))[\Phi(U) - \Phi(\bar U)] + \sum_{\alpha=1}^m \partial_\alpha\theta_\Phi(\Phi(\bar U))[\Psi_\alpha(U) - \Psi_\alpha(\bar U)] \\ - \sum_{\alpha=1}^m \partial_\alpha[\theta_\Phi(\Phi(\bar U))\Omega(\bar U)^\top - \Xi(\bar U)^\top] M_\alpha [U - \bar U] = \sum_{\alpha=1}^m \partial_\alpha \bar U^\top Z_\alpha(U, \bar U).\end{aligned} \tag{5.17}$$

On account of (5.10),

$$\sum_{\alpha=1}^m \partial_\alpha \bar U^\top \mathrm{D}Z_\alpha(\bar U, \bar U) = \left[\sum_{\alpha=1}^m M_\alpha \partial_\alpha \bar U\right]^\top \Gamma(\bar U) = 0. \tag{5.18}$$

Consequently, the right-hand side of (5.17) is of quadratic order in $U - \bar U$.

Upon combining (5.11), (5.12), (5.13) and (5.17), we deduce

$$\begin{aligned}\int_0^T \int_{\mathbb{R}^m} [\partial_t\psi\, h(U, \bar U) + \sum_{\alpha=1}^m \partial_\alpha \psi\, Y_\alpha(U, \bar U)] dx dt + \int_{\mathbb{R}^m} \psi(x, 0)\, h(U_0(x), \bar U_0(x))\, dx \\ \geq \int_0^T \int_{\mathbb{R}^m} \psi \sum_{\alpha=1}^m \partial_\alpha \bar U^\top Z_\alpha(U, \bar U) dx dt.\end{aligned} \tag{5.19}$$

We now fix $t \in (0, T)$ and $\rho > 0$. For any $\sigma \in (0, t]$ and ϵ positive small, we write (5.19) for the test function $\psi(x, \tau) = \chi(x, \tau)\omega(\tau)$, with

$$\omega(\tau) = \begin{cases} 1 & 0 \le \tau < \sigma \\ \epsilon^{-1}(\sigma - \tau) + 1 & \sigma \le \tau < \sigma + \epsilon \\ 0 & \sigma + \epsilon \le \tau < T \end{cases} \tag{5.20}$$

$$\chi(x, \tau) = \begin{cases} 1 & 0 \le \tau < T,\ 0 \le |x| < \rho + \lambda(t - \tau) \\ \epsilon^{-1}[\rho + \lambda(t - \tau) - |x|] + 1 & 0 \le \tau < T, \rho + \lambda(t - \tau) \le |x| < \rho + \lambda(t - \tau) + \epsilon \\ 0 & 0 \le \tau < T,\ \rho + \lambda(t - \tau) + \epsilon \le |x| < \infty \end{cases} \tag{5.21}$$

where λ is the constant appearing in (5.8). This yields

(5.22)

$$\frac{1}{\epsilon}\int_{\sigma}^{\sigma+\epsilon}\int_{|x|<\rho+\lambda(t-\tau)} h(U(x,\tau),\bar{U}(x,\tau))dxd\tau \le \int_{|x|<\rho+\lambda t} h(U_0(x),\bar{U}_0(x))dx$$

$$-\frac{1}{\epsilon}\int_0^{\sigma}\int_{\rho+\lambda(t-\tau)<|x|<\rho+\lambda(t-\tau)+\epsilon}\left[\lambda h(U,\bar{U})+\frac{Y(U,\bar{U})\cdot x}{|x|}\right]dxd\tau$$

$$-\int_0^{\sigma}\int_{|x|<\rho+\lambda(t-\tau)}\sum_{\alpha=1}^{m}\partial_\alpha\bar{U}^\top Z_\alpha(U,\bar{U})dxd\tau+O(\epsilon).$$

On account of (5.8), the second integral on the right-hand side of (5.22) is nonnegative. As shown in [10], since $\Phi(U)$ satisfies (3.9) and $\eta(U)$ satisfies (3.3), in the sense of distributions, there is a subset $\mathcal{T}$ of $[0,t]$, of total measure, such that the functions $\tau\mapsto\Phi(U(\cdot,\tau)),\tau\mapsto\eta(U(\cdot,\tau))$, and thereby also $\tau\mapsto h(U(\cdot,\tau),\bar{U}(\cdot,\tau))$, are weak* continuous in L^∞, for any $\tau\in\mathcal{T}$. Therefore, letting $\epsilon\downarrow 0$ in (5.22) we deduce

$$\int_{|x|<\rho+\lambda(t-\sigma)} h(U(x,\sigma),\bar{U}(x,\sigma))dx \tag{5.23}$$

$$\le\int_{|x|<\rho+\lambda t} h(U_0(x),\bar{U}_0(x))dx-\int_0^{\sigma}\int_{|x|<\rho+\lambda(t-\tau)}\sum_{\alpha=1}^{m}\partial_\alpha\bar{U}^\top Z_\alpha(U,\bar{U})dxd\tau,$$

for any $\sigma\in\mathcal{T}$.

As noted above, $h(U,\bar{U})$ as well as $\sum\partial_\alpha\bar{U}^\top Z_\alpha(U,\bar{U})$ are of quadratic order in $U-\bar{U}$, and $h(U,\bar{U})$ is positive definite. Thus, upon setting

$$u(\tau)=\int_{|x|<\rho+\lambda(t-\tau)}|U(x,\tau)-\bar{U}(x,\tau)|^2dx, \tag{5.24}$$

(5.23) implies

$$u(\sigma)\le au(0)+b\int_0^{\sigma}u(\tau)d\tau, \tag{5.25}$$

for all $\sigma\in\mathcal{T}$, where a depends solely on $\mathcal{D}$ while b also depends on the Lipschitz constant of $\bar{U}$ on $\{(x,\tau):0\le\tau\le t,|x|<\rho+\lambda(t-\tau)\}$. Since $\tau\mapsto U(\cdot,\tau)$ is weak* continuous in $L^\infty, u(\tau)$ is lower semicontinuous so that (5.25) will hold for all $\sigma\in[0,t]$. Then Gronwall's inequality yields $u(t)\le au(0)e^{bt}$, which is (5.3). The proof is complete.

It is clear that the above theorem is directly applicable to the system of elastodynamics (2.7), (2.8), on condition that the strain energy function ε admits the representation (3.12), with $\phi(F,G,\delta)$ having positive definite Hessian, as well as to the Born-Infeld version (2.11), (2.12), (2.13) of Maxwell's equations.

References

[1] BALL, J.M., Convexity conditions and existence theorems in nonlinear elasticity. *Arch. Rational Mech. Anal.* **63** (1977), 337-403.

[2] BENZONI-GAVAGE, S. and D. SERRE, *Multi-dimensional Hyperbolic Partial Differential Equations.* Oxford: Oxford University Press, 2007.

[3] BOILLAT, G., Involutions des systèmes conservatifs. *C.R. Acad. Sci. Paris*, Série I, **307** (1988), 891-894.

[4] BORN, M., and L. INFELD, Foundations of a new field theory. *Proc. Royal Soc. London*, **144A** (1934), 425-451.

[5] BRENIER, Y., Hydrodynamic structure of the augmented Born-Infeld equations. *Arch. Rational Mech. Anal.* **172** (2004), 65-91.

[6] BRESSAN, A., *Hyperbolic Systems of Conservation Laws. The One-dimensional Cauchy Problem.* Oxford: Oxford University Press, 2000.

[7] DAFERMOS, C.M., The second law of thermodynamics and stability. *Arch. Rational Mech. Anal.* **70** (1979), 167-179.

[8] DAFERMOS, C.M., Stability of motions of thermoelastic fluids. *J. Thermal Stresses* **2** (1979), 127-134.

[9] DAFERMOS, C.M., Quasilinear hyperbolic systems with involutions. *Arch. Rational Mech. Anal.* **94** (1986), 373-389.

[10] DAFERMOS, C.M., *Hyperbolic Conservation Laws in Continuum Physics.* (Second Edition). Heidelberg: Springer, 2005.

[11] DEMOULINI, S., STUART, D.M.A. and A.E. TZAVARAS, A variational approximation scheme for three-dimensional elastodynamics with polyconvex energy. *Arch. Rational Mech. Anal.* **157** (2001), 325-344.

[12] DIPERNA, R.J., Uniqueness of solutions to hyperbolic conservation laws. *Indiana U. Math. J.* **28** (1979), 137-188.

[13] FRIEDRICHS, K.O. and P.D. LAX, Systems of conservation laws with a convex extension. *Proc. Natl. Acad. Sci. USA* **68** (1971), 1686-1688.

[14] HOLDEN, H. and N.H. RISEBRO, *Front Tracking for Hyperbolic Conservation Laws.* New York: Springer, 2002.

[15] KATO, T., The Cauchy problem for quasi-linear symmetric hyperbolic systems. *Arch. Rational Mech. Anal.* **58** (1975), 181-205.

[16] LAX, P.D., Hyperbolic systems of conservation laws. *Comm. Pure Appl. Math.* **10** (1957), 537-566.

[17] LAX, P.D., Shock waves and entropy. *Contributions to Functional Analysis*, pp. 603-634, ed. E.A. Zarantonello. New York: Academic Press, 1971.

[18] LEBLANC, V., Post-graduate Memoir, ENS Lyon (2005).

[19] LEFLOCH, P.G., *Hyperbolic Systems of Conservation Laws.* Basel: Birkhäuser, 2002.

[20] MAJDA, A., *Compressible Fluid Flow and Systems of Conservation Laws in Several Space Variables.* New York: Springer, 1984.

[21] QIN, TIEHU, Symmetrizing the nonlinear elastodynamic system. *J. Elasticity* **50** (1998), 245-252.

[22] SERRE, D., *Systems of Conservation Laws*, Vols. 1-2. Cambridge: Cambridge University Press, 1999.

[23] SERRE, D., Hyperbolicity of the nonlinear models of Maxwell's equations. *Arch. Rational Mech. Anal.* **172** (2004), 309-331.

[24] TAYLOR, M.E., *Partial Differential Equations III.* New York: Springer, 1996.

DIVISION OF APPLIED MATHEMATICS, BROWN UNIVERSITY, PROVIDENCE, RHODE ISLAND 02912
E-mail address: dafermos@dam.brown.edu

Titles in This Series

65 **L. L. Bonilla, A. Carpio, J. M. Vega, and S. Venakides, Editors,** Recent advances in nonlinear partial differential equations and applications (Toledo, Spain, June 2006)

64 **Reinhard C. Laubenbacher, Editor,** Modeling and simulation of biological networks (San Antonio, Texas, January 2006)

63 **Gestur Ólafsson and Eric Todd Quinto, Editors,** The radon transform, inverse problems, and tomography (Atlanta, Georgia, January 2005)

62 **Paul Garrett and Daniel Lieman, Editors,** Public-key cryptography (Baltimore, Maryland, January 2003)

61 **Serkan Hoşten, Jon Lee, and Rekha R. Thomas, Editors,** Trends in optimization (Phoenix, Arizona, January 2004)

60 **Susan G. Williams, Editor,** Symbolic dynamics and its applications (San Diego, California, January 2002)

59 **James Sneyd, Editor,** An introduction to mathematical modeling in physiology, cell biology, and immunology (New Orleans, Louisiana, January 2001)

58 **Samuel J. Lomonaco, Jr., Editor,** Quantum computation: A grand mathematical challenge for the twenty-first century and the millennium (Washington, DC, January 2000)

57 **David C. Heath and Glen Swindle, Editors,** Introduction to mathematical finance (San Diego, California, January 1997)

56 **Jane Cronin and Robert E. O'Malley, Jr., Editors,** Analyzing multiscale phenomena using singular perturbation methods (Baltimore, Maryland, January 1998)

55 **Frederick Hoffman, Editor,** Mathematical aspects of artificial intelligence (Orlando, Florida, January 1996)

54 **Renato Spigler and Stephanos Venakides, Editors,** Recent advances in partial differential equations (Venice, Italy, June 1996)

53 **David A. Cox and Bernd Sturmfels, Editors,** Applications of computational algebraic geometry (San Diego, California, January 1997)

52 **V. Mandrekar and P. R. Masani, Editors,** Proceedings of the Norbert Wiener Centenary Congress, 1994 (East Lansing, Michigan, 1994)

51 **Louis H. Kauffman, Editor,** The interface of knots and physics (San Francisco, California, January 1995)

50 **Robert Calderbank, Editor,** Different aspects of coding theory (San Francisco, California, January 1995)

49 **Robert L. Devaney, Editor,** Complex dynamical systems: The mathematics behind the Mandlebrot and Julia sets (Cincinnati, Ohio, January 1994)

48 **Walter Gautschi, Editor,** Mathematics of Computation 1943–1993: A half century of computational mathematics (Vancouver, British Columbia, August 1993)

47 **Ingrid Daubechies, Editor,** Different perspectives on wavelets (San Antonio, Texas, January 1993)

46 **Stefan A. Burr, Editor,** The unreasonable effectiveness of number theory (Orono, Maine, August 1991)

45 **De Witt L. Sumners, Editor,** New scientific applications of geometry and topology (Baltimore, Maryland, January 1992)

44 **Béla Bollobás, Editor,** Probabilistic combinatorics and its applications (San Francisco, California, January 1991)

43 **Richard K. Guy, Editor,** Combinatorial games (Columbus, Ohio, August 1990)

42 **C. Pomerance, Editor,** Cryptology and computational number theory (Boulder, Colorado, August 1989)

41 **R. W. Brockett, Editor,** Robotics (Louisville, Kentucky, January 1990)

40 **Charles R. Johnson, Editor,** Matrix theory and applications (Phoenix, Arizona, January 1989)

TITLES IN THIS SERIES

39 **Robert L. Devaney and Linda Keen, Editors,** Chaos and fractals: The mathematics behind the computer graphics (Providence, Rhode Island, August 1988)

38 **Juris Hartmanis, Editor,** Computational complexity theory (Atlanta, Georgia, January 1988)

37 **Henry J. Landau, Editor,** Moments in mathematics (San Antonio, Texas, January 1987)

36 **Carl de Boor, Editor,** Approximation theory (New Orleans, Louisiana, January 1986)

35 **Harry H. Panjer, Editor,** Actuarial mathematics (Laramie, Wyoming, August 1985)

34 **Michael Anshel and William Gewirtz, Editors,** Mathematics of information processing (Louisville, Kentucky, January 1984)

33 **H. Peyton Young, Editor,** Fair allocation (Anaheim, California, January 1985)

32 **R. W. McKelvey, Editor,** Environmental and natural resource mathematics (Eugene, Oregon, August 1984)

31 **B. Gopinath, Editor,** Computer communications (Denver, Colorado, January 1983)

30 **Simon A. Levin, Editor,** Population biology (Albany, New York, August 1983)

29 **R. A. DeMillo, G. I. Davida, D. P. Dobkin, M. A. Harrison, and R. J. Lipton,** Applied cryptology, cryptographic protocols, and computer security models (San Francisco, California, January 1981)

28 **R. Gnanadesikan, Editor,** Statistical data analysis (Toronto, Ontario, August 1982)

27 **L. A. Shepp, Editor,** Computed tomography (Cincinnati, Ohio, January 1982)

26 **S. A. Burr, Editor,** The mathematics of networks (Pittsburgh, Pennsylvania, August 1981)

25 **S. I. Gass, Editor,** Operations research: mathematics and models (Duluth, Minnesota, August 1979)

24 **W. F. Lucas, Editor,** Game theory and its applications (Biloxi, Mississippi, January 1979)

23 **R. V. Hogg, Editor,** Modern statistics: Methods and applications (San Antonio, Texas, January 1980)

22 **G. H. Golub and J. Oliger, Editors,** Numerical analysis (Atlanta, Georgia, January 1978)

21 **P. D. Lax, Editor,** Mathematical aspects of production and distribution of energy (San Antonio, Texas, January 1976)

20 **J. P. LaSalle, Editor,** The influence of computing on mathematical research and education (University of Montana, August 1973)

19 **J. T. Schwartz, Editor,** Mathematical aspects of computer science (New York City, April 1966)

18 **H. Grad, Editor,** Magneto-fluid and plasma dynamics (New York City, April 1965)

17 **R. Finn, Editor,** Applications of nonlinear partial differential equations in mathematical physics (New York City, April 1964)

16 **R. Bellman, Editor,** Stochastic processes in mathematical physics and engineering (New York City, April 1963)

15 **N. C. Metropolis, A. H. Taub, J. Todd, and C. B. Tompkins, Editors,** Experimental arithmetic, high speed computing, and mathematics (Atlantic City and Chicago, April 1962)

14 **R. Bellman, Editor,** Mathematical problems in the biological sciences (New York City, April 1961)

ISBN 978-0-8218-4211-9

PSAPM/65

DATE DUE
